Springer Series on

Wave Phenomena 3

Springer Series on

Wave Phenomena

Editor: Leopold B. Felsen

Volume 1 **Mechanics of Continua and Wave Dynamics**
By L. Brekhovskikh, V. Goncharov

Volume 2 **Rayleigh-Wave Theory and Application**
Editors: E. A. Ash, E. G. S. Paige

Volume 3 **Electromagnetic Surface Excitations**
Editors: R. F. Wallis, G. I. Stegeman

Electromagnetic Surface Excitations

Proceedings of an International Summer School at the Ettore Majorana Centre, Erice, Italy, July 1–13, 1985

Editors
R. F. Wallis and G. I. Stegeman

With 151 Figures

Springer-Verlag
Berlin Heidelberg New York Tokyo

Professor Richard F. Wallis

Department of Physics, University of California, Irvine, CA 92717, USA

Professor George I. Stegeman

Optical Sciences Center, University of Arizona, Tucson, AZ 85721, USA

Guest Editor:

Theodor Tamir, Ph. D.

981 East Lawn Drive, Teaneck, NJ 07666, USA

Series Editor:

Professor Leopold B. Felsen, Ph. D.

Polytechnic Institute of New York, 333 Jay Street, Brooklyn, NY 11201, USA

ISBN-13:978-3-642-82717-4 e-ISBN-13:978-3-642-82715-0
DOI: 10.1007/978-3-642-82715-0

Library of Congress Cataloging-in-Publication Data. Electromagnetic surface excitation. (Springer series on wave phenomena ; 3) Includes index. 1. Polaritons–Congresses. 2. Electromagnetic waves–Congresses. 3. Surfaces (Physics)–Congresses. 4. Surfaces (Technology)–Congresses. 5. Integrated optics–Congresses. I. Wallis, R.F. (Richard Fisher), 1924-. II. Stegeman, G.I. III. Title: Surface excitations. IV. Series.
QC176.8.P6E425 1986 530.4'1 86-1803

Softcover reprint of the hardcover 1st edition 1986

2153/3150-543210

Preface

This volume is based on lectures and contributed papers presented at the Eighth Course of the International School of Materials Science and Technology that was held in Erice, Sicily, Italy at the Ettore Majorana Centre for Scientific Culture during the period 1–13 July 1985. The subject of the course was "Electromagnetic Surface Excitations". Forty lectures were given by eleven distinguished scientists and engineers from France, Italy, the United Kingdom, and the United States. In addition to the lecturers, approximately fifty participants representing eleven different countries throughout the world took part in the course. Short contributed papers were presented by seventeen participants on the results of their own research.

The subject of the Course is of great importance for both pure science and for practical applications such as telecommunications. A technological revolution is occurring in which the transmission of information by means of electrical currents travelling in copper wires is being replaced by transmission by means of light travelling in objects known as optical wave guides. The manipulation and processing of the light signals prior and subsequent to transmission through the wave guide has resulted in a technology often referred to as integrated optics. Important to the operation of integrated optics devices is the behavior of electromagnetic waves near surfaces and interfaces. One of the goals of the course was to further the dialogue between engineers and physicists in common areas of interest related to the propagation of electromagnetic waves along surfaces.

The subjects covered and the lecturers were the following: Introduction to Electromagnetic Surface Waves, R.F. Wallis; Surface Plasmon Phenomena, F. Abeles; Surface Vibrational Excitations in Noble Metals, V. Bortolani; Surface Wave Electrooptics and Acoustooptics, R.M. De La Rue; Response Functions at Surfaces, R. Del Sole; Fundamentals of Wave Guides and Integrated Optics Technology, E. Garmire; Electromagnetic Surface Excitations on Rough Surfaces, A.A. Maradudin; Surface Brillouin and Raman Scattering, F. Nizzoli; Nonlinear Surface Polariton Interactions: Surface Enhanced Nonlinear Optical Effects, R. Reinisch; Nonlinear Guided Waves, G.I. Stegeman; Surface Magnetic Modes, D.R. Tilley.

The Course was sponsored by a number of organizations. Basic support to the Ettore Majorana Centre was provided by the Italian Ministry of Education, the Italian Ministry of Scientific and Technological Research, and the Sicilian Regional Government. Specific financial contributions were made by the European Physical Society, the European Research Office of the U.S. Army (London), the Science and Engineering Research Council (UK), the National Science Foundation (US), GTE Laboratories (US), IBM Italia, SGS-STES Agrate Brianza, and Unilever Research (UK). We are grateful to all of these organizations for their generous support that made the course possible.

We wish to express our appreciation to the Director of the Ettore Majorana Centre, Professor A. Zichichi, and to the secretaries, Dr. A. Gabriele and Dr. Pinola Savalli, for their excellent handling of administrative matters and whose hospitality and assistance made the experience a very pleasant one indeed. We also would like to thank Professor Minko Balkanski, the Director of the International School of Material Science and Technology, for his kind help in the organization of the Course.

Finally, we extend our gratitude to the lecturers for their outstanding lectures and to all the participants for their penetrating questions and stimulating comments.

December 1985

R.F. Wallis
G.I. Stegeman

Contents

Part I Elementary Surface Excitations

Part II Electromagnetic Probes of Surface Excitations

Part I

Elementary Surface Excitations

Introduction to Electromagnetic Surface Waves

R.F. Wallis

Physics Department, University of California, Irvine, CA 92717, USA

1. Introduction

When an electromagnetic wave propagates through a condensed medium, its properties are modified by the coupling of the electromagnetic wave to the elementary excitations of the medium. The coupled excitation is frequently referred to as a polariton and consists of a photon coupled to a plasmon, optical phonon, magnon, etc. An electromagnetic surface wave or surface polariton is a polariton in which the electromagnetic field is localized at the surface of the medium [1]. In the present discussion we restrict ourselves to planar surfaces.

2. Theory of Surface Polaritons in Isotropic Dielectric Media

We present an elementary treatment of surface polaritons for the case that the media on each side of the surface are isotropic and nonmagnetic. The media occupy the half-spaces $x>0$ and $x<0$ and are characterized by the frequency-dependent dielectric functions $\varepsilon_a(\omega)$ and $\varepsilon_b(\omega)$, respectively. We assume that losses in the media are sufficiently small to permit neglecting the imaginary parts of the dielectric functions. Furthermore, we use the local approximation and neglect spatial dispersion, i.e., the wave vector dependence of the dielectric functions. Under these conditions we derive the surface polariton dispersion relation and then present results for specific cases.

If the magnetic field is eliminated from Maxwell's curl equations, the electric field $\vec{E}$ is found to satisfy the equation

$$\nabla \times \nabla \times \vec{E} + (1/c^2)\partial^2\vec{D}/\partial t^2 = 0 \tag{1}$$

where the displacement $\vec{D}$ is given by $\vec{D}=\varepsilon(\omega)\ \vec{E}$. We seek a solution to (1) in which the electric field is wavelike in the z-direction and decreases exponentially with increasing distance from the interface separating the two media. Explicitly,

$$\vec{E}_a(\vec{r},t) = \vec{E}_a^o \exp(-\alpha_a x)\exp[i(kz-\omega t)],\quad x > 0 \tag{2}$$

$$\vec{E}_b(\vec{r},t) = \vec{E}_b^o \exp(\alpha_b x)\exp[i(kz-\omega t)],\quad x < 0 \tag{3}$$

where k is the wave vector and the decay constants α_a and α_b

must both be positive in order for us to have a properly localized electromagnetic surface wave. Substitution of (2) and (3) into (1) yields the following expressions for the electric fields

$$\vec{E}_a(\vec{r},t) = \left(\frac{ik}{\alpha_a} E^o_{az}, E^o_{ay}, E^o_{az}\right)\exp[-\alpha_a x+i(kz-\omega t)], \quad x>0 \tag{4}$$

$$\vec{E}_b(\vec{r},t) = \left(-\frac{ik}{\alpha_b} E^o_{bz}, E^o_{by}, E^o_{bz}\right)\exp[\alpha_b x+i(kz-\omega t)], \quad x<0 \tag{5}$$

and the decay constants

$$\alpha_a^2 = k^2 - (\omega^2/c^2)\varepsilon_a(\omega) \quad , \tag{6}$$

$$\alpha_b^2 = k^2 - (\omega^2/c^2)\varepsilon_b(\omega) \quad . \tag{7}$$

Also of interest are expressions for the magnetic fields. From the Maxwell equation involving curl $\vec{E}$, we find

$$\vec{B}_a(\vec{r},t) = (ic/\omega)[ikE^o_{ay}, (\omega^2/\alpha_a c^2)\varepsilon_a(\omega)E^o_{az}, \alpha_a E^o_{ay}] \times \exp[-\alpha_a x+i(kz-\omega t)], \; x > 0 \tag{8}$$

$$\vec{B}_b(\vec{r},t) = (ic/\omega)[ikE^o_{by}, -(\omega^2/\alpha_b c^2)\varepsilon_b(\omega)E^o_{bz}, -\alpha_b E^o_{by}] \times \exp[\alpha_b x+i(kz-\omega t)], \; x < 0 \quad . \tag{9}$$

We are now in a position to obtain the dispersion relation by considering the electromagnetic boundary conditions at the surface x=0. The continuity of the tangential components of $\vec{E}$ leads to the relations

$$E^o_{ay} = E^o_{by} \quad , \quad E^o_{az} = E^o_{bz} \quad . \tag{10}$$

The continuity of the normal component $\vec{D}$ yields

$$\frac{\varepsilon_a(\omega)}{\varepsilon_b(\omega)} = -\frac{\alpha_a}{\alpha_b} \tag{11}$$

which constitutes the dispersion relation. An additional relation follows from the continuity of the z-components of $\vec{B}_a$ and $\vec{B}_b$ at x=0, namely,

$$(\alpha_a + \alpha_b)E^o_{ay} = 0 \quad . \tag{12}$$

Since the localization of the surface polariton requires that α_a and α_b be positive, we see from (12) that $E^o_{ay}=0$ and from (10) that $E^o_{by}=0$. The electric field of the surface polariton therefore lies in the sagittal plane, which is defined by the normal to the surface and the direction of propagation. The magnetic field, on the other hand, is seen from (8) and (9) to be normal to the sagittal plane, so the mode is a TM mode.

The dispersion relation given by (11) tells us that $\varepsilon_a(\omega)$ and $\varepsilon_b(\omega)$ must have opposite signs when ω and k correspond to a point on the surface polariton dispersion curve. The medium with negative $\varepsilon(\omega)$ is termed the surface active medium, while that with positive $\varepsilon(\omega)$ is termed the surface inactive medium. Since a negative $\varepsilon(\omega)$ implies an imaginary index of refraction in the absence of damping, the surface active medium is typically highly reflecting in the frequency range where surface polaritons propagate.

3. Surface Plasmon Polaritons

The first specific example that we consider is the case where the surface active medium is a metal or semiconductor containing free current carriers. We can use the Drude form for the dielectric function of a free electron gas,

$$\varepsilon(\omega) = \varepsilon_\infty\left(1 - \frac{\omega_p^2}{\omega^2}\right) \quad , \tag{13}$$

where ε_∞ is the background (high frequency) dielectric constant, $\omega_p^2 = 4\pi ne^2/m^*\varepsilon_\infty$ is the square of the plasma frequency, n is the free carrier concentration, and m^* is the effective mass. If we take the second medium to be vacuum so tht $\varepsilon_b(\omega)=1$, then by using (6), (7), and (13) we can solve the dispersion relation (11) and obtain an explicit result for the frequency of surface plasmon polaritons:

$$\omega_{sp}^2(k) = (1/2\varepsilon_\infty)\{(1+\varepsilon_\infty)c^2k^2+\varepsilon_\infty\omega_p^2 - [((1+\varepsilon_\infty)c^2k^2+\varepsilon_\infty\omega_p^2)^2-4\varepsilon_\infty^2c^2k^2\omega_p^2]^{1/2}\} \quad . \tag{14}$$

A plot of the dispersion relation for n-type InSb is given in Fig. 1. It should be noted that the dispersion curve lies entirely to the right of the light line in vacuum, $\omega=ck$. A simple resonant interaction between the surface polariton and light incident from vacuum is therefore impossible. In the nonretarded limit, $ck\to\infty$, the surface polariton frequency approaches the value

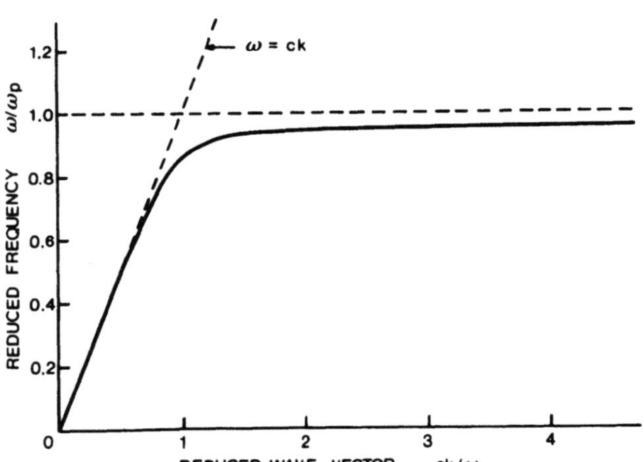

Fig. 1. Dispersion curve for surface plasmon polaritons

$$\omega_{sp}(\infty) = \frac{\omega_p}{(1+\frac{1}{\varepsilon_\infty})^{1/2}} \quad . \tag{15}$$

At small values of the wave vector, the surface polariton is primarily photon-like, whereas at large values of the wave vector, it is primarily plasmon-like.

4. Surface Optical Phonon Polaritons

The second example that we consider concerns a material whose lattice vibrations include one or more optical branches that are electric dipole active. Alkali halides and polar semiconductors fall into this category. For the case of a cubic ionic crystal with two atoms per unit cell, the dielectric function can be taken to have the form

$$\varepsilon(\omega) = \varepsilon_\infty\left(1 + \frac{\omega_{LO}^2-\omega_{TO}^2}{\omega_{TO}^2-\omega^2}\right) \tag{16}$$

where ω_{LO} and ω_{TO} are the long wavelength longitudinal and transverse optical phonon frequencies, respectively. If we again take the second medium to be vacuum so that $\varepsilon_b(\omega)=1$ and substitute this value together with (16) into (11), we obtain the following result specifying the frequency of surface optical phonon polaritons:

$$\omega_{so}^2(k) = (1/2\varepsilon_\infty)\{(1+\varepsilon_\infty)c^2k^2+\varepsilon_\infty\omega_{LO}^2 - [((1+\varepsilon_\infty)c^2k^2+\varepsilon_\infty\omega_{LO}^2)^2 - 4\varepsilon_\infty c^2k^2(\omega_{TO}^2+\varepsilon_\infty\omega_{LO}^2)]^{1/2}\} \quad . \tag{17}$$

In Fig. 2 is plotted $\omega_{so}(k)$ versus k for intrinsic InSb. The dispersion curve starts at the vacuum light line at $\omega=\omega_{TO}$, approaches the asymptotic value

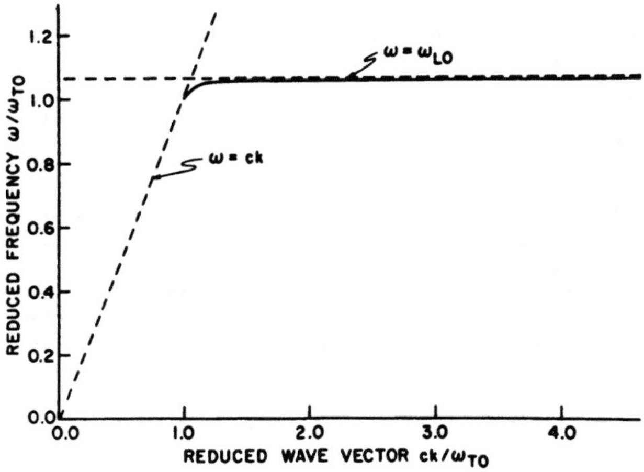

Fig. 2. Dispersion curve for surface optical phonon polaritons

$$\omega_{so}(\infty) = \left(\omega_{TO}^2 + \frac{\omega_{LO}^2-\omega_{TO}^2}{1+\frac{1}{\varepsilon_\infty}}\right)^{1/2} \tag{18}$$

as $ck\to\infty$, and lies in the frequency region where the dielectric function is negative and the crystal is highly reflecting (the reststrahlen region).

5. Guided Waves

Let us consider a slab of medium a bounded by the planes x=d/2 and x=-d/2 and surrounded on both sides by medium b. We seek solutions to Maxwell's equations that are oscillatory in medium a, but that decay exponentially in medium b going away from the slab. Two solutions, which are, respectively, symmetric and antisymmetric with respect to the midplane of the slab, can be written down. For symmetric TM modes, the electric fields take the form

$$\vec{E}_a^S(\vec{r},t) = \left(-\frac{ik}{q}(\sin qx)E_{az}^{So}, 0, (\cos qx)E_{az}^{So}\right)e^{i(kz-\omega t)} \tag{19}$$

$$\vec{E}_b^S(\vec{r},t) = \left(\mp\frac{ik}{\alpha_b}E_{bz}^{So}, 0, E_{bz}^{So}\right)e^{\pm\alpha_b x}e^{i(kz-\omega t)} \tag{20}$$

where the lower and upper signs refer to x>d/2 and x<-d/2, respectively. For antisymmetric TM modes, the electric fields take the form

$$\vec{E}_a^A(\vec{r},t) = \left(-\frac{k}{q}(\cos qx)E_{az}^{Ao}, 0, i(\sin qx)E_{az}^{Ao}\right)e^{i(kz-\omega t)} \tag{21}$$

$$\vec{E}_b^A(\vec{r},t) = \pm\left(\mp\frac{ik}{\alpha_b}E_{bz}^{Ao}, 0, E_{bz}^{Ao}\right)e^{\pm\alpha_b x}e^{i(kz-\omega t)} \quad . \tag{22}$$

The paramater q satisfies the equation

$$k^2 + q^2 = (\omega^2/c^2)\varepsilon_a(\omega) \quad . \tag{23}$$

By imposing the boundary conditions at $x=\pm d/2$, one can obtain the dispersion relations for the guided waves. They are

$$\varepsilon_a(\omega)/\varepsilon_b(\omega) = -\frac{q}{\alpha_b}\cot(qd/2) \quad \text{(symmetric)} \tag{24}$$

$$\varepsilon_a(\omega)/\varepsilon_b(\omega) = \frac{q}{\alpha_b}\tan(qd/2) \quad \text{(antisymmetric)} \quad . \tag{25}$$

It is clear from (23) that $\varepsilon_a(\omega)$ must be positive. Solving (23) and (24) or (25) for given k typically yields a finite number of values of q and ω corresponding to guided waves that are localized in the slab [1]. One also finds that TE guided modes can exist in the slab. These modes are characterized by the fact that only E_y is nonzero. The dispersion relation is [2]:

$$\frac{2\alpha_b q}{q^2-\alpha_b^2} = \tan qd \quad . \tag{26}$$

6. Surface Polaritons in Anisotropic Dielectric Media:

We now turn to systems with a single interface, and consider the case where one of the media is anisotropic. A particularly simple example is that in which the interface is perpendicular to a principal axis of the anisotropic dielectric and the wave vector is parallel to a second principal axis. If $\varepsilon_\perp(\omega)$ and $\varepsilon_\parallel(\omega)$ are the principal values of the dielectric tensor along axes perpendicular to the surface and parallel to the wave vector, respectively, the dispersion relation can be written as [3]

$$\varepsilon_\parallel(\omega)/\varepsilon_b(\omega) = -\alpha_a/\alpha_b \tag{27}$$

where

$$\alpha_a^2 = [\varepsilon_\parallel(\omega)/\varepsilon_\perp(\omega)][k^2 - (\omega^2/c^2)\varepsilon_\perp(\omega)] \tag{28}$$

and α_b is given by (7). Medium b is assumed to be isotropic. Equations (27) and (28) yield surface polaritons similar to those found when medium a is isotropic, provided that $\varepsilon_\parallel(\omega)<0$ and $\varepsilon_\perp(\omega)<0$. However, it is possible to have a second surface polariton branch, called "type II," when $\varepsilon_\parallel(\omega)<0$ and $\varepsilon_\perp(\omega)>0$, provided that k lies in the range

$$(\omega/c)\varepsilon_b^{1/2} < k < (\omega/c)[\varepsilon_\perp(\omega)]^{1/2} \quad . \tag{29}$$

References

1. A. A. Maradudin, R. F. Wallis, and L. Dobrzynski: Handbook of Surfaces and Interfaces, Vol. 3 (Garland Publishing, New York, 1980)

2. D. Marcuse: Theory of Dielectric Optical Waveguides (Academic, New York 1974)

3. V. V. Bryksin, D. N. Mirlin and I. I. Reshina: Zh. Eksp. Teor. Fiz. Pis. 16, 445 (1972); Sov. Phys. JETP Lett. 16, 315 (1972)

Surface Plasmon (SEW) Phenomena

F. Abeles

Laboratoire d'Optique des Solides, Université Pierre et Marie Curie, F-75252 Paris Cedex O5, France

1. Mathematical description of SEW

As it can be shown that SEW are always p-polarized, we consider only this type of wave. Let ε_0 and ε_s be the dielectric functions of the two media in contact, and the z-direction be normal to the interface (plane z = 0). The sagittal plane is the plane xoz . From MAXWELL's equations, it follows that the component of the wave-vector along the interface (k_z) is the same in both media. Let $q_i=(k_x^2-\varepsilon_i\omega^2/c^2)^{1/2}$ be the component of the wave-vector along the normal to the interface (i=0,s). From MAXWELL's equations, it follows that if

$$\vec{H}_i=\frac{\omega}{c}E_i(0,1,0)e^{\pm q_i z}e^{i(k_x x-\omega t)} \quad \text{then}$$

$$\vec{E}_i=E_i(\mp iq_i,0,k_x)e^{\pm q_i z}e^{i(k_x x-\omega t)}$$

(upper sign for z<0, lower sign for z>0)

and the boundary conditions at z=0 (continuity of the x-component of E and of the y-component of H) lead to

$$-q_0E_0=q_sE_s$$

$$\varepsilon_0E_0=\varepsilon_sE_s$$

thus

$$\frac{\varepsilon_0}{\varepsilon_s}=-\frac{q_0}{q_s}. \qquad (1)$$

q_0 and q_s being positive quantities, it follows that $\varepsilon_0\,\varepsilon_s<0$, thus ε_0 and ε_s must be of opposite signs. The medium with negative dielectric function is usually called "active medium". This expresses the fact that the coupling between the EM field and the dipoles giving rise to the SEW occurs in this medium. The surface charge σ_s oscillates like (ε_0- ε_s) exp ($ik_x.x$).

In optics, one usually uses the optical admittance of the medium corresponding to the propagating wave, namely $Z_i=\varepsilon_i/(c/\omega)q_i$ and eq. (1) then becomes

$$Z_0+Z_s=0. \qquad (2)$$

If we replace k_x by a normalized quantity (without dimensions, a pure number) S via $k_x=(\omega/c)S$, then eqs. (1) or (2) lead to

$$S^2 = \frac{\varepsilon_0 \varepsilon_s}{\varepsilon_0 + \varepsilon_s} \tag{3}$$

or, if we keep k_x,

$$k_x^2 = \left(\frac{\omega}{c}\right)^2 \frac{\varepsilon_0 \varepsilon_s}{\varepsilon_0 + \varepsilon_s}. \tag{4}$$

Eqs. (3) or (4) are the dispersion relations for SEW. We found that $\varepsilon_0 \varepsilon_s<0$; it follows that $\varepsilon_0+\varepsilon_s<0$. If $\varepsilon_s<0$ (and $\varepsilon_0>0$), then $|\varepsilon_s|>\varepsilon_0$. For $\varepsilon_0+\varepsilon_s=0$, k_x (or S) become infinite.

Having found the necessary condition for existence of SEW, we must answer a few questions : (a) are there active media and, if there are such media, (b) what is the propagation length along the surface and (c) how do the fields decay away from the surface (penetration depths of the fields in the two media)?

(a) Even if $\varepsilon_0>0$ is real, ε_s is usually a complex quantity : $\varepsilon_s=\varepsilon_{sr}+i\varepsilon_{si}$, i.e. energy can be dissipated in the active medium. The ideal situation would be $\varepsilon_{sr}<0$ and $\varepsilon_{si}=0$. SEW do exist even when $\varepsilon_{si}>0$, provided it is relatively small, i.e. $\varepsilon_{si}\ll|\varepsilon_{sr}|$, so that they are not too much damped. There are essentially two kinds of materials fulfilling this condition :

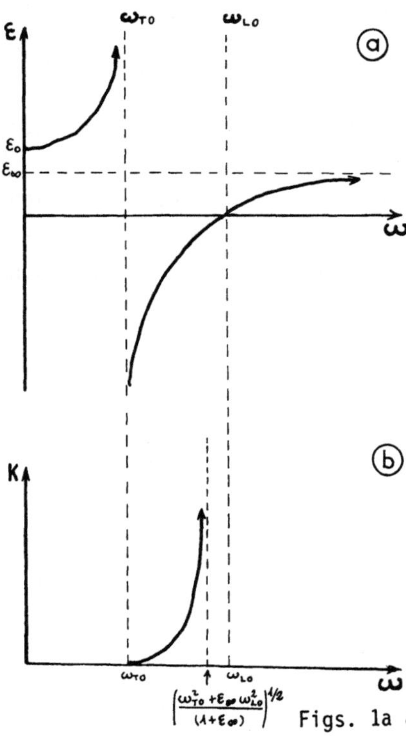

Figs. 1a and 1b

(1) Ionic crystals in the infrared, between roughly 10 and 100 μm, with a dielectric function which (neglecting absorption) can be written as

$$\varepsilon(\omega)=\varepsilon_\infty \frac{\omega_L^2-\omega^2}{\omega_T^2-\omega^2},$$

ω_L and ω_T being the frequencies corresponding to the excitation of longitudinal and transverse optical phonons respectively and ε_∞ the value of ε for $\omega\gg\omega_L$. Moreover, $\omega_L>\omega_T$. In the spectral region between ω_T and ω_L, $\varepsilon<0$ and these materials can be active media for SEW. If $\varepsilon_0=1$, k_x becomes infinitely large for

$$\omega=\left(\frac{\omega_T^2+\varepsilon_\infty\omega_L^2}{1+\varepsilon_\infty}\right)^{1/2}.$$

Figs. 1a and 1b represent $\mathcal{E}(\omega)$ and $k_x(\omega)$ respectively.

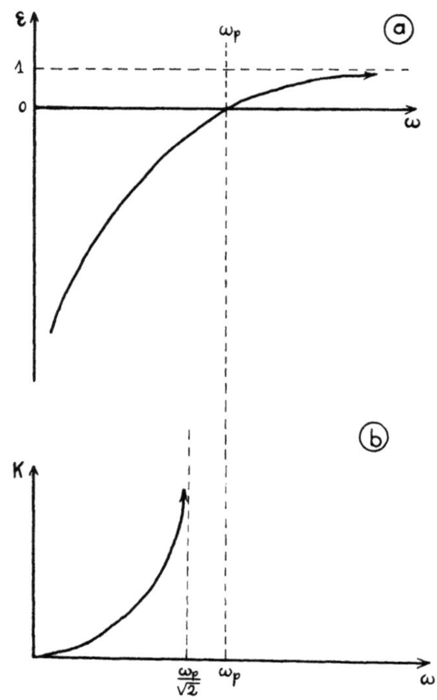

Figs. 2a and 2b

(2) Free electron metals over an extended spectral region (far infrared to ultraviolet, where $\mathcal{E}(\omega)$=-1). In the IR, in fact for all the metals, the photon energy is too small to allow interband transitions, and the dielectric function is practically that of a free electron gas

(neglecting absorption)

$$\mathcal{E}(\omega) = 1 - \left(\frac{\omega_p}{\omega}\right)^2$$

where ω_p is the plasma frequency :

$$\omega_p^2 = \frac{4\pi N e^2}{m}$$

with N = electron density, e = electron charge, m = electron (effective) mass. For $k_x \to \infty$, the SEW (surface plasmons) frequency reaches the value $\omega_s = \omega_p/(1+\mathcal{E}_0)^{1/2}$.

Figs. 2a and 2b represent $\mathcal{E}(\omega)$ and $k_x(\omega)$ respectively.

In fact, other materials, such as some semiconductors, can also be "active media" in a given spectral range.

(b) If $S = S_r + iS_i$ is a solution of the eq. (3), the propagation length L along the surface is given by $L = \lambda/4\pi S_i$. L is defined as being the distance along the surface corresponding to a 1/e reduction of the intensity of the SEW. If we have $\mathcal{E}_0$ real and $|\mathcal{E}_{sr}| \gg \mathcal{E}_{si}$, then

$$S_i = \frac{1}{2S_r} \frac{\mathcal{E}_0^2 \mathcal{E}_{si}}{(\mathcal{E}_0 + \mathcal{E}_{sr})^2 + \mathcal{E}_{si}^2}$$

and

$$S_r^2 = \frac{\mathcal{E}_0(\mathcal{E}_{sr}^2 + \mathcal{E}_{si}^2 + \mathcal{E}_0\mathcal{E}_{sr})}{(\mathcal{E}_0 + \mathcal{E}_{sr})^2 + \mathcal{E}_{si}^2}.$$

In practice, two different situations can occur :

(1) metals and ionic crystals in the infrared : $|\mathcal{E}_{sr}| \gg \mathcal{E}_{si} \gg \mathcal{E}_0$ (taking now absorption in the active medium into account) :

$$S_r = \sqrt{\mathcal{E}_0}, \qquad S_i = \frac{\mathcal{E}_0^{3/2}\mathcal{E}_{si}}{2\mathcal{E}_{sr}^2},$$

and L increases with λ^2. For highly conducting metals (Ag, Au, Cu, Al) it is the order of a few centimeters at $\lambda = 10\ \mu$ m.

(2) free electron gas with small damping in the visible region of the spectrum : $|\varepsilon_{sr}| \gg \varepsilon_{si}$, $|\varepsilon_{sr}| \gg \varepsilon_0$, $\varepsilon_{si} \ll \varepsilon_0$. Here

$$s_r^2 = \frac{\varepsilon_0 \varepsilon_{sr}}{\varepsilon_0 + \varepsilon_{sr}}, \qquad s_i = \frac{1}{2 s_r} \frac{\varepsilon_0 \varepsilon_{si}}{(\varepsilon_0 + \varepsilon_{sr})^2}.$$

L is now much smaller, of the order of μm.

(c) Let us call δ_i the penetration depth of the field in the medium (i=0 or s). It is defined as the distance in the Z direction from the surface over which the electromagnetic energy is divided by e. It is easy to see that $\delta_0/\delta_s \simeq \varepsilon_{sr}/\varepsilon_0$. δ is much smaller in the active medium than in the ε_0 medium. For a metal, which can be represented by a free electron gas for $\omega < \omega_p$ in contact with vacuum (ε_0=1), we have $\delta_s \simeq \lambda_p/4\pi$ (with $\lambda_p = 2\pi c/\omega_p$). This shows that the penetration depth in the metal is practically frequency independent and is essentially determined by the free electrons density. For Al ($\lambda_p \simeq 800$ A) and $\delta_s \simeq$ 60 A. In contradistinction, $\delta_0 \simeq \lambda^2/4\pi\lambda_p$ is strongly wavelength dependent and can be quite large in the infrared.

2. Optical excitation of SEW

SEW are non-radiative waves which cannot be excited directly by light waves propagating in one of the adjacent media. This can be easily understood if we notice that their phase velocity is v= $\omega/\mathrm{Re}(k_x)$, which, according to eq. (3) shows that

$$v = \frac{c}{\sqrt{\varepsilon_0}} \cdot \mathrm{Re}\sqrt{\frac{\varepsilon_0 + \varepsilon_s}{\varepsilon_s}}$$

Thus v is smaller than the phase velocity of light in the medium ε_0, namely $c/\sqrt{\varepsilon_0}$. If k_0 is the component of the wave vector along the surface for light propagating in this medium under incidence θ, then $k_0 = (\omega/c)\sqrt{\varepsilon_0}\sin\theta$ and $k_0 < k_x$. It is thus clear that SEW require some special arrangement for their excitation. We consider this problem now.

1) If the surface is sinusoidally modulated with period d (the grating lines being parallel to the y direction), FANO [1] has shown that, if the modulation amplitude is small with respect to the wavelength of light, the effect of the modulation is the modification of k_0 by addition of terms $2\pi n/d$ (n=0, 1, 2...), i.e. we have now instead

$$k_n = k_0 + \frac{2\pi n}{d}.$$

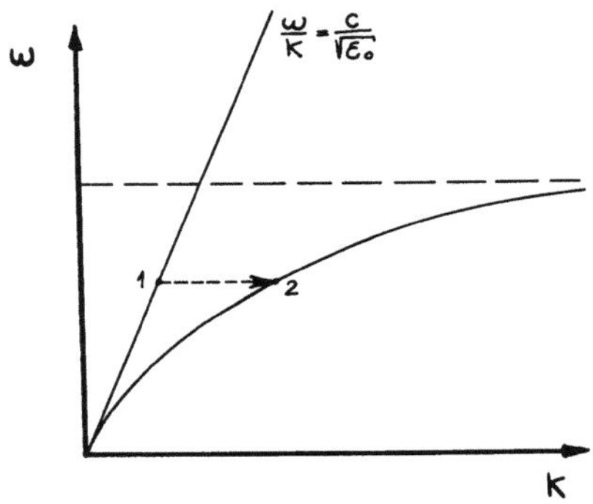

Fig. 3

If, for a given value of d, an integer n exists such that $k_n=(\omega/c)S_n$, SEW can be excited. Fano used his considerations to explain Wood's anomalies for gratings, which were well known by the experimentalists. Fig.3 shows schematically this process. Theory and experiment have shown that the modulation depth h of the grating plays a significant part, the dispersion curve being a function of both d and h. For further reading we refer to the review papers by RAETHER [2] and MARADUDIN [3].

(2) A randomly rough surface, when Fourier analyzed, can be described by a set of wavevectors k_α (and amplitudes A_α), each of them being able to give rise to other wavevectors according to $k_{n\alpha}=k_0+nk_\alpha$. This shows that SEW can always be excited at a rough surface. We refer again to the review papers already mentioned by RAETHER [2] and MARADUDIN [3].

3) For smooth plane surfaces, one uses methods based on attenuated total reflection (ATR). Evanescent waves with the k_x corresponding to the real part of the solution of eq. (4), i.e. with $k_x > k_0=(\omega/c)\,\varepsilon_0^{1/2}$, can be generated by the use of a total reflection prism with dielectric constant ε_p working at an angle of incidence θ_p larger than the critical angle $\theta_c=\sin^{-1}(\varepsilon_p/\varepsilon_0)^{1/2}$. The wave propagating in the ε_0 medium being evanescent (the field amplitude decaying exponentially) the surface $\varepsilon_0/\varepsilon_s$ of interest should not be too far from the prism surface (at a distance of the order of the wavelength of light in the ε_0 medium). This method of generating SEW was first suggested by OTTO [4]. Fig. 4 shows this configuration.

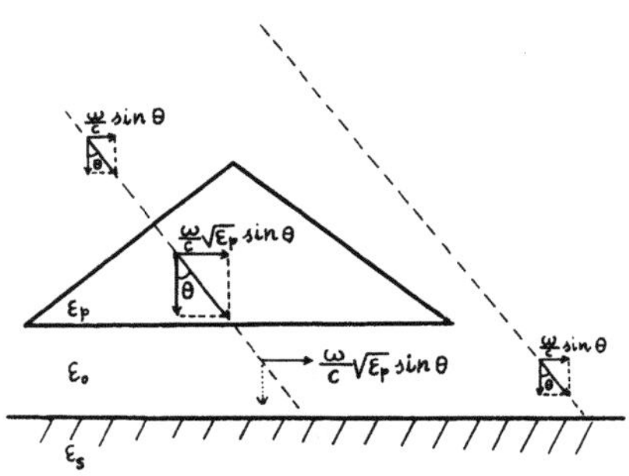

Fig. 4 Principle of the optical excitation of SEW by attenuated total reflection in the Otto configuration.

Eqs. (3) and (4) show that ε_0 and ε_s play symmetrical roles, as could be expected. Therefore, it is possible to have the ε_s medium

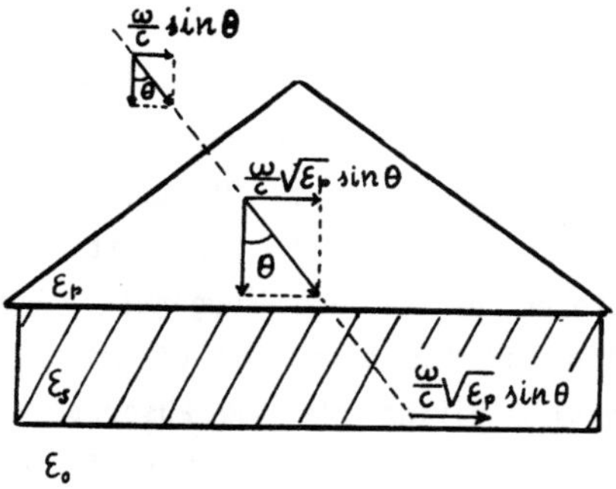

Fig. 5 Principle of the optical excitation of SEW by attenuated total reflection in the Kretschmann configuration.

in contact with the prism. Being absorbing, the thickness of this medium should be definitely smaller than the wavelength of light in vacuum. For metals, in the visible region of the spectrum, it should be smaller than $\lambda/10$ approximately. This indicates that the active medium is now a thin film deposited on the prism surface. The method has been first suggested by KRETSCHMANN [5] and this configuration is depicted in fig. 5.

3. Detection and observation of SEW

It can be performed by various techniques and we will mention now some of them.

1) The most immediate is the detection of the reflected light in a ATR configuration (Otto or Kretschmann). The incident light being p-polarized, one measures the reflectance R_p either at fixed frequency and variable incidence or at fixed incidence and varying wavelength. For a perfect coupling, one would have $R_p=0$, the incident energy being totally transferred to the SEW. With no radiative de-excitation (via surface roughness, for instance), the transmitted energy is null and the optical absorption is $A_p = 1-R_p$. In practice, there is never a complete transfer to the SEW (except in very special occasions) and A_p reaches a maximum. This will be discussed more in detail below.

A more sophisticated method of detection consists in measuring both the amplitude and the phase of the reflected wave. This is performed in an ellipsometric measurement using the same configuration, which provides the two quantities $\Psi = \tan^{-1}(R_p/R_s)^{1/2}$ and $\Delta = \delta_p - \delta_s$. The subscript s refers to s-polarized light and the δ are the phase shifts at reflection. R_s and δ_s being practically constant when either θ or ω vary in the region of SEW excitation, $\tan\Psi$ is proportional to $R_p^{1/2}$ and Δ varies like δ_p. The simultaneous measurement of the amplitude ($R_p^{1/2}$) and phase (δ_p) is an advantage of the ellipsometry as compared to photometry. This will also be discussed more in detail below.

2) Another method is based on the detection of the light emitted because of the scattering of SEW by surface roughness in a ATR experiment. We have already noticed that surface roughness can concur to the excitation of SEW. Reciprocally, SEW can be scattered by surface roughness. Two processes are possible : a) a modification of the direction of the SEW wave vector with conservation of its modulus giving rise to absorption; b) a modulus change, bringing it in the radiative region. In this latter case, there will be light emission, which can be detected and measured. Scattering of SEW can also occur due to volume inhomogeneities in the active medium or to dielectric function fluctuations. All the real surfaces present some "roughness" (either superficial or bulk) and are therefore more or less scattering the SEW. The emitted (scattered) intensity is $I(\theta)=|A(\theta)|^2 X$, where θ is the emission angle, A the intensity of the SEW field at the surface and X is a (rather complicated) function describing the surface roughness. In the vicinity of S given by eq. (3), $A(\theta)\propto[S(\theta)-S]^{-1}$, where $S(\theta)=\varepsilon_p^{1/2}\sin\theta$ and $|A(\theta)|^2=|A[S(\theta)]|^2$ is a lorentzian. The angular position of the maximum of the emitted light, θ_m, is related to the real part of S via $\varepsilon_p^{1/2}\sin\theta_m=\mathrm{Re}\,S$ and the width $\Delta\theta$ of the emission curve is related to the imaginary part of S via $\varepsilon_p^{1/2}\cos\theta_m\Delta\theta=\mathrm{Im}\,S$ (WEBER et al. [6]). For a more thorough discussion, we refer again to the review article by RAETHER [2].

3) SEW can be detected from the inspection of the surface photoelectric yield (BOSENBERG [7], MACEK et al. [8], HINCELIN and SEPTIER [9]). For clean surfaces, the method is limited to energies greater than the work function. The SEW de-excited in emitted photoelectrons are detected by this method : a maximum of the yield Y is observed when the excitation conditions are satisfied. The shape of the resonance curve $Y(\theta)$ is similar to $I(\theta)$ discussed above. It is related to the intensity of the electromagnetic field of the SEW in a region corresponding to the extraction depth for photoelectrons.

4) Last but not least we indicate the method of observation based on the property of SEW to propagate along the surface. This is very convenient especially at long wavelengths, i.e. in the infrared. SEW are excited by any of the methods used in integrated optics (half-prisms, gratings, end-fire coupling). It has been first demonstrated by SCHOENWALD

et al. [10] and used intensively by Sievers and coworkers (SIEVERS and SCHLESINGER [11]) and by Zhizin and coworkers (ZHIZHIN et al. [12]). This method has proved to be extremely effective for the investigation of adsorbed molecules on metals.

4. Excitation of SEW by attenuated total reflection (ATR)

We consider now in detail the problem in both the Otto and the Kretschmann configurations described in the preceding chapter. We will write S_m for the solution of eq.(3). In the Otto configuration, we have the ε_0 film embedded between the semi-infinite media ε_p and ε, whereas in the Kretschmann configuration the ε medium is embedded between the semi-infinite media ε_p and ε_0. Let us call r_1 and r_2 the amplitude reflection coefficients (Fresnel) at the first and second surface of the film. Remember that we are working with a p-polarized wave. If $S=\varepsilon_p^{1/2}\sin\theta$, θ being the angle of incidence in the first semi-infinite medium, then, in the vicinity of S_m, it can be shown that

$$r_2=\frac{A}{S-S_m} \qquad \text{with} \qquad A=\frac{2S_m^3}{\varepsilon_0-\varepsilon} \tag{5}$$

in the Krestschmann configuration, with a change of sign in A for the Otto configuration. Eq.(5) shows that $|r_2|\gg 1$ in this angular region. We assume, of course, a fixed frequency. The reflected amplitude r is given by

$$r=r_1\,\frac{1+(r_2/r_1)\,e^{i\beta}}{1+r_1r_2\,e^{i\beta}} \tag{6}$$

where r_1 and r_2 are the (complex) reflection coefficients (Fresnel) at the first and second interface and $\beta=2(\omega/c)\sqrt{\varepsilon-S^2}$ or $\beta=2(\omega/c)\sqrt{\varepsilon_0-S^2}$ according to the chosen (Kretschmann or Otto respectively) configuration, d being the thickness of the interval between the prism and the surface supporting SEW. Moreover, if $r_1=\rho_1\exp(i\delta_1)$, then $\rho_1\simeq 1$. For the Otto configuration, $\rho_1=1$. β is practically purely imaginary : $\beta\simeq i\alpha$ for both configurations (it is purely imaginary for the Otto configuration). Taking these remarks into account, eq. (6) becomes

$$r\simeq r_1\,\frac{S-S_m+(A/\rho_1)\,e^{-\alpha}\,e^{-i\delta_1}}{S-S_m+A\rho_1\,e^{-\alpha}\,e^{i\delta_1}} \tag{7}$$

or

$$r\simeq r_1\,\frac{S-S_m-\Delta S_n}{S-S_m-\Delta S_d} \tag{8}$$

with

$$\Delta S_n=-\frac{A}{\rho_1}\,e^{i(\beta-\delta_1)}, \qquad \Delta S_d=-A\rho_1 e^{i(\beta+\delta_1)},$$

both being small quantities for the values of d which are of interest to us. With the notations $S_m = S_{mr} + iS_{mi}$, $\Delta S_n = \Delta S_{nr} + i\Delta S_{ni}$, $\Delta S_d = \Delta S_{dr} + i\Delta S_{di}$, the reflectance $R_p = |r|^2$ for both configurations becomes

$$R_p = |r|^2 = \rho_1^2 \frac{(S-S_{mr}-\Delta S_{nr})^2+(S_{mi}+\Delta S_{ni})^2}{(S-S_{mr}-\Delta S_{mr})^2+(S_{mi}+\Delta S_{di})^2}, \quad (9)$$

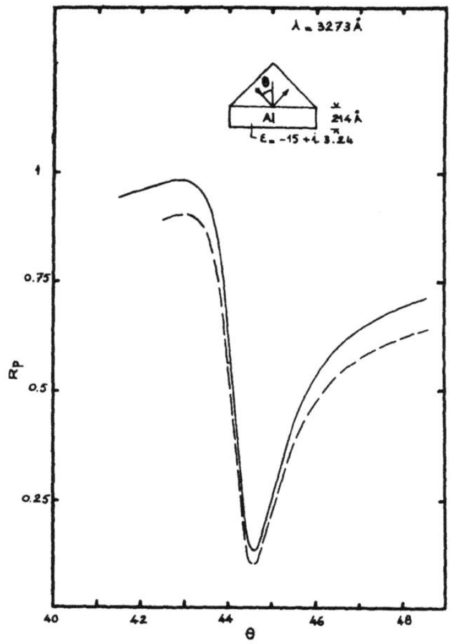

Fig. 6 R_p vs. θ for an Al film (d = 214 Å, ε = -15+i.3.24) deposited on a prism ($\varepsilon_p = 1.48^2$) and in contact with air, ($\varepsilon_0 = 1$) for λ = 3270 A.
Full line : computed with the exact expression, dashed line : computed with the approximate equation (9).

In order to test the validity of this formula, Fig. 6 shows the R_p vs. θ curves computed with the exact and the approximate expressions for an Al film (d = 214 A, $\varepsilon_p^{1/2}$ = 1.48, ε = -15 + i 3.24, ε_0 = 1, λ = 3270 A). Notice the asymmetry of the $R_p(S)$ curves given by eq.(9) reflecting the fact that ΔS_n and ΔS_d are different.

For some metals, like Ag, in the visible $|\varepsilon_r| \gg \varepsilon_i$ and $\rho_1 \simeq 1$, $\beta \simeq i\alpha$. Then $\Delta S_d = \Delta S_n^* = \Delta S_r + i\Delta S_i$, this situation occurring always in the Otto configuration. Eq.(9) becomes now

$$R_p = \frac{(S-S_{mr}-\Delta S_r)^2+(S_{mi}-\Delta S_i)^2}{(S-S_{mr}-\Delta S_r)^2+(S_{mi}+\Delta S_i)^2}. \quad (10)$$

Some remarks are in order.

a) Eqs. (9) and (10) are only approximate and valid in the vicinity of $S = S_m$. They should not be used beyond this region.

b) From eq. (10), the minimum of R_p is

$$R_{p,min} = \left(\frac{S_{mi}-\Delta S_i}{S_{mi}+\Delta S_i}\right)^2 \quad (11)$$

and it occurs for $S = S_{mr} + \Delta S_r$, i.e. it is shifted from the exact value corresponding to SEW excitation, which is $S = S_{mr}$. The shift is small, because $\Delta S_r \simeq 2 S_{mr}^3 e^{-\alpha} \cos\delta_p / (\varepsilon_0 + |\varepsilon_r|)$. It is due to the influence of the coupling prism. In principle, ΔS_r could be made very small by increasing the distance d between the coupling prism and the $\varepsilon_0/\varepsilon$ surface

($e^{-\alpha} \rightarrow 0$ then), but in that case ΔS_i decreases as $e^{-\alpha}$ too and $R_{p,min}$ approaches the value 1, which means that the efficiency of the coupling to SEW waves is negligible and becomes undetectable. This is easy to understand when we notice that the amplitude of the EM wave at the $\varepsilon_0/\varepsilon$ surface is exceedingly small.

c) We will write $d = d_m$ for the value of d for which $R_{p,min} = 0$. Then $A_p = 1 - R_{p,min} = 1$, i.e. all the energy carried by the incident EM wave has been absorbed and converted into a SEW. The "coupling efficiency" is now maximum and equal to unity. The two dampings, namely the "surface" (via S_i) and the "volume" (via ΔS_i) ones are then equal. For under- or over-damping $R_{p,min}>0$ and, in fact, by a mere inspection of the R_p vs. S curve, one cannot guess which is the actual situation. Nevertheless a curve fitting procedure with the exact formula for R_p vs. S can provide both ε and d (i.e. $\varepsilon_r, \varepsilon_i$ and d) assuming that ε_p and ε_0 are known. It is clear that S and Δ S can be varied either by keeping ω = const and changing θ or by keeping θ = const and varying ω (i.e. the wavelength of the incident light).

d) Last, but not least, do not forget that we have assumed plane waves in all the discussion, i.e. light beams with cross-section much larger than the squares of the other lengths entering our calculations and of constant intensity (not narrow gaussian beams, for instance).

5. Ellipsometry

Let the complex reflected amplitudes be

$$r_s = \sqrt{R_s}\, e^{i\delta_s}, \qquad r_p = \sqrt{R_p}\, e^{i\delta_p}$$

In the photometric measurements discussed above, one was measuring R_p. In an ellipsometric experiment, one measures (as indicated above):

$$\tan\Psi = \sqrt{\frac{R_p}{R_s}} \qquad \text{and} \qquad \Delta = \delta_p - \delta_s .$$

The point to be noticed is that R_s and δ_s are slowly varying functions of S in the region of resonance (maximum of $A_p = 1-R_p$). Therefore one can write with a good approximation

$$\tan\Psi \propto \text{const.}\sqrt{R_p}, \qquad \Delta \simeq \delta_p - \text{const.}$$

Thus ellipsometry provides information about both the amplitude and the phase of the reflected p-polarized light in the vicinity of the region where SEW are excited.

Eqs. (9) or (10) give practically $\tan^2\Psi$, because $R_s \simeq 1$. In order to have an estimate for Δ, the simplest is to assume that $\Delta S_n = \Delta S_d^*$

$= \Delta S_r + i\Delta S_i$ and to use eq. (8). From it, after some elementary manipulations, one obtains

$$\tan(\Delta - \Delta_m) = \frac{2\Delta S_i (S - S_{mr} - \Delta S_r)}{(S - S_{mr} - \Delta S_r)^2 + S_{mi}^2 - \Delta S_i^2} \qquad (12)$$

where, clearly, Δ_m is the value of Δ at the resonance, i.e. for $S = S_{mr} + \Delta S_r$.

The interesting point to notice now is that the Δ variations with S are completely different, according to whether there is (in our terminology) under-or overdamping.

a) $d>d_m$. The "volume" damping is larger than the SEW damping, i.e. $S_i > S_i$. It is then easy to see from eq. (12) that Δ reaches a maximum (Δ_{max}) and a minimum (Δ_{min}) in the vicinity of the resonance. One finds that

$$\tan(\Delta_{max} - \Delta_m) = \tan(\Delta_m - \Delta_{min}) = \tan\frac{\Delta_{max} - \Delta_{min}}{2} = \frac{\Delta S_i}{\sqrt{S_{mi}^2 - \Delta S_i^2}} \qquad (13)$$

the positions of the maximum and of the minimum being given by

$$S_{1,2} = S_{mr} + \Delta S_r \mp \sqrt{S_{mi}^2 - \Delta S_i^2} \qquad (14)$$

Eqs. (13) and (14) show that knowing the positions and the values of Δ_{max} and Δ_{min} one can obtain the essential information about the active medium and its distance from the prism via

$$\frac{S_1 + S_2}{2} = S_{mr} + \Delta S_r, \qquad \frac{S_1 - S_2}{2} = \sqrt{S_{mi}^2 - \Delta S_i^2}$$

and

$$\frac{S_2 - S_1}{2} \tan\frac{\Delta_{max} - \Delta_{min}}{2} = \Delta S_i .$$

Measurements of $\tan\Psi$ simply serve as a check of the quantities obtained from the use of the Δ vs. S (or θ) values.

b) $d < d_m$. The "volume" damping is now less than the SEW damping. Eq. (12) shows that Δ has a monotonous variation without any maximum or minimum and covering an angular range larger than 180°. In the vicinity of the resonance, $\Delta = \Delta_m \pm 90°$ for

$$S_{1,2} = S_{mr} + \Delta S_r \mp \sqrt{\Delta S_i^2 - S_{mi}^2} \qquad (15)$$

and the above considerations concerning the essential information about the active medium and its distance from the coupling prism are again valid, with obvious replacement of Δ_{max} and Δ_{min} by $\Delta_m \pm 90°$.

Figures 7 and 8 show Δ vs. λ and $\tan\Psi$ vs. λ for Ag films of different thicknesses d in the Kretschmann configuration with $\theta = 33.3°$ on a prism with $\varepsilon_p^{1/2} = 1.89$. The Ag dielectric function is given by

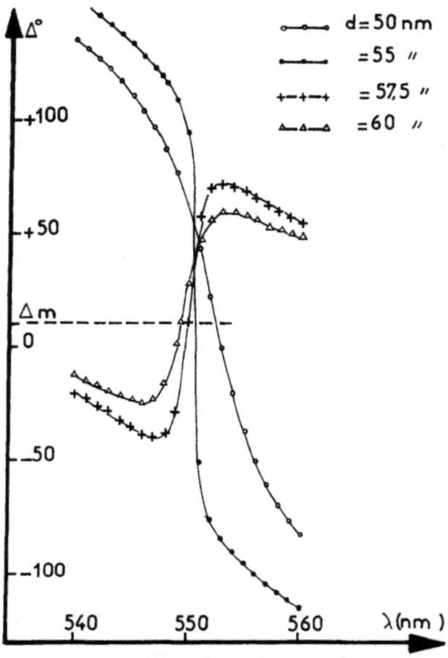

Fig. 7 (degrees)vs. (nm) for Ag films (thickness d) in the Kretschmann configuration (after 13). See text for data used in the computations.

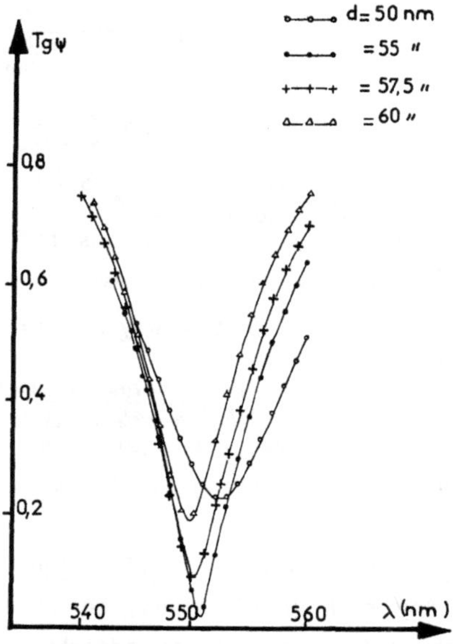

Fig. 8 tan vs. (nm) for Ag films (thickness d) in the Kretschmann configuration (after 13). See text for data used in the computations.

$\varepsilon = 4-(\lambda/\lambda_p)^2[1+i(\lambda/\lambda_\tau)]^{-1}$ with λ_p= 129 nm and λ_τ= 3×10^4 nm. It is clearly apparent from the Δ vs. λ curves that 55 nm < d_m < 57.5 nm.

Additional remarks. Although all the preceding mathematical considerations were in terms of $S=\sqrt{\varepsilon_p}\sin\theta$ as a variable, figs. 8 and 9 considered a fixed θ and variable λ (or $\omega=2\pi c/\lambda$). Eqs. (7) and (8) can be rewritten in r vs. ω notation, but using now a complex ω . In the same way, we saw that $S_m = S_{mr} + i\, S_{mi}$ and S_{mi} was related to the angular broadening (damping) of the resonance, we will find now that if $\omega=\omega_m$ is a solution of the dispersion relation

$$Z_o(\omega_m)+Z(\omega_m)=0,$$

this solution is a complex quantity, the imaginary part of which gives the frequency broadening of the resonance (or its finite life-time in the time language).

If we assume that $\partial Z_o/\partial\omega=0$, which is usually a very good approximation, then the eq. corresponding to the previous eq. (5) is now

$$r_2 \simeq \frac{4\varepsilon}{(\varepsilon_0-\varepsilon)\,\frac{\partial\varepsilon}{\partial\omega}\,(\omega-\omega_m)} \qquad (5')$$

and

$$r \simeq r_1 \frac{\omega-\omega_m-\Delta\omega_n}{\omega-\omega_m-\Delta\omega_d} \qquad (8')$$

with

$$\Delta\omega_n = \frac{4\varepsilon^2}{(\varepsilon_0-\varepsilon)(\partial\varepsilon/\partial\omega)}\frac{1}{\rho_1}\,e^{i(\beta-\delta_1)}, \qquad \Delta\omega_d = \Delta\omega_n\cdot\rho_1^2\cdot e^{i2\delta_1}$$

both being small quantities as before. Again when $|\varepsilon_r| \gg \varepsilon_i$, $\rho_1 \simeq 1$ and $\beta \simeq i\alpha$ (with α real), we have

$$\Delta\omega_n = \Delta\omega_d^* = \Delta\omega = \frac{4\varepsilon^2}{(\varepsilon_0-\varepsilon)(\partial\varepsilon/\partial\omega)}\,e^{-\alpha}e^{-i\delta_1}$$

and R_p becomes

$$R_p = \frac{(\omega-\omega_{mr}-\Delta\omega_r)^2+(\omega_{mi}-\Delta\omega_i)^2}{(\omega-\omega_{mr}-\Delta\omega_r)^2+(\omega_{mi}+\Delta\omega_i)^2} \qquad (10')$$

The R_p vs. ω curves are again symmetrical for ω in the vicinity of the resonance corresponding to $\omega = \omega_{mr}+\Delta\omega_r$.

6. Influence of slight modifications of the $\varepsilon_0/\varepsilon$ surface

Any such modification, be it due to a change of the charge density in the active medium, an adsorbed layer, surface roughness, etc., could be phenomenologically described by the addition of small terms both in the numerator and in the denominator of r_2, which becomes now

$$r_2 = \frac{Z-Z_0+a}{Z+Z_0+b}$$

with $|a|$ and $|b|$ much smaller than unity. Working in the region where $Z + Z_0 = 0$, it becomes clear that the effect of the b term will be quite important, whereas a will be negligible with respect to $Z - Z_0 \simeq 2Z$. This is the mathematical reason for the great sensitivity of SEW to surface modifications. Physically, one can understand this property of SEW from the fact that they are bound to the surface and travelling along it.

Let us now be more specific, and assume the presence of a very thin layer characterized by its dielectric function ε_f and thickness d_f, between ε and ε_0. It can be shown, by using the first order approximation for thin films, i.e. assuming $2\pi d_f/\lambda \ll 1$ and neglecting higher order terms, that

$$\begin{matrix} a \\ b \end{matrix} = -i\frac{\omega}{c}d_f\left[ZZ_0\left(1-\frac{S^2}{\varepsilon_f}\right) \mp \varepsilon_f\right] \qquad (16)$$

at least for a homogeneous layer. If the surface layer is inhomogeneous (selvedge region) and is a function of the normal (z) coordinate, the following substitutions should be made in the above expressions :

$$\varepsilon_f \to \bar{\varepsilon}_f = \frac{1}{d_f}\int_0^{d_f}\varepsilon_f(z)\,dz \quad \text{and} \quad \frac{1}{\varepsilon_f} \to \overline{\left(\frac{1}{\varepsilon_f}\right)} = \frac{1}{\bar{\bar{\varepsilon}}_f} = \frac{1}{d_f}\int_0^{d_f}\frac{dz}{\varepsilon_f(z)},$$

showing that an inhomogeneous layer is equivalent to an anisotropic (uniaxial) layer with $\varepsilon_{fx}=\varepsilon_{fy}\neq\varepsilon_{fz}$. More sophisticated treatments of the selvedge region of the active material would lead to the same result except that the estimate of ε_{fx} and ε_{fz} will require the microscopic description of the region.

We will assume again that the approximations leading to eq. (10) are valid in order to obtain the simplest formula displaying the essential features related to the presence of the surface layer (assumed now to be homogeneous) :

$$r = \frac{S - S_m - \Delta S - \Delta S_f}{S - S_m - \Delta S^* - \Delta S_f}$$

with

Notice that eq. (17) is similar to eq. (10), the only difference being that S_m has been replaced by $S_m + \Delta S_f$, showing that ΔS_f is related to the modification Δk_f of the SEW wave-vector due to the superficial layer ($\Delta k_f = (\omega/c)\Delta S_f$). ΔS_f is independent of the coupling and its imaginary part gives the change of the intrinsic damping of the SEW.

For highly reflecting and low absorbing active materials, $\varepsilon \simeq \varepsilon_r < 0$ and ε_i very small, ΔS_f is essentially real for ε_f real (e.g. a non-absorbing oxide layer on top of a metal). Then the only effect of such a surface layer in an ATR photometric or ellipsometric measurement will be a ΔS_{fr} translation of the resonance curve. If, moreover, (and this is often the case) $\varepsilon_f > \varepsilon_o$, then $\Delta S_{fr} > 0$.

Figs. 9 and 10 show computed curves for tan Ψ and Δ vs.S (or θ) for an Ag layer (d = 580 A) in the Kretschmann configuration (λ = 5620 A). Notice that the superficial layer has ε_f = $(n_c)^2$ = 9 and its thicknesses are d_c.

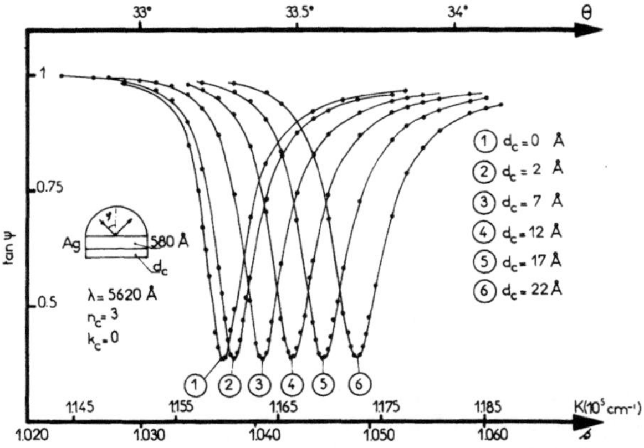

Fig. 9 tan Ψ vs.S (or k or θ , upper horizontal scale) for an Ag layer covered with various thicknesses d_c of a film with ε_c= 9.

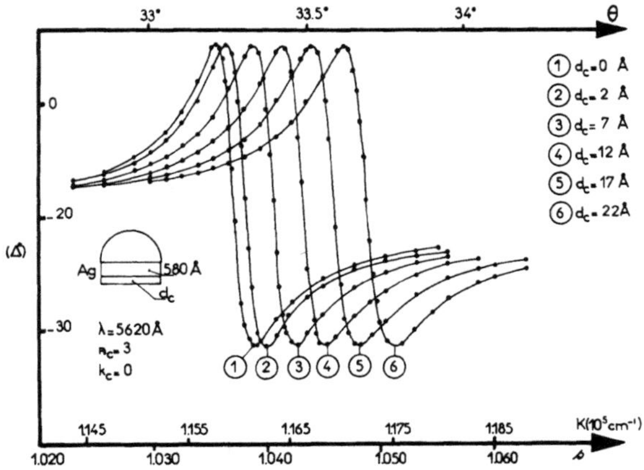

Fig. 10 Δ vs.S (or k or θ , upper horizontal scale) for the same configuration as fig. 10.

Figs. 11 and 12 show experimental results obtained during a SEW investigation of the tarnishing of a silver surface in the Kretschmann configuration. The prism (here a half-cylinder) had $\varepsilon_p = (1.89)^2$ and the experiments were performed at fixed frequency (λ = 5620 A) and variable θ (or S, or k). The tarnish layer was found to have $\varepsilon_f = 2.9 + i.0.4$ and its thicknesses, corresponding to the various curves from left to right, were d_f = 0,2, ..., 30 A. The circles are experimental

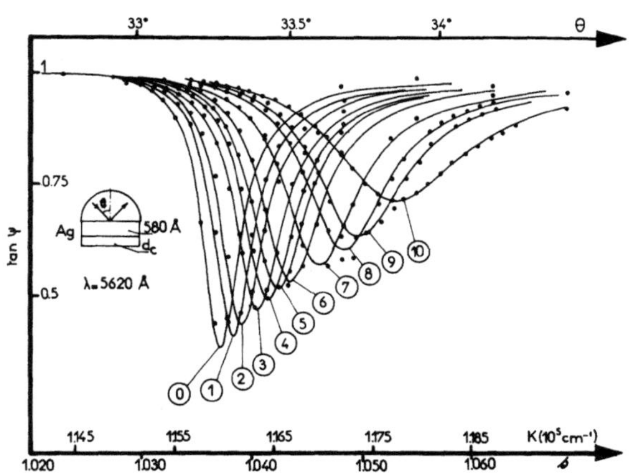

Fig. 11 tanΨ vs.S (or k or θ , upper horizontal scale) at various stages of the tarnishing of a Ag film in the Kretschmann configuration (d_f = 0,2,...., 30 A from left to right). The circles correspond to experimental data and the full curves were computed. (After [14]).

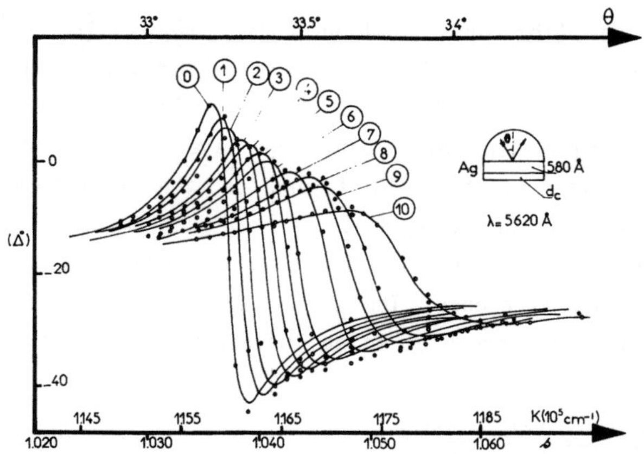

Fig. 12 Same as fig. 12, but with Δ vs.S (or k, or θ). (After [14]).

data and the full curves were computed. Notice that there is both a shift and a broadening of the resonance, and the minimum of tan Ψ increases with d_f, a result which comes out from eq. (17).

When the superficial layer is a metal with $\varepsilon_f < 0$ (ε_{fi} small), the shift of the resonance can be on either side of the resonance for

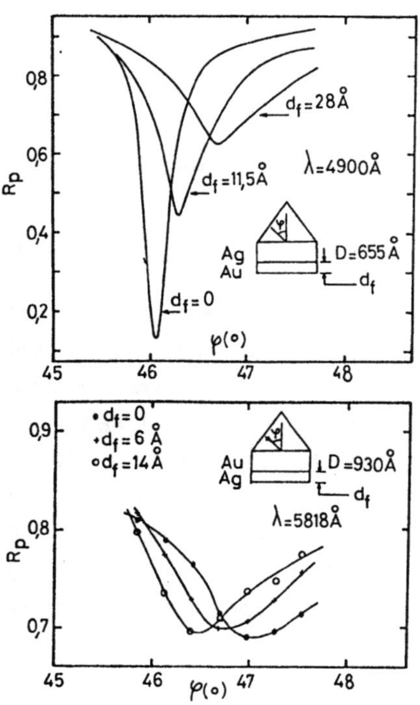

Fig. 13 Effect of a very thin Au layer on a Ag surface ($|\varepsilon_{fr}| < |\varepsilon_r|$, upper curve) and of a very thin Ag layer on a gold surface ($|\varepsilon_{fr}| > |\varepsilon_r|$, lower curve). (After [15]).

the bare $\varepsilon_0/\varepsilon$ surface. Eq. (18) shows that when $|\varepsilon_r| > |\varepsilon_{fr}|$, $\Delta S_{fr} > 0$, the reverse shift occurring when $|\varepsilon_r| < |\varepsilon_{fr}|$. This is demonstrated in fig. 13. On top, one sees the effect of a thin Au layer on Ag ($|\varepsilon_r| > |\varepsilon_{fr}|$), which is the equivalent of a depletion layer on Ag. The bottom curves show the effects of thin Ag layers on the Au/vacuum resonance, ΔS_{fr} being now negative ($|\varepsilon_r| < |\varepsilon_{fr}|$).

We have discussed in this chapter the situation corresponding to experiments performed at fixed frequency. The preceding discussion can now be repeated for experiments performed at fixed incidence and variable frequency. It is found that

$$r = \rho_1 \frac{\omega - \omega_m - \Delta\omega_m - \Delta\omega_f}{\omega - \omega_m - \Delta\omega_d - \Delta\omega_f} \tag{19}$$

with

$$\Delta\omega_f = \frac{2\varepsilon^2(\varepsilon_0 + \varepsilon - \varepsilon_0\varepsilon/\varepsilon_f - \varepsilon_f)}{\sqrt{-(\varepsilon_0+\varepsilon)}\,(\varepsilon - \varepsilon_0)(\partial\varepsilon/\partial\omega)} \frac{\omega}{c} d_f . \tag{20}$$

Here again, the surface layer gives rise to an additional term $\Delta\omega_f$ which is proportional to d_f/λ .

Finally, we must notice that our approximations will break down when $\varepsilon_f \simeq 0$, because of the presence of a term in $1/\varepsilon_f$ in eqs. (18) and (20). Physically, this means that we are exciting plasma oscillations in the layer, which will interact with the SEW. If we write $\varepsilon_f \simeq (\partial\varepsilon_f/\partial\omega).\Delta\omega_f$, it can be show, that eq. (20) becomes

If, moreover, $\varepsilon = 1 - (\lambda/\lambda_p)^2$ and $\varepsilon_f = 1 - (\lambda/\lambda_{pf})^2$ with $\lambda_{pf} \gg \lambda_p$, then eq. (21) becomes (with $\varepsilon_0 = 1$) :

$$\frac{\delta\lambda}{\lambda_{pf}} \simeq \pm\sqrt{\frac{\pi d_f}{\lambda}} . \tag{22}$$

Two points come out from eq. (20), namely that, working at fixed incidence, a) there is a splitting of SEW resonance when $\varepsilon_f = 0$ and b) the splitting may be very important even for a very thin layer, because it is proportional to $(d_f/\lambda)^{1/2}$ and not, as in other cases, to d_f/λ .

Fig. 14 shows computed Rp vs. λ curves for bare Al (d = 250 A) and for the same Al film covered with a K layer 5 A thick and for two values of the plasma damping in K, in a Kretschmann configuration for θ = 43°95. Notice that a 5 A thick K layer would lead to a 800 A splitting of the SEW resonance, the effect of K plasma damping being only a damping of the two resonances in the $R_p(\lambda)$ curve [16].

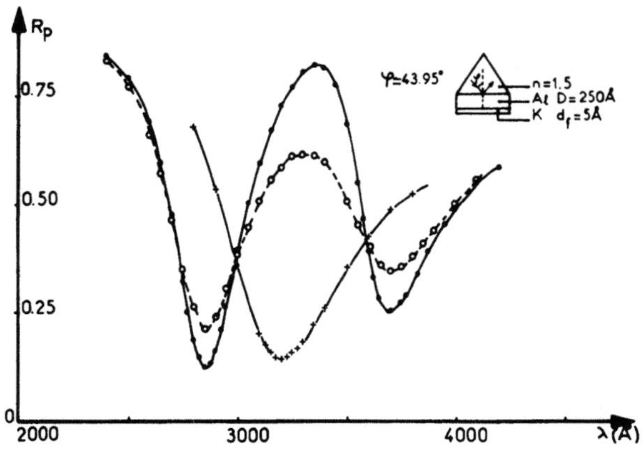

Fig. 14 Computed values of R_p in an ATR configuration schematically shown in the insert for an Al surface covered with a very thin K layer (d_f= 5A) and for different values of the absorption in the layer (open circles : pf f = 10, fullcircles : pf f = 50). The crosses give the values of R_p for a bare Al surface. (After [16]).

7. SEW at the two surfaces of an active film

1) Consider an active film ε surrounded by two inactive media ε_0 and ε_2 . If ε_0 and ε_2 propagate evanescent waves, the dispersion relation for SEW at the two surfaces $\varepsilon_0/\varepsilon$ and $\varepsilon/\varepsilon_2$ is obtained by writing that the Fresnel coefficient for p-polarized light in the three media ($\varepsilon_0/\varepsilon/\varepsilon_2$) configuration has a pole. From the classical thin film formula, one obtains the condition

$$Z_0 + Z_2 + \left(\frac{Z_0 Z_2}{Z} + Z\right) \tanh \alpha = 0 \qquad (23)$$

with $\alpha = (\omega/c)\sqrt{s^2-\varepsilon}\cdot d$, d being the thickness of the active medium and the Z's being the optical admittances already defined. For large thicknesses, tanh α = 1 and eq. (23) becomes

$$(Z_0 + Z)(Z + Z_2) = 0 \qquad (24)$$

We recover the two modes corresponding to SEW at each surface : $\varepsilon_0/\varepsilon$ and $\varepsilon/\varepsilon_2$, which are totally decoupled.

For smaller values of d, we obtain from eq. (23) a dispersion curve with two branches. For $k \to \infty$, their asymptotic values correspond to $\varepsilon_0 + \varepsilon = 0$ and $\varepsilon_2 + \varepsilon = 0$.

Figure 15 depicts the actual situation. There will be two resonances in the R_p curves for scanning either at fixed θ and variable ω (upper left curve) or at fixed ω and variable θ (lower right curve).

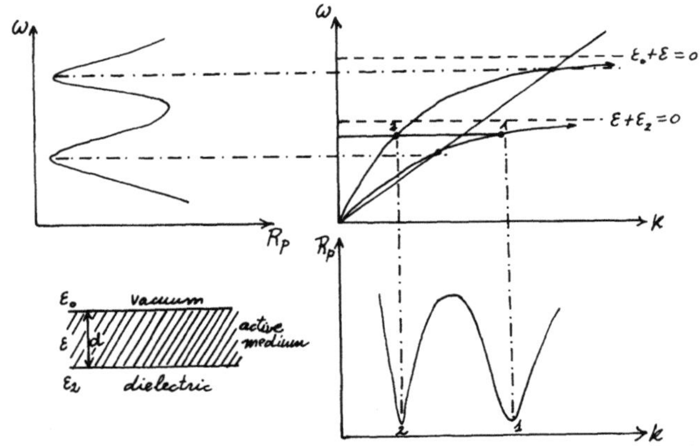

Fig. 15 An active film between two inactive media. Upper right : dispersion curves. Upper left : R_p vs. ω at fixed incidence. Lower right : R_p vs. k at fixed frequency.

The important point to notice is that for not too small values of d/λ , tanh α is not very different from unity and the two resonances are practically decoupled.

Figure 16 shows experimental results for a gold (active) film surrounded by a MgF_2 film ($\varepsilon_2 = 1.384^2$) on one side (d_2 = 1734 A) and by vacuum ($\varepsilon_o = 1$) on the other side. The tan Ψ vs. θ measurements are performed at fixed frequency (λ = 6093 A). The full line shows the experimental results and the dashed curve the computed results. Notice that, as already mentioned above, tan Ψ is proportional to $R_p^{1/2}$. The

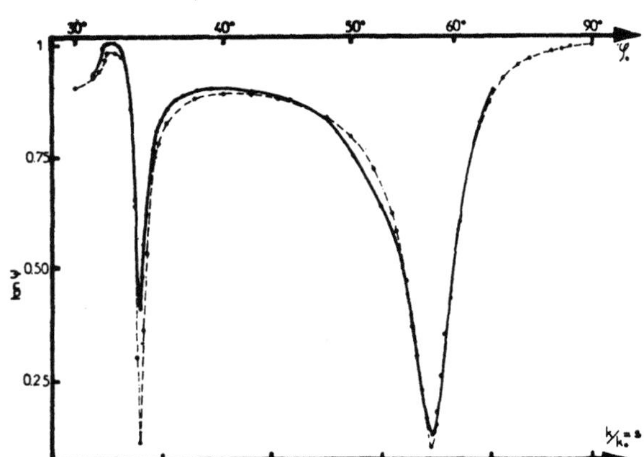

Fig. 16 Experimental results for a gold film surrounded by two inactive media ($\varepsilon_0 = 1$, $\varepsilon_2 = 1.384^2$) : tan Ψ vs. k (or θ , upper abscissas scale). (After [17]).

first resonance from the left corresponds to SEW excitation at the vacuum-gold surface and the second one to the Au-MgF_2 interface. The differences between the experimental and the computed curves are due to experimental uncertainties. The Δ vs. θ curve showed the corresponding resonant behavior.

When the active film is relatively thin (thickness below about 1000 A) and is surrounded by the same medium on both sides ($\varepsilon_0 = \varepsilon_2$), the two resonances are associated with short- and long-range SEW (Sarid [18]).

2) Configuration : metal 2 (ε_2, plasma, ω_{p2})/ metal 1 (ε_1, plasma, ω_{p1})/ vacuum (ε_0). We assume that $\omega_{p1} < \omega_{p2}$. Then for $\omega_{p1} < \omega < \omega_{p2}$, $\varepsilon_1(\omega) > 0$ and $\varepsilon_2(\omega) < 0$. We can excite a SEW at the $\varepsilon_1/\varepsilon_2$ interface. For $\omega < \omega_{p1}/\sqrt{2}$, we can excite a SEW at the $\varepsilon_1/\varepsilon_0$ surface. The dispersion relations for this configuration are obtained again by solving eq. (23). (Notice that we write now ε_1 instead of ε). If d is not too large, there are two branches for the dispersion curve corresponding to the two surfaces ($\varepsilon_2/\varepsilon_1$ and $\varepsilon_1/\varepsilon_0$) of the film. Their starting frequencies (corresponding to S = 1 and ω = ck, light line) are

$\omega = 0$ for the lower branch ($\varepsilon_1/\varepsilon_0$ surface)

$\omega = \sqrt{\dfrac{\omega_{p1}\omega_{p2}(\omega_{p1}+\omega_{p2}\tanh\alpha)}{\omega_{p2}+\omega_{p1}\tanh\alpha}}$ for the upper branch ($\varepsilon_2/\varepsilon_1$ interface).

Their asymptotic values (for $k \to \infty$) are

$\dfrac{\omega_{p1}}{\sqrt{1+\varepsilon_2}} = \dfrac{\omega_{p1}}{\sqrt{2}}$ for the lower branch ($\varepsilon_1/\varepsilon_0$ surface).

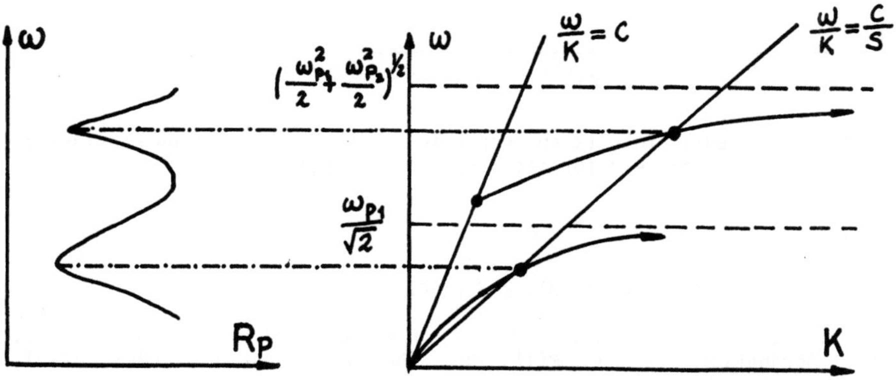

Fig. 17 Dispersion curves and R_p vs. ω for the configuration : metal 2 (plasma)/metal 1 (plasma)/vacuum (ε_0). It is assumed that $\omega_{p1} < \omega_{p2}$.

$\sqrt{\frac{\omega_{p1}^2+\omega_{p2}^2}{2}}$ for the upper branch ($\varepsilon_2/\varepsilon_1$ interface).

Because $0<\omega<\omega_{p1}/\sqrt{2}$ for the lower branch and $\omega>\omega_{p1}$ for the upper branch, it is clear that ATR measurements at fixed ω will show only one resonance, while ATR measurements at fixed θ (and variable ω) will display two resonances. This is shown schematically in fig. 17. Fig. 18 shows the experimental confirmation of these considerations. The two metals are : Al(ε_2) and Ag(ε_1) with $\omega_{p2}>\omega_{p1}$. The R_p^2 vs. λ curves show two resonances. The dashed curve is for the prism/Al/Ag/ vacuum configuration, the minimum at λ = 4080 A corresponding to the excitation of SEW at the Ag/vacuum surface. The dot-dashed curve is obtained after the deposition of a d_f = 9 A layer on Ag. The R_p minimum at 4080 A is strongly displaced towards higher frequencies. After oxidation of the Al layer, we recover the full line curve. The crosses give computed values for d_f = 0. The experimental data clearly show the decoupling of the resonance at λ = 3200 A (SEW at the Al/Ag interface) from the one corresponding to the excitation of a SEW at the Ag/vacuum surface.

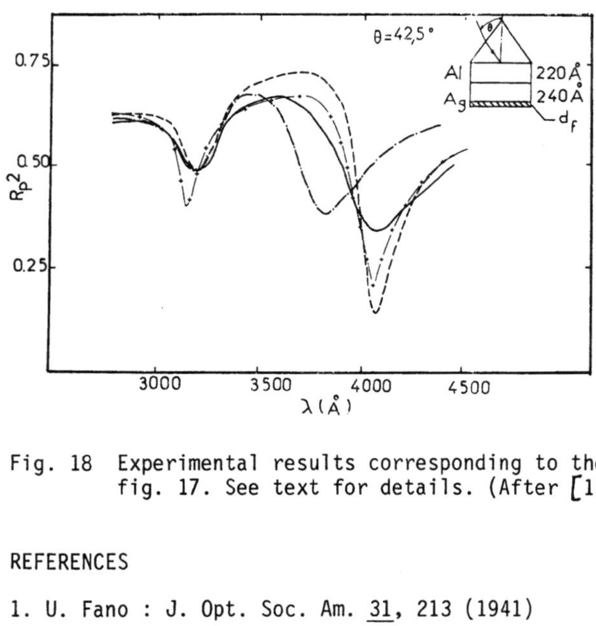

Fig. 18 Experimental results corresponding to the configuration considered in fig. 17. See text for details. (After [15]).

REFERENCES

1. U. Fano : J. Opt. Soc. Am. 31, 213 (1941)

2. H. Raether : "Surface plasmons and roughness", in Surface Polaritons, V.M. Agranovich and D.L. Mills eds. (North-Holland, Amsterdam, 1982), pp. 331-403.

3. A.A. Maradudin : "Interaction of surface polaritons and plasmons with surface roughness", in Surface Polaritons, V.M. Agranovich and D.L. Mills eds. (North-Holland, Amsterdam, 1982), pp. 405-510.

4. A. Otto : Z. Physik 181, 232 (1965) and Z. Physik 216, 398 (1968).

5. E. Kretschmann : Z. Physik 241, 313 (1971).

6. W.H. Weber and S.L. McCarthy : Phys. Rev. Lett. 39, 153 (1977).

7. J. Bösenberg : Phys. Lett. 37A, 439 (1971).

8. C.H. Macek, A. Otto and W. Steinmann : Phys. Status Solidi B51, K59 (1972).

9. G. Hincelin and A. Septier, in Proc. 7th Int. Vac. Congr. and 3rd Int. Conf. on Solid Surfaces, R. Dobrozemsky, F. Rüdenauer, F.P. Viehböck and A. Breth eds. (F. Berger and Söhne, Austria, 1977) vol. II, p. 1269.

10. J. Schoenwald, E. Burstein and J.M. Elson : Solid State Commun. 12, 185 (1973).

11. A.J. Sievers and Z. Schlesinger : J. Physique C10, 13 (1983).

12. G.N. Zhizhin, M.A. Moskalova, E.V. Shomina and V.A. Yakovlev : "Surface electromagnetic wave propagation on metal surfaces", in Surface Polaritons, V.M. Agranovich and D.L. Mills eds. (North-Holland, Amsterdam, 1982), pp. 93-144.

13. F. Abelès : J. Physique 38 C5, 67 (1977)

14. F. Abelès and T. Lopez-Rios, in Proc. Taormina Conf. on Structure of Matter : Polaritons, E. Burstein and F. de Martini eds. (Pergamon, New York, 1974), p. 241.

15. T. Lopez-Rios, Ph. Thesis, Paris 1980.

16. T. Lopez-Rios : Opt. Commun. 17, 342 (1976).

17. F. Abelès and T. Lopez-Rios : Opt. Commun. 11, 169 (1974).

18. D. Sarid : Phys. Rev. Lett. 47, 1927 (1981).

Surface Magnetic Modes

D.R. Tilley

Department of Physics, University of Essex, Colchester C04 3S0, U.K.

1 Introduction

The interactions within a system of spins are of two kinds: short-range *exchange interaction* and long-range *electromagnetic interactions*. The former determine short-wavelength modes while the latter are important at long wavelengths. For very long wavelengths, of order c/ω_R, where ω_R is the appropriate (ferromagnetic or antiferromagnetic) resonance frequency, retardation of the electromagnetic interaction is important, and the bulk or surface modes are then called magnetic polaritons. In fact most observations have been of modes of somewhat shorter wavelength, for which electromagnetic effects can be treated in a magnetostatic approximation. The exchange-dominated short-wavelength modes, or spin waves, have been the subject of much theoretical study, but there has been very little experimental work. Partly for this reason, we restrict discussion of surface spin waves to some brief comments. The division into three wavelength régimes is not hard and fast. For example, important effects arise because of the modification of the magnetostatic modes by exchange interactions.

A general introduction to the subject of surface magnetic modes is found in two recent book chapters. SARMENTO and TILLEY [1] discuss the polariton and magnetostatic regions, whilst MILLS [2] deals with magnetostatic and spin-wave modes.

2 Surface Ordering and Surface Spin Waves

Consider a system of spins on a lattice, coupled by ferromagnetic nearest-neighbour exchange. The lattice is truncated at a plane surface, so that we are dealing with a semi-infinite specimen. The Hamiltonian is

$$\mathcal{H} = -\tfrac{1}{2} \sum_{\underset{\sim}{R},\underset{\sim}{\delta}} J(\underset{\sim}{R},\underset{\sim}{\delta}) \underset{\sim}{S}_{\underset{\sim}{R}} \cdot \underset{\sim}{S}_{\underset{\sim}{R}+\underset{\sim}{\delta}} - g\mu_B H \sum_{\underset{\sim}{R}} S^z_{\underset{\sim}{R}} \qquad (1)$$

The sums are over site positions $\underset{\sim}{R}$ and nearest-neighbour vectors $\underset{\sim}{\delta}$ and H is an external static magnetic field in the z direction. There is no reason to suppose that the exchange interactions between spins in or near the surface layer are the same as between spins in the bulk. In addition, a site in the surface layer has a smaller number of nearest neighbours. Thus the presence of a surface leads to important static effects:

(1) The spin averages $\langle S^z \rangle$ are a function of distance from the surface; this is called surface reconstruction. If J is larger at the surface, the surface spins can order at a higher temperature than the bulk spins, as observed recently on the Gd(0001) surface by WELLER et al. [3].

(2) The specific heat is altered from its bulk value; a discussion is given by MILLS [2].

The spectrum of surface spin waves can be found from eqn. (1) with the same approximations as are used for bulk spin waves. Equations of motion for the spin-deviation operators $S^{\pm} = S^x \pm iS^y$ are linearized by the approximation $S^z \to \langle S^z \rangle$, which is best at low temperatures. Since eqn. (1) has translational invariance in the plane of the surface, surface spin waves are characterized by a 2-D wave vector $\underset{\sim}{k}_{/\!/}$. For the (001) surface of a spin-S ferromagnet on a simple-cubic lattice, with exchange constants J_s between spins in the surface layer and J otherwise, the low-temperature surface spin wave frequency is

$$\omega_s(\underset{\sim}{k}_{/\!/}) = g\mu_B H + 2SJ(2-\cos k_x a-\cos k_y a) - SJ(\gamma-1)^2/\gamma \qquad (2)$$

where $\gamma = \exp(-\lambda a) = [1 + 2(1-J_s/J)(2-\cos k_x a-\cos k_y a)]^{-1}$ and $S^+_{\underset{\sim}{R}} = S_o \exp(-\lambda z)$ $\exp(ik_x x+ik_y y-i\omega t)$. Eqn. (2) may be compared with the expression for bulk spin waves

$$\omega_B(\underset{\sim}{k}_{/\!/},k_z) = g\mu_B H + 2SJ(3-\cos k_x a-\cos k_y a-\cos k_z a) \qquad (3)$$

On a ω - $\underset{\sim}{k}_{/\!/}$ plot, eqn. (3) is represented by a bulk continuum, since k_z may lie anywhere within the first Brillouin zone. For $J_s < J$ the surface mode lies below the bulk continuum.

Up to now there are no experimental results concerning these surface spin-wave modes. However the availability of spin-polarized sources, as described by PIERCE and CELOTTA [4], for example, means that in principle inelastic scattering of spin-polarized electrons might be used to investigate their properties.

3 Magnetic Susceptibility Tensor

We now turn to our main theme, the magnetostatic and polariton-type modes associated with magnetic surfaces. These modes are at wavelengths long compared with interatomic distances, and their properties can therefore be derived from macroscopic equations. The specimen is characterized by a magnetic susceptibility tensor, and we assume that the bulk susceptibility characterises the whole specimen up to the surface. That is, surface reconstruction, which takes place within a few atomic layers of the surface, is assumed to be irrelevant to the properties of the long-wavelength modes. In this section, therefore, we find the magnetic susceptibility tensor of a bulk specimen. We deal with a ferrimagnet; the results for a ferromagnet and an antiferromagnet follow as special cases.

We consider a two-sublattice ferrimagnet with uniaxial anisotropy such that the equilibrium values of the sublattice magnetizations $\underset{\sim}{M}_1$ and $\underset{\sim}{M}_2$ are directed in the +z and -z directions respectively; we write M_1, M_2 for the equilibrium values of M_1^z and M_2^z; $M_1 > 0$ and $M_2 < 0$. The susceptibility is the linear response of $\underset{\sim}{M}_1 + \underset{\sim}{M}_2$ to an applied field $\underset{\sim}{H}$ of frequency ω. We describe the motion of $\underset{\sim}{M}_1$ and $\underset{\sim}{M}_2$ by the usual torque equations of magnetic resonance theory [5]

$$d\underset{\sim}{M}_i/dt = \gamma_i(\underset{\sim}{M}_i \times \underset{\sim}{H}_i^{eff}) \qquad i = 1,2 \tag{4}$$

with effective field H_1^{eff}

$$\underset{\sim}{H}_1^{eff} = \underset{\sim}{H}_o + \underset{\sim}{H}_{A1} - \lambda\underset{\sim}{M}_2 + {}^1/_3(\underset{\sim}{M}_1+\underset{\sim}{M}_2) + \underset{\sim}{H} \tag{5}$$

$\underset{\sim}{H}_2^{eff}$ is given by a similar expression with 1 and 2 interchanged. The gyromagnetic ratios γ_1 and γ_2 are taken as negative and damping is neglected. $\underset{\sim}{H}_o$ is a static field in the z-direction and $\underset{\sim}{H}_{Ai}$ are static anisotropy fields in the ±z-direction. The term $\lambda\underset{\sim}{M}_2$ is the interlattice exchange; intralattice exchange is not included, since it gives a zero contribution to eqn. (4). ${}^1/_3(\underset{\sim}{M}_1+\underset{\sim}{M}_2)$ is a Lorentz term describing local-field corrections [6].

Eqn. (6) can be derived from the quantum mechanical equations of motion for spins $\underset{\sim}{S}_i$ that follow from an appropriate Hamiltonian. The macroscopic limit is taken with $\underset{\sim}{M}_i \propto \underset{\sim}{S}_i$. $\underset{\sim}{H}_{Ai}$ appears as an approximate representation of the effects of a single-ion anisotropy term $D_A(S_i^z)^2$. We do not go into the details of this macroscopic averaging procedure here. Note, however, that eqn. (5) involves a long-wavelength treatment of the exchange, in that the exchange field on sublattice 1 is taken simply proportional to $\underset{\sim}{M}_2$. For example, if the exchange Hamiltonian is $-\frac{1}{2}J \sum_{R,\delta} \underset{\sim}{S}_{\underset{\sim}{R}}^{(1)}.\underset{\sim}{S}_{\underset{\sim}{R}+\underset{\sim}{\delta}}^{(2)}$ where the superscripts denote the sublattices and the notation otherwise is as in eqn. (1), then in the first step of the macroscopic averaging $\underset{\sim}{S}_{\underset{\sim}{R}+\underset{\sim}{\delta}}^{(2)}$ is taken as the local value of a continuum field $\underset{\sim}{S}^{(2)}(\underset{\sim}{r})$. The expansion $S_{R+\delta}^{(2)} = \underset{\sim}{S}^{(2)}(\underset{\sim}{R}) + \underset{\sim}{\delta}.\nabla S^{(2)}(\underset{\sim}{R}) + \frac{1}{2}\delta_\ell\delta_m\nabla_\ell\nabla_m\underset{\sim}{S}^{(2)}(\underset{\sim}{R}) + \ldots$ is then applied. The assumption in eqn. (5) is that only the first term of this expansion contributes. If the wavelength of the mode under consideration is sufficiently short, then higher-order terms must be taken into account. The linear term vanishes by symmetry when summed over nearest neighbours, so the first correction comes from the quadratic term. Thus, a more general form of eqn. (5) includes a term $D\nabla^2\underset{\sim}{M}_2$. The *exchange stiffness constant* D is proportional to Ja^2, where a is the nearest-neighbour distance. The inclusion of this term is important for interpretation of some experiments on Brillouin scattering from ferromagnetic films. For the remainder of this section and for the next section, we ignore the exchange stiffness; the corrections due to its inclusion will be discussed qualitatively in §5.

For the derivation of the susceptibility eqn. (4) is linearized in the r.f. field $\underset{\sim}{H}$. The components M_i^x, M_i^y are of first order, but the deviations of M_i^z from their static values are of second order. Thus, only the transverse components M_i^x, M_i^y enter the linearized theory, and it is convenient to use a rotating wave representation $M_i^\pm = M_i^x \pm iM_i^y$, $H^\pm = H^x \pm iH_y$. The equations of motion then take the form

$$(\omega+\omega_{12})M_1^+ + \omega_{11}M_2^+ = \omega_{M11}H^+$$

$$\omega_{22}M_1^+ - (\omega-\omega_{21})M_2^+ = \omega_{M22}H^+ \tag{6}$$

The equations for M_1^-, M_2^- are obtained from these by changing the sign of ω. The constants are $\omega_{12} = \gamma_1\lambda M_2 - \gamma_1 H_{A1} - \gamma_1 H_o - {}^1/_3\gamma_1 M_2$, $\omega_{21} = -\gamma_2\lambda M_1 + \gamma_2 H_{A2} + \gamma_2 H_o + {}^1/_3\gamma_2 M_1$, $\omega_{ii} = -\gamma_1\lambda M_i + {}^1/_3\gamma_i M_i$, $\omega_{Mii} = \mp\, \gamma_i M_i$, the upper sign referring to $i = 1$.

Some implications of eqn. (6) are discussed by SARMENTO and TILLEY [1]. The susceptibilities $\chi^\pm$ are defined by $M_1^\pm + M_2^\pm = \chi^\pm H^\pm$ since it is obvious from eqn. (6) that the susceptibility tensor is diagonal in the rotating-wave approximation. Since eqn. (6) is a 2 x 2 matrix equation, there are two resonance frequencies ω_1 and ω_2, called the exchange resonance frequency and the ferromagnetic resonance frequency [5]. Typical orders of magnitude are 10-100 cm^{-1} for the former and 1 cm^{-1} for the latter.

Since the susceptibility tensor is diagonal in the rotating-wave representation, it is not diagonal in the original (x,y) coordinate frame. In fact,

$$\chi = \begin{pmatrix} \chi_a & i\chi_b \\ -i\chi_b & \chi_a \end{pmatrix} \tag{7}$$

This is often called the gyromagnetic form.

It is convenient to quote explicitly the forms that apply to ferromagnets. The torque equation (4) still applies, but there is only one spin lattice. Hence terms of eqn. (5) that are proportional to $\underset{\sim}{M}$ can be dropped, since they do not contribute to the vector product of eqn. (4). Thus eqn. (5) is simplified to $\underset{\sim}{H}^{eff} = \underset{\sim}{H}_o + \underset{\sim}{H} + D\nabla^2\underset{\sim}{M}$.

For later use, we include here the exchange-stiffness term.

The susceptibility tensor for a ferromagnet may be derived by specialisation of the general results quoted; it still has the general form of eqn. (7). With exchange stiffness neglected, the constants are $\chi_a = \omega_m\omega_o/(\omega_o^2-\omega^2)$, $\chi_b = \omega_m\omega/(\omega_o^2-\omega^2)$, $\chi^\pm = \omega_m/(\omega_o\pm\omega)$ where $\omega_m = -\gamma M$ and $\omega_o = -\gamma(H_o+H_A)$. Very often $H_A \ll H_o$, so that the resonance frequency ω_o is simply $-\gamma H_o$.

The discussion of magnetic surface and interface modes in the polariton and magnetostatic regimes is a matter of solving Maxwell's equations and boundary conditions when the magnetic regions are specified by the susceptibility tensor of eqn. (7). The interest is not restricted to geometries in which surfaces are present; even the solutions in a bulk medium are of some complexity [1].

4 Magnetostatic Modes of a Slab

This section is concerned with purely magnetostatic modes; the modifications required when exchange stiffness is incorporated are the subject of §5. The definitive calculation was presented long ago [7]. We consider a slab of thickness s with applied field and static magnetization in the surface plane. The z axis is taken along the field and the x axis normal to the slab. In the magnetostatic approximation, curl $\underset{\sim}{H} = 0$, so that $H = \nabla\psi$. $\underset{\sim}{M}$ is related to $\underset{\sim}{H}$ by the susceptibility tensor of eqn. (7), and we introduce a background permeability [8] to write $\underset{\sim}{B} = \mu\mu_o(\underset{\sim}{H}+\underset{\sim}{M})$.

The equation div $\underset{\sim}{B} = 0$ becomes

$$(1+\chi_a)(\partial^2\psi/\partial x^2+\partial^2\psi/\partial y^2) + \partial^2\psi/\partial z^2 = 0 \quad \text{for } s/2 < x < s/2$$

$$\nabla^2\psi = 0 \quad \text{for} \quad x < -s/2 \quad \text{and} \quad x > s/2 \tag{8}$$

We seek solutions of the form

$$\begin{aligned} \psi &= c \exp(-\alpha x)\exp[i(k_y y+k_z z)] \quad &&\text{for} \quad x > s/2 \\ &= d \exp(\alpha x)\exp[i(k_y y+k_z z)] \quad &&\text{for } x < -s/2 \\ &= [a \exp(ik_x x) + b\exp(-ik_x x)] \exp[i(k_y y+k_z z)] &&\text{for } -s/2 < x < s/2 \end{aligned} \tag{9}$$

The functions must all have the same y- and z-dependence, in order that boundary conditions can be satisfied over all the surfaces. Substitution of eqn. (9) into eqn. (8) yields $\alpha^2 - k_y^2 - k_z^2 = 0$ and $(1+\chi_a)(k_x^2+k_y^2) + k_z^2 = 0$. It is seen from eqn. (9) that α must be real, in order that the mode be localised in the slab, while k_x can be either real or imaginary. With k_x real, the dependence on x within the slab is oscillatory, while with k_x imaginary, the mode is more or less localized at one or other surface, and may be called a surface mode. Real values of k_x are possible only when $1 + \chi_a < 0$.

The dispersion equation is found by substituting eqn. (9) into the boundary conditions ψ and B_x continuous at each interface. Thus one finds four homogeneous linear equations for a, b, c and d; the condition for a solution is

$$(\alpha/\mu)^2 + 2(\alpha/\mu)k_x(1+\chi_a)\cot(k_x s) - k_x^2(1+\chi_a)^2 - \chi_b^2 k_y^2 = 0 \tag{10}$$

This equation gives the dispersion curves ω versus $\underset{\sim}{k}_{//} = (k_y,k_z)$.

The above considerations hold for any material with a gyromagnetic susceptibility of the form of eqn. (7). For the particular case of a ferromagnet, χ_a and χ_b may be written $\chi_a = \Omega_H/(\Omega_H^2-\Omega^2)$ and $\chi_b = \Omega/(\Omega_H^2-\Omega^2)$ with $\Omega = \omega/\omega_m$ and $\Omega_H = \omega_0/\omega_m$.

We now give a summary of the main features of the dispersion curves for ferromagnets, taking $\mu = 1$ [7]. There are two independent variables, k_y and k_z, and it is convenient to define $\phi = \arctan(k_z/k_y)$. A general picture of the dispersion curves is shown in Fig. 1. The guided modes with real k_x, called volume modes by DAMON and ESHBACH [7], occupy the frequency interval $\Omega_H < \Omega < (\Omega_H^2+\Omega_H)^{\frac{1}{2}}$. As the thickness s increases, these guided modes become more numerous and therefore more closely packed in frequency, until as $s \to \infty$ they constitute the bulk continuum.

The modes with imaginary k_x are surface modes with amplitude decaying exponentially with distance from the surface. We discuss in detail the special case $k_z = 0$ and $\phi = 0$, which is propagation at right angles to the static magnetic field. For this case eqn. (10) becomes

$$\Omega^2 = \Omega_H^2 + \Omega_H + [\Omega_H(\mu-\mu^{-1})+\mu]/[\mu+\mu^{-1}+2\coth(k_y s)] \tag{11}$$

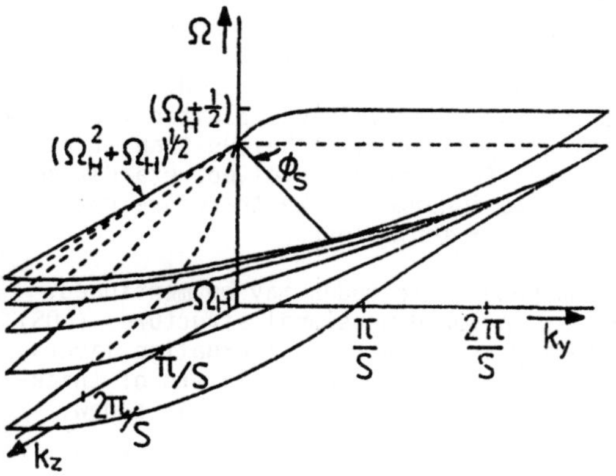

Fig.1 Dispersion curves of magnetostatic modes of a ferromagnetic slab [7]

The limiting values are $\Omega \to (\Omega_H^2+\Omega_H)^{\frac{1}{2}}$ as $k_y \to 0$, and $\Omega \to \Omega_s$ as $k_y \to \infty$, where $\Omega_s = \Omega_H + \mu/(\mu+1)$ may be called the surface frequency for this problem. For the particular case $\mu = 1$ shown in Fig. 1, $\Omega_s = \Omega_H + \frac{1}{2}$. As shown in Fig. 1, then, for $k_z = 0$ the surface-mode frequency rises from the top of the bulk continuum, $\Omega = (\Omega_H^2+\Omega_H)^{\frac{1}{2}}$, at $k_y = 0$ to $\Omega = \Omega_s$ as $k_y \to \infty$. As ϕ increases from zero, the limiting frequency for $|k\ | \to \infty$ decreases, and as shown in Fig. 1, the surface mode runs along the top of the bulk continuum for $\phi = \phi_s$, where $\tan^2\phi_s = [(\mu^2-1)\Omega_H^2 + \mu^2\Omega_H]/\Omega_H^2$. In the particular case $\mu = 1$, $\tan^2\phi_s = 1/\Omega_H$. For $|\phi| > \phi_s$, there simply is no surface mode.

The result is a distinguishing characteristic of the Damon-Eshbach (DE) mode. For a semi-infinite slab, the surface mode propagates in one sense of k_y and in directions from $-\phi_s$ to ϕ_s about this k_y direction; it does not propagate in the opposite direction. This is referred to as non-reciprocal propagation, and is generally explained by invoking the absence of time-reversal invariance in a magnetic system. Ultimately this is true, but the argument for non-reciprocity needs careful formulation [2].

Another characteristic of the DE mode is that, as might be expected, the magnetic field pattern is concentrated on one surface of the slab. The field pattern can be found once the coefficients a, b, c and d are determined; it is depicted elsewhere [1,7].

Figure 1 and the discussion were concerned with ferromagnets. For these, ω_0 and ω_m are typically in the microwave region. This has a number of consequences. First, applications of magnetostatic surface waves are to signal processing at microwave frequencies [9]. Second, observation by attenuated total reflection (ATR) would at best be cumbersome, because of the large specimen and gap sizes that would be required, and in fact it has not yet been achieved. On the other hand, the frequencies are very suitable for Brillouin scattering of light, and that technique has proved very successful.

It is a curious feature of the literature that although eqn. (10) applies to antiferromagnets, there was no discussion of its implications for nearly twenty years. Eventually, CAMLEY [10] provided an account of the magnetostatic mode on a semi-infinite antiferromagnet, and later STAMPS and CAMLEY

[11] dealt with an antiferromagnetic slab. There has been no experimental work reported, but the development is important in principle, because antiferromagnetic resonance frequencies are typically in the far infrared. Thus, potential applications are in a different and relatively unexploited frequency region. Likewise, ATR is a more promising experimental technique than it is for ferromagnets, but because of the higher frequencies involved, retardation of the electromagnetic field is likely to be important. We therefore return to antiferromagnets in the discussion of polariton modes.

One further calculation concerning magnetostatic modes may be mentioned. There is considerable interest at present in low-dimensional structures (LDS), which are prepared by epitaxy or sputtering as a stack of alternating layers of different materials. A magnetic LDS consists of magnetic layers of thickness d_1 alternating with nonmagnetic layers of thickness d_2 so that a new, one-dimensional period $D = d_1 + d_2$ is introduced. The technology is such that structures with D of order tens or hundreds of Angstroms are possible. Theoretical studies of magnetostatic modes in a magnetic LDS were first given recently [12,13]. For an infinite LDS, it is clear that the introduction of the period D leads to a Brillouin zone edge at π/D, with stop-bands at the zone edges. As well as studying this effect in an infinite LDS, the papers cited also deal with a semi-infinite specimen. They consider the possible existence of a DE-type surface mode, localised at the vacuum interface in the sense that its amplitude decays exponentially through successive layers from the surface. The result is striking: there is such a surface mode separate from the bulk continuum provided $d_1 > d_2$. Where the mode exists, its frequency is equal to that of the DE mode on the surface of the magnetic material alone. This prediction, which has been confirmed by light-scattering experiments, is illustrated in Fig. 2 by the results of numerical calculations for the Ni/Mo LDS system.

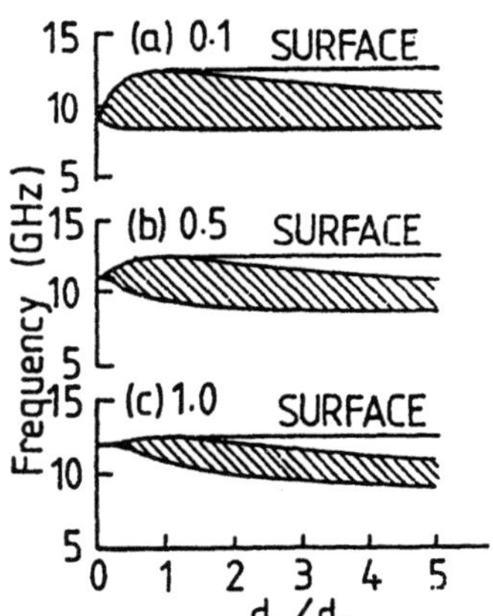

Fig. 2 Surface modes (lines) and bulk continuum (shaded) for a magnetic LDS of ferromagnetic Ni and nonmagnetic Mo [12]. Propagation perpendicular to magnetic field; values of $k_y d_1$ marked. Applied field 0.1T.

5 Magnetostatic Modes: Inclusion of Exchange Stiffness

We give here a largely qualitative account of the way in which the inclusion of exchange stiffness modifies the results of the previous section. It is convenient to consider semi-infinite and slab geometries separately. For bulk modes, the exchange stiffness simply adds a term Dk^2 to the magnetostatic frequency at sufficiently high wave number k [14].

For a semi-infinite specimen, both the bulk continuum and the surface mode are modified. In the exchange-dominated region, the bulk mode has frequency $\omega = D(k_{//}^2 + k_x^2)$ where $k_{//}$ and k_x have the same meaning as in §4. Since k_x can take any value, the top of the bulk continuum moves up, and the DE mode now runs through the continuum. This means that the DE mode can lose energy by radiation into the continuum; it is 'leaky' [15]. For a leaky mode the definition of a dispersion curve is somewhat indistinct. Putting this refinement on one side for the moment, a simple argument [16,17] is that in the exchange-dominated region, the DE dispersion curve should be $\omega \sim D(k_{//}^2 + \alpha^2)$, where α is the spatial decay parameter. Since $\alpha^2 = k_{//}^2$, one expects $\omega \sim 2Dk_{//}^2$. This is in fair agreement with the result $\omega = 1.76\ D\ k_y^2$ obtained analytically [15] and with numerical results [18].

Exchange stiffness introduces a term in $\nabla^2\underset{\sim}{M}$ into the equation of motion for $\underset{\sim}{M}$. A quantitative description of a semi-infinite specimen therefore requires a boundary condition on $\underset{\sim}{M}$. This can be found by following through the process by which the continuum limit of the microscopic equations of motion is taken. The equations of motion of spins in the surface layer lead to the boundary condition [19]. In particular if the surface anisotropy field is neglected, the so-called RW boundary condition is $dM/dx = 0$. To allow for surface anisotropy and possible surface reconstruction, CAMLEY and MILLS [18] use the more general form $d\underset{\sim}{M}/dx - \lambda\underset{\sim}{M} = 0$, where λ is an adjustable parameter.

In drawing a dispersion curve,it is assumed that both ω and k are real; for a leaky mode which radiates energy and is therefore damped this is not the case. A better description is given by evaluation of appropriate Green functions from which spectral densities can be found. The Green functions for the leaky DE mode are calculated in the definitive paper of CAMLEY and MILLS [18]. They present numerical results for the spectral densities and go on to calculate the cross-section for Brillouin scattering.

For a slab, the inclusion of exchange stiffness and the RW boundary condition greatly modifies the real-k_x guided modes shown for the pure magnetostatic case as the lower part of Fig. 1. The RW boundary condition produces the quantization condition $k_x = n\pi/s$ with $n = 1,2, \ldots$. As s decreases, therefore, the exchange energy $D(k_x^2 + k_y^2)$ increases, and eventually the guided-mode frequency exceeds that of the DE surface mode, which is unaffected by the boundary condition. The results of calculations exhibiting this effect, together with light-scattering experimental results, are shown in Fig. 3. A more exact calculation, which involves diagonalisation of a 6 x 6 matrix,[16] shows that the RW boundary condition leads to mixing of the DE and guided mode near the crossing point. Thus "mode repulsion" is observed, as shown in the inset to Fig. 3.

6 Magnetostatic Modes: Brillouin Scattering

The first experimental work on the Damon-Eshbach mode is due to Brundle and Freedman [21], who investigated microwave propagation on the surface of an yttrium-iron garnet (YIG) slab. They measured delay times for a pulse to travel from one electrode to the other, and confirmed that the mode was of the DE type,by observing that the transmitted pulse disappeared when a metal plate was placed against one face of the slab, but was unaffected when it was placed against the other face. More recent experimental work has been by

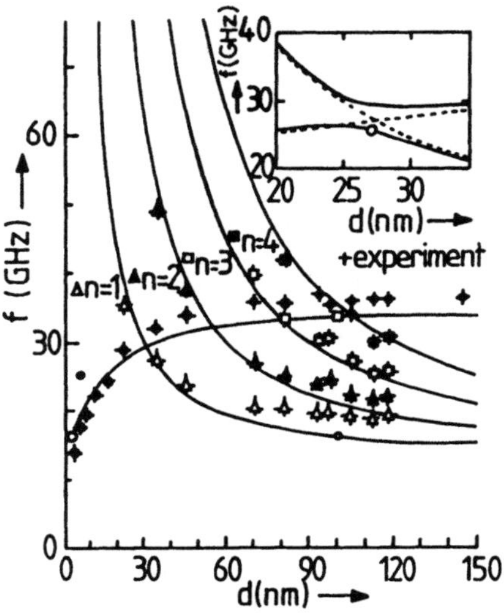

Fig. 3 Frequencies of guided-wave modes and DE mode versus film thickness s calculated for Fe [20]. Applied field $B_0 = 0.1T$, $k_{//} = 1.82 \times 10^5\ cm^{-1}$. Full lines, calculations, crosses experimental points. Insert shows mode repulsion between n = 1 and DE mode for $B_0 = 0.02T$, $k_{//} = 2.42 \times 10^5\ cm^{-1}$.

means of Brillouin scattering, which was applied from about the middle of the 1970's.

The availability of Brillouin-scattering results was a result of Sandercock's achievement in developing the multipass Fabry-Perot interferometer to provide a highly sensitive spectrometer for frequency shifts around 10^{-1} to $10\ cm^{-1}$ ($1\ cm^{-1}$ = 30 GHz = 0.12 meV). [22]

General discussions of light-scattering spectroscopy are given in many books. [23,24] for example. In scattering off a bulk mode, incident light of frequency ω_I and wave number $\underset{\sim}{k}_I$ is scattered into ω_S and $\underset{\sim}{k}_S$ with creation (Stokes process) or absorption (anti-Stokes) of an excitation quantum $(\omega,\underset{\sim}{q})$. The kinematic relations are $\omega_S = \omega_I \mp \omega$, $\underset{\sim}{k}_S = \underset{\sim}{k}_I \mp \underset{\sim}{q}$, the upper sign applying for Stokes scattering. In general $\omega \ll \omega_I$; consequently $|\underset{\sim}{k}_S| \approx |\underset{\sim}{k}_I|$ and a scan over ω_S essentially probes the excitation spectrum at fixed $|\underset{\sim}{q}| = 2n(\omega_I/c)\sin\frac{1}{2}\theta$. Here n is the refractive index at frequency ω_I and θ is the angle between $\underset{\sim}{k}_I$ and $\underset{\sim}{k}_S$.

For magnetic modes, the coupling energies $U_1 = a^{ijk}E_I^iE_S^jM^k$ and $U_2 = b^{ijkl}E_I^iE_S^jM^kM^l$ both contribute to the light-scattering process [25]. U_1 must appear since it is linear in all three fields $\underset{\sim}{E}_I$, $\underset{\sim}{E}_S$ and $\underset{\sim}{M}$; the necessity for U_2 arises from the fact that M^z has a zero-order, static value, so that U_2 also contains terms linear in the three time-dependent fields. Coefficients a^{ijk} and b^{ijkl} are related to the classical coefficients for the magneto-optic Faraday and Voigt effects. Because these two terms contribute, the anti-Stokes to Stokes ratio I_{As}/I_S for scattering off magnetic systems generally departs from the value $n_B(\omega)/[n_B(\omega) + 1]$ ($n_B(\omega)$ = Bose factor) generally found for non-magnetic systems.

The light scattering is due to thermal fluctuations of the scattering field M, the relevant quantity being the power spectrum $\langle M^i(\underset{\sim}{r})M^j(\underset{\sim}{r}')\rangle_\omega$. This is conveniently found by means of the fluctuation dissipation theorem from the Green, or linear response function [26] $\langle\langle M^i(\underset{\sim}{r});M^j(\underset{\sim}{r}')\rangle\rangle_\omega$.

In light scattering off surface modes, the scattered light is observed in reflection from the (generally opaque) specimen surface; typically, the angles $\theta_I(\theta_s)$ between the direction of the incident (scattered) light and the normal to the surface are fixed. Wave-vector components k_I^x and k_s^x normal to the surface are complex, and the bulk kinematic relations are replaced by $\omega_s = \omega_I \mp \omega$, $\underset{\sim}{k}_s'' = \underset{\sim}{k}_I'' \mp \underset{\sim}{q}''$ where $\underset{\sim}{k}_s'' = (k_s^y, k_s^z)$ etc. In consequence, the experiment consists of a scan over ω at fixed $\underset{\sim}{q}''$; surface modes show up as more-or-less sharp peaks, while the bulk continuum generates broad structure. The detailed theory involves calculation of surface-related Green functions.

We now turn to the experimental results, reviewed also in [16] and [20]. Figure 4 shows the first observation of the DE mode [27]. Note that results from a Fabry-Perot are periodic with period $\lambda_I/2$; consequently, the results marked for frequency shift -80 to 0 GHz are identical to those from 0 to 80 GHz. Note also that parasitically scattered light at ω_I shows up as a strong instrumentally-broadened central peak.

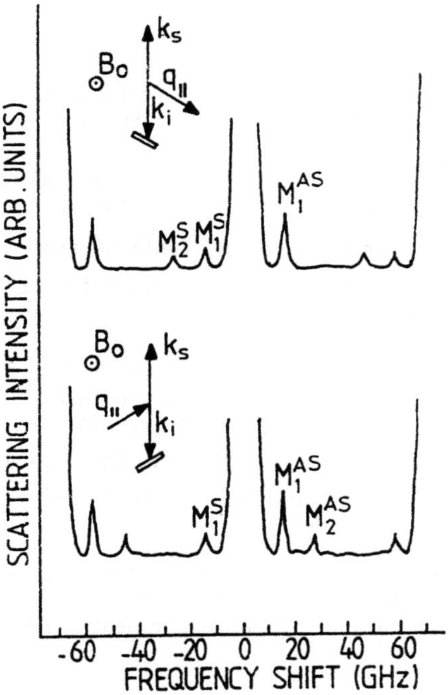

Fig. 4 Light-scattering spectrum from EuO at 4.5K in applied field 0.3T [27]. For the lower curve, the specimen orientation was changed so as to reverse the sign of $\underset{\sim}{q}''$.

The two peaks marked M_1^S and M_1^{AS} are due to scattering off the bulk modes. Of more interest are the peaks marked M_2^S and M_2^{AS}, which are indeed due to the DE mode. In the scattering geometry used $\underset{\sim}{q}''$ was transverse to $\underset{\sim}{M}$ and since $\underset{\sim}{q}''$ has opposite signs for Stokes and anti-Stokes scattering, the DE mode shows up on only one side, as is indeed the case for the data shown.

Results for scattering off Fe were presented by SANDERCOCK and WETTLING [28]. Fe has a much shorter optical penetration depth than EuO. Consequently bulk modes with a wider range of q^x values contribute to the scattering, and the bulk scattering feature becomes much broader, extending into the frequency region where DE mode scattering is expected. Partly because of this, and partly because D is much larger than in EuO, exchange corrections become important. CAMLEY and MILLS [18] gave the full theory for the light scattering, based on their exchange-corrected Green functions. Their results agree well with the data.

Scattering off thin films exhibits some novel features. Figure 1 shows that the bulk continuum is replaced by discrete modes; this is confirmed by the experimental results [29], shown in Fig. 5. For thinner films, it was shown already (Fig. 3) that because of the exchange term Dk^2 and the RW boundary condition,the standing-wave modes increase in frequency. It is also found [20] that for very thin films, s $\sim$ 10 nm, the DE-mode scattering intensity increases with decreasing s; the reasons are now understood [30].

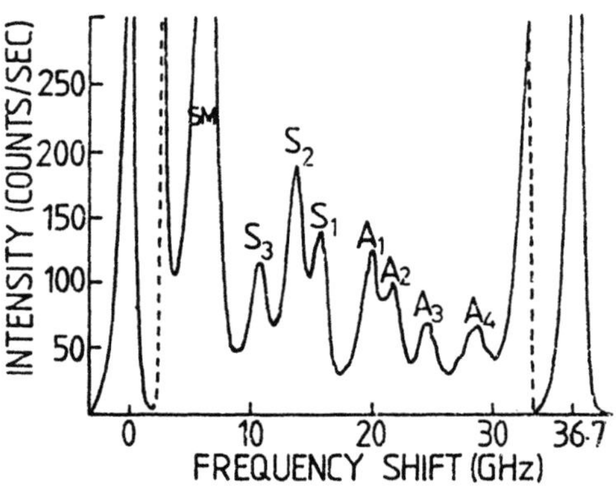

Fig. 5 Brillouin scattering off an amorphous sample of $Fe_{80}B_{20}$ of thickness 106 nm in an applied field 0.26 TESLA [29]. Peak SM is the DE mode. Peaks S_1, S_2, S_3 (Stokes side) and A_1 to A_4 (anti-Stokes side) are due to the discrete modes.

At the end of §4 mention was made of theoretical investigations of layered magnetic materials (LDS). These led to the prediction, illustrated in Fig. 2, that provided $d_1 > d_2$ there is a surface mode at the DE frequency. This was confirmed in Brillouin scattering experiments [31]. For $d_1 > d_2$ there is a mode with dependence of frequency on H exactly as predicted by the DE theory. Equally important, for $d_1 < d_2$ there is no such mode.

All the experimental work described concerns ferromagnets. Although calculations of the dispersion curves for semi infinite and thin-film antiferromagnets have been presented [10,11], there are as yet no experimental reresults on antiferromagnets.

7 Surface Magnetic Polaritons

Sections 4 to 6 were concerned with modes in which the coupling to the electromagnetic field can be described within the magnetostatic approximation. As seen in §6, the experimental results concern magnetostatic modes,or sometimes magnetostatic modes modified by exchange stiffness. We now go beyond the magnetostatic approximation and include effects due to retardation of the electromagnetic field. Although a considerable amount of theoretical work has been done,there are as yet no experimental results; prospects for experiment will be discussed. We treat, in turn, magnetic polaritons in a bulk specimen, on the surface of a semi-infinite specimen, and in a thin film.

The dispersion equation for a bulk magnetic polariton is found by solving Maxwell's equations for a medium characterized by the gyromagnetic susceptibility of eqn. (7). It is convenient to use a permeability tensor $\underset{\sim}{\mu}$:

$$\underset{\sim}{\mu} = \begin{pmatrix} \mu_{xx} & \mu_{xy} & 0 \\ -\mu_{xy} & \mu_{xx} & 0 \\ 0 & 0 & \mu_{zz} \end{pmatrix} \tag{12}$$

where $\mu_{xx} = \mu(1+\chi_a)$, $\mu_{xy} = i\mu\chi_b$, $\mu_{zz} = \mu$, and we include a background permeability μ. Maxwell's equations then yield the wave equation

$$\nabla^2\underset{\sim}{H} - \nabla(\nabla.\underset{\sim}{H}) - (\varepsilon/c^2)\underset{\sim}{\mu}\,\ddot{\underset{\sim}{H}} = 0 \tag{13}$$

and divergence condition $\nabla.(\underset{\sim}{\mu}\underset{\sim}{H}) = 0$. Note that $\underset{\sim}{\mu}\underset{\sim}{H}$ and $\underset{\sim}{\mu}\ddot{\underset{\sim}{H}}$ are tensor products. Substitution of $\underset{\sim}{H} \propto \exp(i\underset{\sim}{k}.\underset{\sim}{r} - i\omega t)$ leads, eventually, to the dispersion equation ω versus $\underset{\sim}{k}$. The dispersion curves for a ferromagnet were first calculated by AULD [14]. More discussion of theoretical results for ferromagnets, and also for antiferromagnets and ferrimagnets, is given in [1]. Polaritons in a bulk antiferromagnet have been observed by means of their influence on antiferromagnetic resonance [32,33].

The algebraic problems involved in a theoretical discussion of surface magnetic polaritons on a semi-infinite medium are fairly severe, and full accounts for all relative orientations of field, wavevector and surface have not been given. So far, attention has mainly been restricted to the so-called Voigt geometry, in which $\underset{\sim}{k}$ is at right angles to $\underset{\sim}{H}_0$ and $\underset{\sim}{H}_0$ is in the surface plane, i.e. the special case $k_z = 0$ and $\phi = 0$ of §4. We therefore discuss mostly the Voigt geometry. We take the interface between the magnetic medium and vacuum as the plane $x = 0$ with the medium in the half-space $x < 0$. The mathematical problem is to solve eqn. (13) in the medium and the ordinary wave equation $\nabla^2\underset{\sim}{H} - (1/c^2)\ddot{\underset{\sim}{H}} = 0$ in the vacuum, together with the usual electromagnetic boundary conditions on the plane $x = 0$.

A mode localized at the surface has the form $\underset{\sim}{H} = \underset{\sim}{H}_1\exp(iky)\exp(\alpha_1 x)\exp(-i\omega t)$ for $x < 0$ and $\underset{\sim}{H} = \underset{\sim}{H}_0\exp(iky)\exp(-\alpha_0 x)\exp(-i\omega t)$ for $x > 0$ where k is real and α_0 and α_1 are real and positive. Substitution into the z-component of eqn. (13) gives $H_{1z} = 0$ and the boundary condition on $\underset{\sim}{H}$ then gives $H_{0z} = 0$. Thus the surface polariton is a TE mode.

The x- and y-components of eqn. (13) give

$$\alpha_1^2 - k^2 + (\varepsilon\omega^2/c^2)\mu_v = 0 \tag{14}$$

where $\mu_v = \mu_{xx} + \mu_{xy}^2/\mu_{xx}$ is called the Voigt permeability. Likewise

$$\alpha_0^2 - k^2 + \omega^2/c^2 = 0 \tag{15}$$

The boundary conditions at the interface take the form $\mu_{xx}H_{1x}/H_{1y} + \mu_{xy} = H_{0x}/H_{0y}$. The ratios are found from the divergence equations to be $H_{1x}/H_{1y} =$

$(ik\mu_{xx} + \alpha_1\mu_{xy})/(ik\mu_{xy} - \alpha_1\mu_{xx})$ and $H_{0x}/H_{0y} = ik/\alpha_0$, so that

$$\alpha_1 = -\alpha_0\mu_v + ik\mu_{xy}/\mu_{xx} \tag{16}$$

As long as the resonance modes are undamped, μ_{xy} is pure imaginary, and eqn. (16) involves purely real quantities.

Equations (14), (15) and (16) are simultaneous equations for the three quantities α_0, α_1 and k; elimination of α_0 and α_1 yields an equation for the dispersion function $k(\omega)$ [8]. The dispersion curves for a ferromagnet found by numerical solution are shown in Fig. 6. It is seen that propagation is non-reciprocal. The +k mode is the generalisation of the DE mode to include retardation. For large k the curve coincides with the DE one, being asymptotic to the surface frequency, which in non-reduced units is $\omega_s = [\omega_0 + \mu(\omega_0 + \omega_m)]/(\mu+1)$. For small k the DE curve (the $k_z = 0$ section of Fig. 1) has the unphysical feature that it crosses the vacuum photon line and reaches $k_y = 0$ at the frequency $\omega_v = (\omega_0^2 + \omega_0\omega_m)^{\frac{1}{2}}$. The calculation with retardation corrects this; the mode still terminates at ω_v, but at the non-zero number $k_v = (\omega_v/c)[(\omega_0+\omega_m)/\omega_m]^{\frac{1}{2}}$ (for $\varepsilon = 1$).

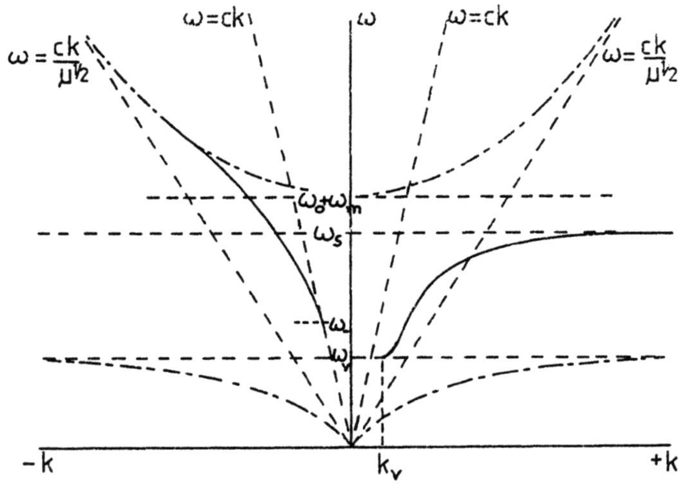

Fig. 6 Surface polariton dispersion curves (——) for the +k and -k directions on a ferromagnet in the Voigt geometry. Also shown (-·-·-) is the bulk dispersion curve $\varepsilon = 1$. After HARTSTEIN et al. [8]

In addition to the +k mode, Fig. 6 shows a mode propagating in the -k direction. It exists only for $\mu > 1$; it starts on the vacuum photon line at frequency $\omega_- = (\mu-1)^{-\frac{1}{2}}[\omega_0^2 + \mu\omega_0(\omega_0+\omega_m)]^{\frac{1}{2}}$ and merges with the bulk polariton dispersion curve at a finite frequency.

HARTSTEIN et al. [8] give a brief qualitative account of the surface polariton as k moves away from the Voigt direction, restricting attention to

$\varepsilon = 1$. As for the magnetostatic surface mode, the surface mode exists only for $\phi < \phi_s$. This is qualified by FUKUI et al. [34] who allow $\varepsilon \neq 1$. They find that there is a mode propagating along $\underset{\sim}{H}_o$, called the Faraday direction. It is a reciprocal TM mode, with $k^2 = (\omega^2/c^2)\varepsilon(\varepsilon-\mu_{yy})/(\varepsilon^2-1)$.

There has not been any direct observation of these surface ferromagnetic polaritons. The frequencies are of order the ferromagnetic resonance frequency ω_o, typically around 1 GHz, with wave numbers of order ω_o/c. In light scattering, therefore the scattering angle would have to be very small, in fact impractically small, to get down from the magnetostatic into the polariton region. Surface polaritons on dielectrics are investigated by attenuated total reflection (ATR) [35]. In principle ATR might be applied to surface ferromagnetic polaritons [36]. However, since the wavelengths are of order 1 cm or greater, specimens would have to be very large.

The difficulties of observation of surface magnetic polaritons are in principle reduced for antiferromagnets, since resonance frequencies are in the far-infrared. Light scattering might apply, although even detection of surface dielectric polaritons is a difficult experiment [37]. Surface polaritons on unaxial antiferromagnets have been considered by CAMLEY and MILLS [38]; they conclude that ATR is viable.

The magnetic polariton modes of a ferromagnetic film were first investigated by KARSONO and TILLEY [39] and by MARCHAND and CAILLE [40]. The first paper considers only modes in which k_x is imaginary. These are the generalisation to include retardation of the surface modes occupying the frequency interval $(\Omega_H^2+\Omega_H)^{\frac{1}{2}} < \Omega < \Omega_H + \frac{1}{2}$ in Fig. 1, and the generalisation to finite thickness of the surface modes of Fig. 6; Fig. 7 shows an example of the dispersion curves. Marchand and Caillé deal with modes of real k_x. These guided modes are the generalisation of those in the frequency interval $\Omega_H < \Omega < (\Omega_H^2+\Omega_H)^{\frac{1}{2}}$ of Fig. 1; in a semi-infinite medium they go over into the bulk modes.

Fig. 7 Dispersion curves (——) for a ferromagnetic slab of thickness s for $\mu = 1.25$ and $\Omega_H = 2.0$. The curves are labelled by values of the dimensionless parameter $|\gamma|M_o s/c$. The +k mode of Fig. 6 is shown (- - -); this is the $s_r \to \infty$ limit. Also shown are the magnetostatic modes. The vacuum light line and the horizontal asymptote are indicated (........). After KARSONO and TILLEY [39].

The possibilities for experiments may be assessed as for semi-infinite media. FUKUI et al. [34] calculate ATR spectra for the Karsono-Tilley type modes, but just as with a single-surface mode the long wavelength implies a specimen of very large surface area. The outlook is presumably more hopeful for antiferromagnets, but to date no theoretical work has been published on dispersion curves or ATR spectra for polariton modes in an antiferromagnetic film.

To conclude, we point out how theoretical work on magnetic surface polaritons falls short of that on magnetostatic modes. The experimental work lags well behind, since no results have been published. As far as theory is concerned:

1. Theoretical studies of dispersion curves, for both films and semi-infinite media, have mainly been concerned with the Voigt configuration. There has been no full study for all directions of propagation.

2. The only calculation for antiferromagnets is for a semi-infinite medium in the Voigt configuration [38].

3. Although extensive calculations of Green functions have been carried out for magnetostatic modes [26] and for bulk magnetic polaritons in ferromagnets [41] and antiferromagnets [42], no such calculations have been reported for surface magnetic polaritons.

References

1 E.F. Sarmento and D.R. Tilley: Chapter 16 of *Electromagnetic Surface Modes*, ed. A.D. Boardman (Wiley, New York 1982)

2 D.L. Mills: Chapter 3 of *Surface Excitations*, ed. V.M. Agranovich and R. Loudon (North-Holland, Amsterdam 1984)

3 D. Weller, S.F. Alvarado, W. Gudat, K. Schröder and M. Campagna: Phys. Rev. Lett. *54*, 1555 (1985)

4 D.T. Pierce and R.J. Celotta: Chapter 6 of *Optical Orientation*, ed. F. Meier and B. Zakharchenya (North-Holland, Amsterdam 1984)

5 S. Foner: *Magnetism* (ed. G.T. Rado and H. Suhl), Vol. 1, 393 (1963)

6 A. Brooks Harris: Phys. Rev. *143*, 353 (1966)

7 R.W. Damon and J.R. Eshbach: J. Phys. Chem. Solids *19*, 308 (1961)

8 A. Hartstein, E. Burstein, A.A. Maradudin, R. Brewer and R.F. Wallis: J. Phys. C *6*, 1266 (1973)

9 W.S. Ishak and K-W Chang: Hewlett-Packard Journal *36*, No 2, 10 (1985)

10 R.E. Camley: Phys. Rev. Lett. *45*, 283 (1980)

11 R.L. Stamps and R.E. Camley: J. Appl. Phys. *56*, 3497 (1984)

12 R.E. Camley, T.S. Rahman and D.L. Mills: Phys. Rev. B*27*, 261 (1983)

13 P. Grünberg and K. Mika: Phys. Rev. B*27*, 2955 (1983)

14 B.A. Auld: J. Appl. Phys. *31*, 1642 (1960)

15 T. Wolfram and R.E. Dewames: Prog. Surf. Sci. *2*, 233 (1972)

16 P. Grünberg: to be published in Progress in Surface Science (1985)

17 C.E. Patton: Phys. Reports *103*, 251 (1984)

18 R.E. Camley and D.L. Mills: Phys. Rev. B*18*, 4821 (1978)

19 G.T. Rado and J.R. Weertman: J. Phys. Chem. Solids *11*, 315 (1959)

20 P. Grünberg, M.G. Cottam, W. Vach, C. Mayr and R.E. Camley: J. Appl. Phys. *53*, 2078 (1982)

21 L.K. Brundle and N.J. Freedman: Electronics Lett. *4*, 132 (1968)

22 J.R. Sandercock: *Topics in Applied Physics, Vol 51 - Light scattering in solids II* (Springer-Verlag, Berlin 1982)

23 W. Hayes and R. Loudon: *Scattering of Light by Crystals* (Wiley, New York 1978)

24 M.G. Cottam and D. Lockwood: forthcoming monograph

25 W. Wettling, M.G. Cottam and J.R. Sandercock: J. Phys. C*8*, 211 (1975)

26 M.G. Cottam and A.A. Maradudin: Chapter 1 of *Surface Excitations*, ed. V.M. Agranovich and R. Loudon (North Holland, Amsterdam 1984)

27 P. Grünberg and F. Metawe: Phys. Rev. Lett. *39*, 1561 (1977)

28 J.R. Sandercock and W. Wettling: J. Appl. Phys. *50*, 7784 (1979)

29 M. Grimsditch, A. Malozemoff and A. Brunsch: Phys. Rev. Lett. *43*, 711 (1979)

30 M.G. Cottam: J. Phys. C *16*, 1573 (1983)

31 M. Grimsditch, M.R. Kahn, A. Kueny and I.K. Schuller: Phys. Rev. Lett. *51*, 498 (1983)

32 R.W. Sanders, V. Jaccarino and S.M. Rezende: Solid State Comm. *28*, 907 (1978)

33 R.W. Sanders, R.M. Belanger, M. Motokawa, V. Jaccarino and S.M. Rezende: Phys. Rev. B*23*, 1190 (1981)

34 M. Fukui, H. Dohi, J. Matsuura and O. Tada: J. Phys. C *17*, 1783 (1984)

35 A. Otto: Festkörperprobleme *XIV*, 1 (1974)

36 J. Matsuura, M. Fukui and O. Tada: Solid State Comm. *45*, 157 (1983)

37 S. Ushioda and R. Loudon: Chapter 12 of *Surface Polaritons*, ed. V.M. Agranovich and D.L. Mills (North Holland, Amsterdam 1982)

38 R.E. Camley and D.L. Mills: Phys. Rev. B*26*, 1280 (1982)

39 A.D. Karsono and D.R. Tilley: J. Phys. C *11*, 3487 (1978)

40 M. Marchand and A. Caillé: Solid State Comm. *34*, 827 (1980)

41 E.F. Sarmento and D.R. Tilley: J. Phys. C *9*, 2943 (1976)

42 E.F. Sarmento and D.R. Tilley: J. Phys. C *10*, 795 (1977)

Surface Vibrational Excitations in Noble Metals

V. Bortolani and G. Santoro

Department of Physics and G.N.S.M., University of Modena,
I-41100 Modena, Italy

1. Introduction

The increasing sophistication of the experimental techniques recently reached in surface physics has put in evidence new elementary excitations. One of the more exciting developments is related to the mesure of surface vibrational properties [1]. Inelastic atomic scattering [2], electron energy-loss spectrocscopy [3], Brillouin scattering [4] have been used successfully to study surface phonons of alkali halides, noble metals, semiconductors and chemisorbed systems. In particular, the high resolution (≈0.3 meV) obtained with inelastic atomic scattering has shown the existence of new excitations. These localised and resonant vibrational modes are not only related to the lack of translational invariance normal to the surface, as one would expect , but also to the variation of the interatomic forces occurring in the surface region. The difference between the surface interactions and the bulk ones, particularly important in noble metals, is responsible for the occurence of new phonon branches in the continuous part of the spectrum. These lectures are devoted to the interpretation of the experimental inelastic atom-surface reflection coefficients of noble metals through realistic calculations of the surface phonon spectra. At present, a microscopic theory which fully accounts for the termination of the electron gas exists only for simple metals [5,6] but is not applicable to noble and transition metals. Therefore for such materials we will use a phenomenological approach based on the force constant parametrisation of the bulk dynamics.

We will first discuss the theory of the He-metal interaction. For metals having a small corrugated surface we use the distorted wave Born approximation (DWBA) [7]. The atom surface interaction is treated in Sec. 2 where explicit formulae for the cross-section are derived in terms of the phonon displacement field. The interaction potential which enters in the cross-section is discussed in Sec. 3. The bulk dynamical problem is presented in Sec. 4. In the last section (5) we determine the surface displacement field and the surface phonons by solving the dynamical problem of a slab. We will consider the force constants in the surface region different from those of the bulk because of the different role played by the conduction electrons in the surface region. We will show that the change of the surface-force constants allows us to explain the peaks observed in the time-of-flight experiments in silver and gold.

2. Cross-Section in the Distorted Wave Born Approximation

Explicit formulae for the inelastic one-phonon scattering cross-section in the DWBA have been derived some years ago by CABRERA et al. [8]. In this section we present the essential points of the derivation. For a two-body potential $V(\mathbf{r}-\mathbf{r}_l-\mathbf{u}_l)$ between the gas atom at $\mathbf{r}$ and the atoms in the solid at $\mathbf{r}_l+\mathbf{u}_l$ the interaction is:

$$\sum_l \mathbf{u}_l \cdot \nabla_{\mathbf{r}_l} V(\mathbf{r}-\mathbf{r}_l) \tag{1}$$

to first order in the phonon-induced displacements $\mathbf{u}_l$. For the repulsive part of the interaction, only the atoms in the surface layer contribute appreciably to the sum in (1). The l-dependence of the $\mathbf{u}_l$, for a phonon of lateral momentum $\mathbf{Q}$, is then given by $\exp(i\mathbf{Q}.\mathbf{R}_l)$ (capital letters denote vectors in the plane of the surface). The matrix element of the interaction in (1) must be taken between eigenstates of the static potential:

$$\sum_l V(\mathbf{r}-\mathbf{r}_l). \tag{2}$$

Only the zero Fourier component of the potential $V_0(z)$ is appreciable for smooth metallic surfaces such as the (111) surface of the noble and transition metals. The eigenfunctions of $V_0(z)$ are then of the form $\chi_{k_z}(z)\ \exp(i\mathbf{K}.\mathbf{R})$ where, for an atom of energy E and mass m:

$$k_z{}^2 + K^2 = 2mE. \tag{3}$$

The matrix elements for transitions from $(\mathbf{K}_i, k_{iz})$ to $(\mathbf{K}_f, k_{fz})$ contain the integral:

$$\sum_l e^{i\mathbf{Q}.\mathbf{R}_l}\ \nabla_{\mathbf{r}_l} \int_S d^2R\, e^{i(\mathbf{K}_i-\mathbf{K}_f).\mathbf{R}}\, V(\mathbf{r}-\mathbf{r}_l) \tag{4}$$

which vanishes unless $\mathbf{Q}=\mathbf{K}_f-\mathbf{K}_i+\mathbf{G}$ where $\mathbf{G}$ is a reciprocal lattice vector. Ultimately then, we need to compute:

$$N \int_{-\infty}^{0} dz\, \chi_{iz}(z)\, \chi_{fz}(z) \int_S d^2R\, e^{-i(\mathbf{Q}+\mathbf{G}).\mathbf{R}}\, \nabla_{\mathbf{r}_l} V(\mathbf{r}-\mathbf{r}_l)\Big|_{z_l=0} \tag{5}$$

where N is the number of atoms for unit surface area and $\chi_{iz}(z)$, $\chi_{fz}(z)$ are eigenfunctions of the laterally averaged potential:

$$V_{AV}(z) \equiv V_0(z) = N \int_S d^2R\, V(\mathbf{R},z). \tag{6}$$

The Fourier components of the potential then determine both the dynamic interaction and the static potential. ∇_l can be replaced by $[i(\mathbf{Q}+\mathbf{G}), \partial/\partial z]$. By taking the square modulus of (5) and by summing all processes in the volume $d^2K_f dk_{fz}=(2mE_f)^{1/2}m\ dE_f$ and dividing by $dE_f d\Omega_f$ we obtain the differential reflection coefficient or cross-section per unit surface area. For flat surfaces only the $\mathbf{G}=0$ term can be retained, so that we obtain:

$$\frac{d^2R}{dE_f d\Omega_f} = \frac{1}{NM} \sum_{\mathbf{Q},j} \frac{(2mE_f)^{1/2} k_{iz}}{\omega(\mathbf{Q},j)}\, n[\omega(\mathbf{Q},j)]\, e^{-2W} \times$$

$$\left| \mathbf{e}(\mathbf{Q},j).\langle \chi_{iz} | (i\mathbf{Q}, \frac{\partial}{\partial z}) V(\mathbf{Q},z) | \chi_{fz} \rangle \right|^2 \delta(\mathbf{K}_f-\mathbf{K}_i-\mathbf{Q})\ \delta[E_f-E_i-\omega(\mathbf{Q},j)]. \tag{7}$$

Here we have introduced the phonon frequencies $\omega(\mathbf{Q},j)$ and the polarization vectors $e(\mathbf{Q},j)$. $n(\omega)$ is the Bose factor, phonon creation (energy loss) corresponds to $\omega(\mathbf{Q},j)<0$; phonon annihilation (energy gain) corresponds to $\omega(\mathbf{Q},j)>0$. $V(\mathbf{Q},z)$ is the 2-dimensional Fourier transform of the potential where now $\mathbf{Q}$ is considered in the extended zone scheme. The Debye Waller factor 2W is given by:

$$2W = \sum_{\mathbf{Q},j} \left[|k_{iz}|+k_{fz}\right]^2 <|\mathbf{e}(\mathbf{Q},j)|^2>. \tag{8}$$

It must be remarked that, due to the form of the matrix element in (7), only the displacements that are polarised in the sagittal plane (defined by $\mathbf{Q}$ and by the surface normal) can be detected in scattering experiments. As a consequence, for in-plane scattering the shear horizontal (SH) modes (polarised perpendicular to the sagital plane) do not contribute to the scattering intensity.

3. Atom-Surface Potential

The atom surface potential is generally divided as follows:

$$V(\mathbf{r}) = V_R(\mathbf{r}) + V_A(z). \tag{9}$$

The short-range repulsive part V_R arises from the overlap of the electronic cloud of the atom and the surface electrons, while the long-range attractive polarization potential V_A has the asymptotic form:

$$V_A(z) = -\frac{C}{z^3}. \tag{10}$$

To determine the repulsive part, which gives the major contribution to the matrix elements in (7) we use the ESBJERG and NORSKOV [9] approximation:

$$V_R(z) = \alpha\, \overline{\rho(\mathbf{r})} \tag{11}$$

where α can be determined from electron-He atom scattering data. $\rho(r)$ is the surface charge density averaged with the atom electrostatic potential. We have found that the averaging effects of the He atom on the surface charge density at the typical incident energies used in atom-scattering experiments (E_i =8-30 meV) are a minor correction,so that in (11) we will take directly the charge density $\rho(r)$. We then evaluate $\rho(r)$ with a superposition of atomic charges. The reason for using a simple superposition of atomic charges is that the impinging He atom experiences the charge density at large distances from the surface, where the surface states and the bonding effects are negligible.

Equation 11 is evaluated at the classical turning point for which:

$$V_R(\mathbf{R},z_T) = E_i. \tag{12}$$

Around this region the potential can be written as:

$$V_R(\mathbf{R},z) = E_i\, e^{-\beta[z-z_1-D(\mathbf{R})]} \tag{13}$$

where β is the softness parameter and $D(\mathbf{R})$ is the corrugation function [10]. The corrugation is measured with respect to z_T evaluated at R=0 so that D(0)=0. To determine β and D we have evaluated $\rho(r)$ with a thirty atom cluster. In Fig. 1 is plotted the corrugation function $D(\mathbf{R})$ for Ag(111) with $\mathbf{R}$ along the $[11\bar{2}]$ surface direction.

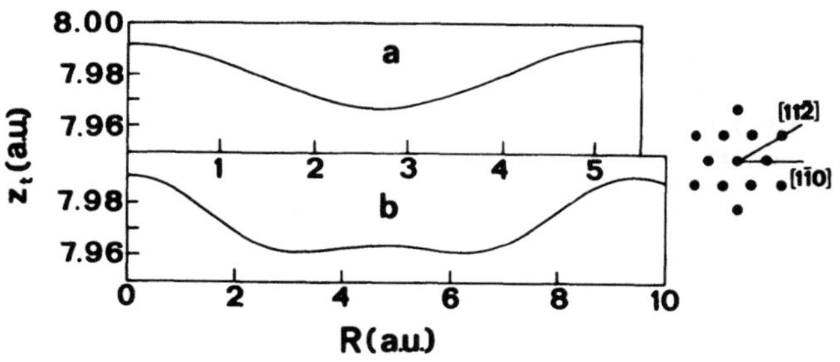

Fig.1 Corrugation profile D(**R**) for Ag (111) with R along the Σ [11$\bar{2}$] and T [1$\bar{1}$0] directions. E_i=17.5 meV

We have found an analytic form of the potential which reproduces very accurately the results of the full calculation:

$$V_R(\mathbf{R},z) = V \sum_l e^{-\beta[(\mathbf{R}-\mathbf{R}_l)^2+(z-z_l)^2]^{1/2}}. \tag{14}$$

We have considered only surface atoms at z_l=0. From our calculations for noble metals,it turns out [10] that the classical turning point z_t is very large (z_t>4 Å). For this reason, if the coordinate R happens to be in the l-th surface cell $|\mathbf{R}-\mathbf{R}_l|<<z_t$. In this cell we can expand the potential as:

$$e^{-\beta[(\mathbf{R}-\mathbf{R}_l)^2+z^2]^{1/2}} \simeq e^{-\beta z}\, e^{-\frac{\beta}{2z_t}(\mathbf{R}-\mathbf{R}_l)^2}. \tag{15}$$

This expansion has to be applied to all the nearest neighbors. For all the other neighbors,the contribution of the two-body potential to the total potential is negligible,so that this approximation can be performed without altering the accuracy of the total potential. V_R can thus be written as:

$$V_R(\mathbf{R},z) = V e^{-\beta z} \sum_l e^{-\frac{\beta}{2z_t}(\mathbf{R}-\mathbf{R}_l)^2}. \tag{16}$$

The 2-dimensional Fourier transform of the potential has the form:

$$e^{-Q^2/Q_c^2}. \tag{17}$$

This lateral cut-off gives a quantitative explanation of the Armand effect i.e. of the reduction in the cross-section produced by the cooperative movement of neigboring surface atoms. The cut-off factor reduces the scattering intensities for large lateral momentum transfer. We want to point out that this cut-off is just associated with the deformation of charge produced by a single moving atom and exists because of the softness of the repulsive potential.

We turn now to discuss the long-range attractive dipolar interaction. As shown by ZAREMBA and KOHN [11] an approximate form of V_A is given by:

$$V_A(z) = -\frac{C_{VW}}{(z-z_{VW})^3} \tag{18}$$

where for Au C_{VW}=0.27 eV Å^3 and z_{VW}=1.33 Å. The divergence of V_A at $z=z_{VW}$ has been recently considered by NORLANDER and HARRIS [12]. These authors found that the

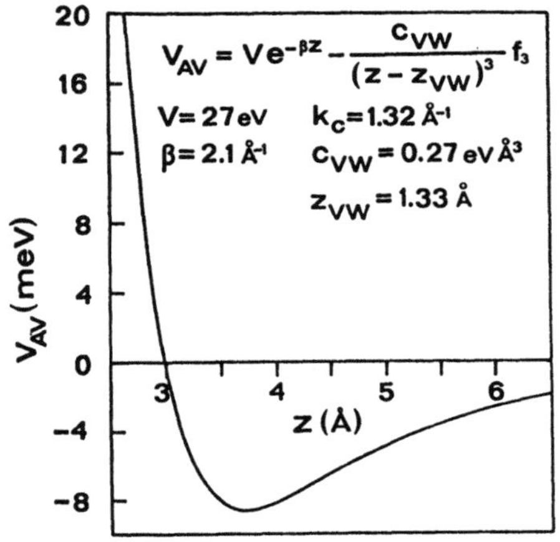

Fig.2 The laterally averaged potential for He:Au(111)

inclusion of exchange and correlation between the He atom and the surface regularize the divergence. In fact according to these authors V_A has to be multiplied by the function:

$$f_3(x) = [1-2x(1+x)+1]e^{-2x}\,; \qquad x=k_c(z-z_{VW}) \tag{19}$$

that eliminates the divergence. k_c is a free parameter of the order of the inverse of the atomic radius. The laterally averaged total potential can then be written as:

$$V_{AV}(z) = V\,e^{-\beta z} - \frac{C_{VW}}{(z-z_{VW})^3}\, f_3[k_c(z-z_{VW})]\,. \tag{20}$$

The value of k_c can be obtained by fitting for a given surface the physisorption well or the bound states of the He-metal potential. In Fig. 2 is drawn this potential for He/Au(111).

4. Bulk Phonons

The pseudopotential theory for metals indicates that the effective interaction between atoms contains at least two-body and three-body contributions. The last contribution becomes particularly important in the case of noble and transition metals,where the d interaction produces directional forces. To take into account of the functional dependence of these interactions,we consider an effective potential of the form:

$$v = \sum_{i,j} v^{(2)}(r_{ij}) + \sum_{i,j,k} v^{(3)}(\cos\theta_{ijk})\,. \tag{21}$$

Here $r_{ij} = |\mathbf{r}_i - \mathbf{r}_j|$ and θ_{ijk} is the angle centred on atom i formed by the vectors $\mathbf{r}_i-\mathbf{r}_j$ and $\mathbf{r}_i-\mathbf{r}_k$. The tangential ($\alpha$) and radial ($\beta$) force constans are defined by:

$$\alpha_{ij} = \frac{1}{r_{ij}}\frac{dv^{(2)}}{dr_{ij}}\,; \qquad \beta_{ij} = \frac{d^2v^{(3)}}{dr_{ij}^2}\,. \tag{22}$$

We also introduce a second-order angular force constant by:

$$\delta_{ijk} = \frac{d^2V^{(3)}}{d\cos\theta_{ijk}^2} \frac{1}{3a_0^2}. \tag{23}$$

The bulk dynamical matrix depends linearly on the force constants. With a weighted least square fitting procedure, the values of the force constants are determined from the experimental data. We fit the phonon spectra measured with neutron scattering experiments and the elastic constants. The fitting procedure is carried out by considering angular forces up to second nearest neighbors (n.n.), because of their short range nature. The range of the central forces is chosen to reproduce all the experimental data to within a few percent. In the case of Au, where is present a Kohn anomaly in the Σ direction, it is necessary to consider interactions up to four nearest neighbors. We want to stress that with our procedure,the leading force constants (those relative to the first and second shells of neighbors) converge to a definite values when the interaction length is taken sufficiently large. In the case of Pt, where the Kohn anomaly is more pronounced, the procedure gives convergent results only when 6 n.n. are considered in the fitting [13]. In this case,the overall agreement between theory and experiments is excellent,and the Kohn anomaly is reproduced accurately.

5. Surface Phonons

To determine the surface phonons,we study the dynamics of a slab of N planes. The calculations presented here refer to an ideal surface without relaxation or reconstruction. By solving the 3Nx3N dynamical problem we obtain, for a given phonon lateral momentum **Q**, the 3N frequencies $\omega(j,Q)$ of the slab and the corresponding 3N dimensional polarization vectors $\mathbf{e}(j,Q)$ with j=1,3N. The elements of the dynamical matrix which depends on the force constants defined in the previous section are too lenghtly [13] and are not reported here. The value of N is taken large enough (N>45) in order to eliminate interference effects between the two surfaces. The splitting in energy due to the modes of the second surface is only of importance in a region of $Q<Q_{BZ}/N$. Here Q_{BZ} is the zone border lateral momentum. A part this small region,our results reproduce the phonon spectrum of a semiinfinite crystal over the entire 2-D Brillouin zone. We firstly discuss the results obtained by using in the dynamical matrix of the slab the force constants derived from the bulk fitting without considering any modification of the interactions in the surface region.

The comparison with the recent experimental surface phonons measured by DOAK et al. [14] is very puzzling and similar for all the noble metals. The localized Rayleigh wave (RW) is only well represented in the first half of the Brillouin zone. However, the most striking result is the absence in the theoretical calculations [15] of a phonon branch located in the continuum of the bulk projected branches, which instead gives a well-defined peak in the experimental spectra. We want to note that the discrepancies between theory and experiment are larger for Au than for Ag and Cu. The origin of this disagreement is imputed to the different screening properties of the electron gas at the surface with respect to the bulk region. The inverse dielectric function has been evaluated perturbatively for a confined free electron gas [5,6]. The main result of these calculations proves that the screening properties of the surface differ from those of the bulk only for atoms lying in the surface plane. A qualitative explanation of the surface screening has been given by HEINE and MARKS [16] in terms of an attractive multi-atom s-p and s-d electron interaction against a two boby d-d electron repulsion. For flat surfaces the s-p attraction increases in the topmost layer, giving rise to a tangential compressive stress. The net force produces a lateral surface contraction and a lowering of the surface force constants. However, we can introduce in our phenomenological approach these effects by considering the surface force constants as free parameters. Of course

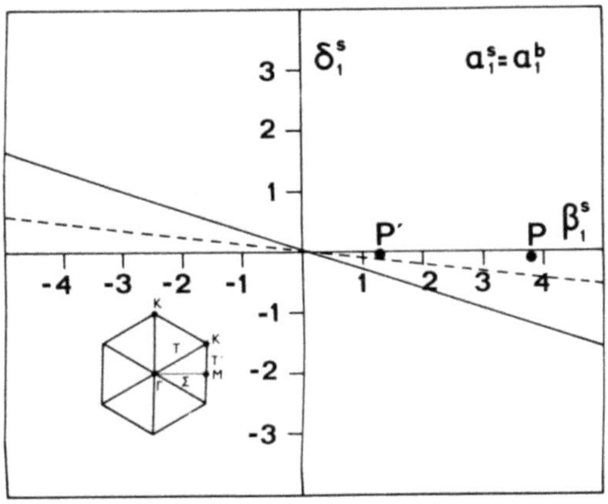

Fig.3 Stability diagram for the Ag(111) surface with $\alpha_1{}^S = \alpha_1{}^b$. The 2-D Brillouin zone is drawn in the inset

the choice of the surface force constants is not arbitrarly: only values that preserve the stability of the surface are acceptable [15].

We study now the stability problem of the Ag(111) surface. By varying arbitrarly the radial and angular surface force constants $\beta_1{}^S$ and $\delta_1{}^S$, we determine the region of surface stability by drawing the lines along which the frequency of a phonon of a given symmetry is zero [15]. The full line in Fig.3 refer to the instability of an M phonon polarized in the sagittal plane,and the dashed line corresponds to a SH M surface phonon, polarized perpendicularly to the sagittal plane. The stability region lies above the two lines where all the phonon frequencies remain real. A small lowering of the angular surface force constant $\delta_1{}^S$ makes the surface unstable for M SH modes. The most important result is obtained by reducing the radial surface force constant $\beta_1{}^S$. In this case M modes located just above the longitudinal threshold decraese their frequencies

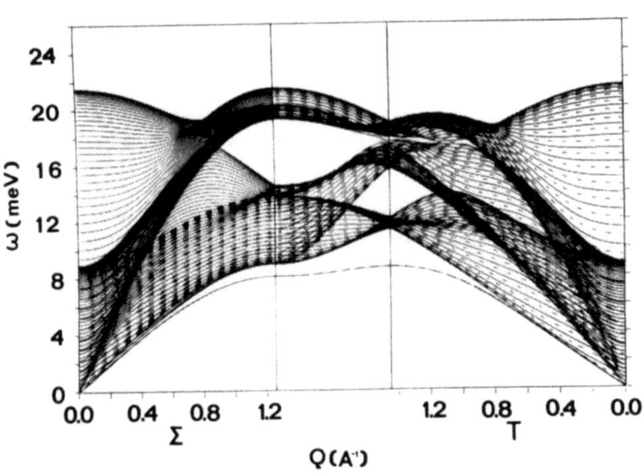

Fig.4 Surface phonons of Ag(111) with ideal surface force constants ($\beta_1{}^S = \beta_1{}^b$). In the Σ direction the dashed lines refer to pure SH modes. In the other two directions the dashed lines refer to modes odd with respect to the center of the slab. Triangles: experimental data from [14]

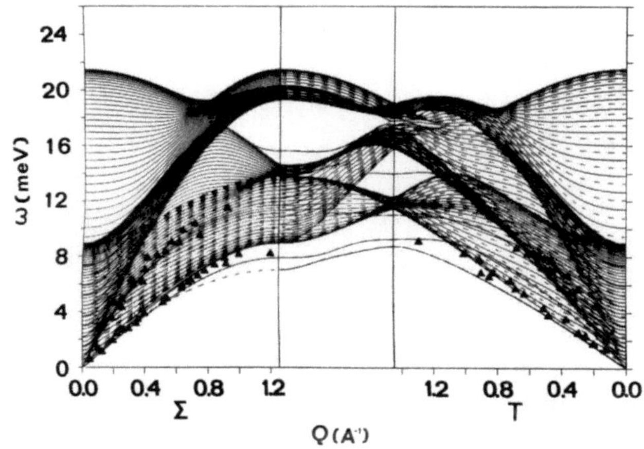

Fig.5 As in Fig.4 but with the modified force constants ($\beta_1{}^s = 0.48\beta_1{}^b$)

and become very localized. These resonant modes in M enhance the surface phonon density of states, which is observable with atomic scattering. This is illustrated in Figs. 4 and 5.

The figure 4 refers to the ideal calculation,which corresponds to the point P in the stability diagram. The figure 5 corresponds to $\beta_1{}^s=0.48\beta_1{}^b$ (point P' in Fig.3). We first discuss the phonon branches in the Σ direction.

In the Σ direction the dynamical matrix decouples in a part containing only sagittal modes and in a part relative to the SH modes. The lowering of $\beta_1{}^s$ produces a resonant mode of longitudinal character (LR) polarized along **Q**, which explains the experimental branch observed in the continuous part of the spectrum. The situation is more complicated for the T [1$\bar{1}$0] direction. With **Q** along T,the symmetry of the slab does not allow the decoupling between the sagittal and SH components of the displacement. The coupling between the two components produces a Pseudo Rayleigh Wave (PRW) [17]. This mode, for $\beta_1{}^s=\beta_1{}^b$, becomes a sharp resonance in the acoustic limit [17] and broadens at large **Q**. The reduction of $\beta_1{}^s$ produces a localization of the PRW over the entire zone and lowers considerably its frequency for $Q>0.2Q_{BZ}$. The PRW and the RW hybridize and there is an exchange in their polarization, as shown in Fig.6. The PRW acquires a vertical component (in the intermediate Q region $0.6Q_{BZ}<Q<0.8Q_{BZ}$) which makes detectable this mode. At the zone boundary the PRW is circularly polarized on the surface and the RW becomes a Shear Vertical (SV) mode. As in the case of the Σ direction, the lowering of the force constants is responsible for the appearing of the longitudinal resonance which is observable in the reflection coefficient for $Q<0.8Q_{BZ}$.

The surface phonons of Au(111) are presented in Fig.7. The calculations are performed by considering central forces up to 4 n.n. and angular forces up to 2 n.n.. To fit the frequency of the surface phonons, we have to reduce the surface force constants more than in the case of Ag. We belive that this strong reduction is related to the repulsive d-d overlapping interaction which enters the total effective potentials. This overlap energy is greater for the 5-d wavefunction of Au than for the 4-d wavefunction of Ag. At the surface, the attractive s-p interaction is strongly cancelled by the repulsive d-d interaction more for Au than for Ag. The large lowering of the force constants introduces new features in the phonon spectra. In the Σ direction the LR decreases in frequency and becomes

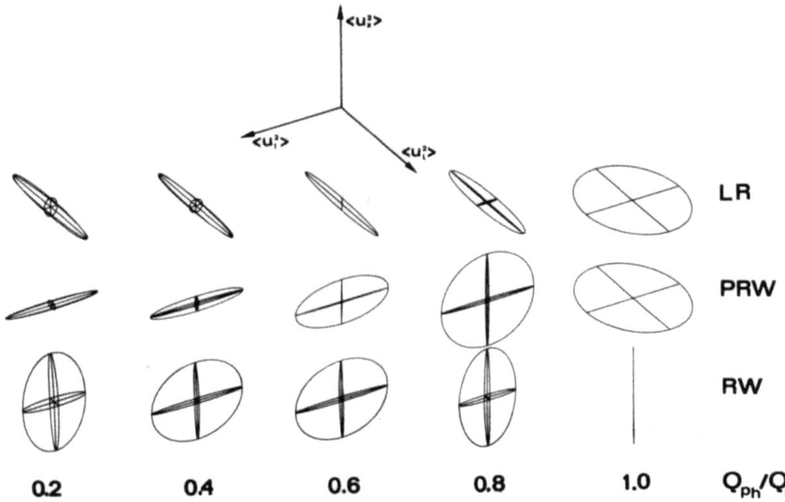

Fig.6 Polarization ellipses of the three relevant localized modes of Ag(111) with **Q** along T[1$\bar{1}$0]. The vectors $\langle u_l \rangle$ and $\langle u_z \rangle$ define the sagittal plane.

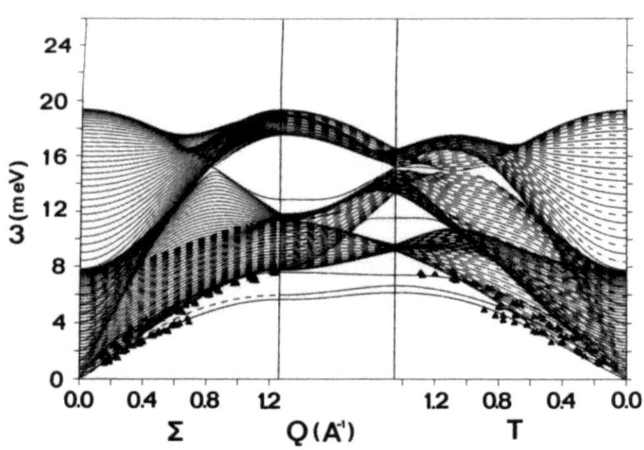

Fig.7 Surface phonons of Au(111) with modified force constants ($\beta_1^s = 0.3\beta_1^b$; $\delta_1^s = 0.5\delta_1^b$). Continuous and dashed lines as in Fig.4. The experimental data (triangles) are from [18].

a localized mode at the zone boundary. The strong hybridization between the LR and RW modes makes the LR mode detectable over the entire Brillouin zone, while the scattering intensity of the RW mode decreases at large **Q**. However, the disappearance of the experimental Rayleigh branch in the second half of the BZ can be fully explained by including in the calculations the solitonic reconstruction of the Au(111) surface. In the T direction there is a strong hybridization among the RW, PRW and LR modes. At large Q, the LR acquires a large SV polarization and becomes the only observable mode, as illustrated by the polarizations depicted in Fig.8. In the intermediate Q region, all the modes are detectable, while at small Q the PRW mode becomes almost SH, so that one observe the RW and LR modes.

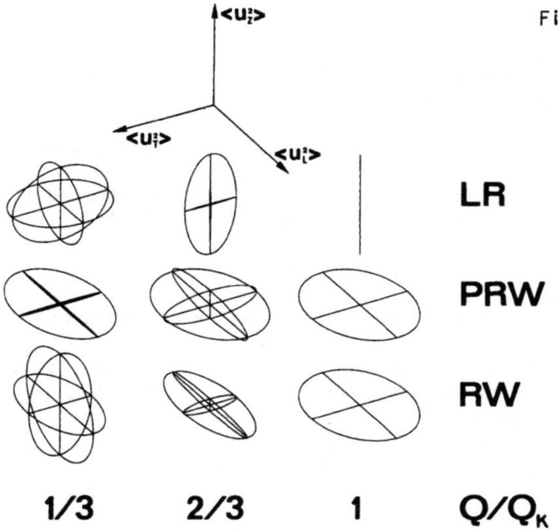

Fig.8. As in Fig.6 for Au(111)

To make a more direct comparison with the experiments, we present in Fig.9 the calculated and experimental time-of-flight spectra of Au(111). The experimental spectra have been obtained with the tilt angle between the incident and scattered beams fixed at 90°. θ_i represents the incidence angle with respect to the surface normal. The presence both in the theory and in the experiments of the second peak in addition to the Rayleigh one gives the convincing proof of the existence of the LR. Furthermore, by comparing the calculated and the experimental spectra, one can determine the relative amplitude of the SV and longitudinal displacements of the relevant modes that produces the peaks in the cross-section. In many cases this is necessary, because there is a very delicate balance between the polarizations entering the square modulus of (7). This also shows that inelastic scattering experiments can be useful in studying surface reconstruction. In particular, as already mentioned, it is possible [19] to explain the disagreement between the calculated and the experimental intensities of the RW peak at large **Q** (Fig.9b) in terms of the surface reconstruction.

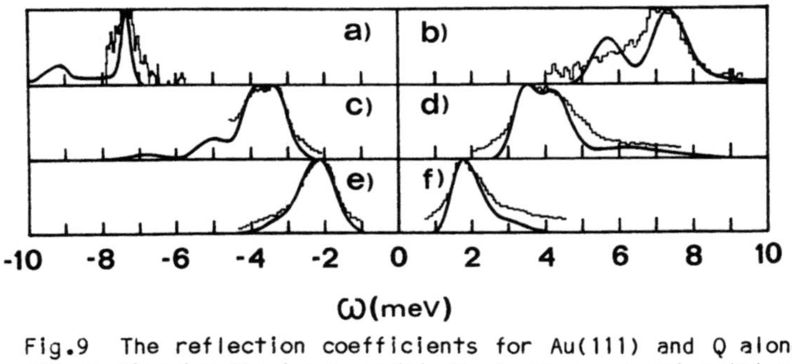

Fig.9 The reflection coefficients for Au(111) and Q along Σ. Full lines: calculation for the 4 n.n. model. Hystograms: experimental spectra at T=100°K [18]. a) θ_i=19° and E_i=11.2 meV; b) θ_i=63° and E_i=8.2 meV; c) θ_i=36° and E_i=12.1 meV; d) θ_i=55° and E_i=8.1 meV; e) θ_i=40° and E_i=12.3 meV; f) θ_i=50° and E_i=8.1 meV

References

1. G.I. Stegeman and F. Nizzoli: "Surface Vibrations" in **Surface Excitations**, Ed. V.M. Agranovich and R. Loudon (North-Holland, Amsterdam, 1984) chap. 4
2. M.G. Brusdeylins, R.B. Doak and J.P. Toennies: Phys.Rev.Lett. **16**, 937 (1981)
3. S. Lehwald, J.M. Szeftel, H. Ibach, T.S. Rahman and D.L. Mills: Phys. Rev. Lett. **5**, 518 (1983).
4. J.R. Sandercock:"Trends in Brillouin Scattering" in **Light Scattering in Solids III**, Ed. M. Cardona and G. Guntherodt (Springer, Berlin, 1982) p. 173
5. C. Beatrice and C. Calandra: Phys. Rev. **B29**, 6139 (1983); C. Calandra, A. Catellani and C. Beatrice: Surface Sci. **148**, 90 (1984)
6. A. Eguiluz, A.A. Maradudin and R.F. Wallis: to be published
7. V. Celli: "Theory of atom surface scattering" in **Dynamics of Gas-Surface Interaction**, Ed. G.Benedek and U.Valbusa (Springer, Berlin, 1982) p. 1
8. N. Cabrera, V. Celli and R. Manson: Phys. Rev. Lett. **22**, 346 (1969); N. Cabrera, V. Celli, F.O. Goodman and R. Manson: Surface Sci. **19**, 67 (1970); R. Manson and V. Celli: Surf. Sci. **26**, 695 (1971)
9. N. Esbjerg and J.K. Norskov: Phys. Rev. Lett. **45**, 807 (1980)
10. V. Bortolani, A. Franchini, N. Garcia, F. Nizzoli and G. Santoro: Phys. Rev. **B28**, 7358 (1983)
11. E. Zaremba and W. Kohn: Phys. Rev.**B15**, 1769 (1977)
12. P. Norlander and J. Harris: J. Phys. C: Solid St. Phys. **17**, 1141 (1984)
13. V. Bortolani, A. Franchini and G. Santoro: "Surface phonon calculation in metals and comparison with the experimental techniques" in **Electronic Structure,Dynamics and Quantum Structural Properties of Condensed Matter** (Plenum, New York 1985) p. 401
14. R.B. Doak, U. Harten and J.P. Toennies: Phys. Rev. Lett. **51**, 578 (1983)
15. V. Bortolani, A. Franchini, F. Nizzoli and G. Santoro: Phys. Rev. Lett. **52**, 429 (1984)
16. V. Heine and L.D. Marks: to be published
17. G.W. Farnell: "Properties of elastic surface waves" in **Physical Acoustics**, vol. VI, Ed. W.P.Mason and R.N. Thurston (Academic Press, New York, 1970) p. 109
18. V. Bortolani, G. Santoro, U. Harten and J.P. Toennies: Surface Sci. **148**, 82 (1984) and to be published
19. V. Bortolani, G. Santoro, U. Harten and J.P. Toennies: to be published

Electromagnetic Surface Excitations on Rough Surfaces

A.A. Maradudin

Department of Physics, University of California, Irvine, CA 92717, USA*
and
Max-Planck-Institut für Festkörperforschung, Heisenbergstrasse 1,
D-7000 Stuttgart 80, F. R. G.

I. Introduction

All solid surfaces, even very carefully prepared ones, possess some degree of roughness. It becomes a matter of interest, then, to examine the consequences of that roughness for physical phenomena associated with surfaces. The propagation of surface waves across a rough surface has a particular interest in this regard, because the fields associated with such waves can be localized to the surface within a distance comparable with the lengths characterizing the surface roughness. The interaction between a surface wave and surface roughness can be very strong as a result.

For example, when a surface wave propagates across a rough surface, it can be scattered by the hills and valleys on the surface into other surface waves and into bulk waves. The surface wave is attenuated thereby, even if it is propagating on the surface of a lossless medium. It can be important for applications of surface waves to know what degree of surface roughness is tolerable before the attenuation to which it gives rise reaches an unacceptable level. The propagation of surface waves across periodically corrugated surfaces give rise to pass bands and stop bands in their dispersion curves that can be useful in the design of filters for surface waves, to the possibility of the controlled conversion of surface waves into bulk waves, and to the phenomenon of wave slowing. In addition, surface roughness gives rise to new types of surface waves that cannot exist in the absence of roughness.

From a comparison of theoretical and experimental results for the propagation characteristics of surface waves across a rough surface, it should also be possible to obtain useful information concerning the nature of the surface, the extent to which it is ideal and the ways in which it departs from ideality.

We will study all of these consequences of surface roughness on the propagation of surface waves for the particular case of surface electromagnetic waves (surface polaritons). These are electromagnetic waves that propagate in a wavelike fashion along the surface of a dielectric medium and whose

---

*Permanent address.

amplitudes decay exponentially with increasing distance from the surface into the dielectric and into the vacuum in contact with it.

In these lectures I will consider three types of rough surfaces, each of which would be planar in the absence of the roughness. The first type of rough surface is characterized by a surface profile that is deterministic but not periodic. It is realized, for example, by an isolated protuberance or indentation on an otherwise planar surface. The second type of rough surface is characterized by a surface profile that is deterministic but periodic in one or two directions. It corresponds to a classical grating or bigrating, respectively. The third type of rough surface is characterized by a stochastic surface profile, and is a randomly rough surface.

To present a discussion of the effects of all three types of surface roughness on surface polaritons from a unified point of view, I will base it on the Rayleigh hypothesis [1], the vectorial equivalent of the Kirchhoff integral [2], and on the extinction theorem [3]. The Rayleigh hypothesis is the assumption that the expressions for the electromagnetic fields that are valid in the regions of space above the maximum vertical excursion of the surface profile from flatness, and below the minimum excursion, can be continued in to the surface itself and used in satisfying the boundary conditions there. It is known [4] that the Rayleigh hypothesis is invalid for a surface profile that is nonanalytic, e. g. for one that has corners or cusps. It is also known [5-8] that this hypothesis has a limited applicability to periodically corrugated surfaces described by analytic profiles: crudely put, the ratio of the maximum amplitude of such a profile to its period cannot be too large for it to be valid, and a definite prescription exists for determining what "too large" is for a given profile [7,8]. However, against all these limitations on the validity of the Rayleigh hypothesis, one can set empirical evidence that, particularly in the numerical determination of the dispersion curves of surface waves propagating across periodically corrugated surfaces, its use leads to convergent results in the case of nonanalytic profiles, and in the case of analytic profiles to convergent results for such large amplitude surface roughness that the Rayleigh hypothesis must surely be invalid [9-13]. It has been argued [7,14] that this numerical, as opposed to mathematical, convergence may be asymptotic in nature, in a sense that will be described later, and there is now some evidence in support of that suggestion [12].

No rigorous information is available concerning the conditions under which the Rayleigh hypothesis is valid for a randomly rough surface, but it is reasonable to assume that they are comparable to those for periodically corrugated surfaces. In what follows, we will assume that the use of the Rayleigh hypothesis is valid for the randomly rough surfaces we consider.

As was indicated above, surface polaritons are electromagnetic surface waves, i.e., they are solutions of Maxwell's equations in which the effects of retardation--the finiteness of the speed of light--are included. An important sub-class of surface polaritons are surface plasmons [15]. These can be viewed as the limiting case of surface polaritons when the speed of light is allowed to become infinitely large. Alternatively, and equivalently, they are solutions of Laplace's equation that propagate along the surface of a dielectric medium, and whose amplitudes decay exponentially with increasing distance from the surface into the dielectric and into the vacuum adjoining it. They are therefore electrostatic surface waves.

Despite the fact that surface plasmons are a limiting case of surface polaritons, in an operational sense they can be said to have an existence of their own, independently of the existence of surface polaritons, since there are experiments that measure them directly. These involve, for example, external probes that have a longitudinal electric field associated with them, such as a beam of electrons, whose wavelength is shorter than the vacuum wavelength of the dipole active excitation in the dielectric medium with which the surface plasmon is associated.

Surface roughness can affect surface plasmons just as it affects surface polaritons. It is often easier, however, to treat the effects of surface roughness on surface plasmons by starting directly from Laplace's equation, rather than by passing to the limit of an infinite speed of light in the corresponding results for a surface polariton. It is the former approach to the interaction of surface plasmons with surface roughness that will be taken in these lectures.

It is my purpose in these lectures to survey existing theoretical work on the interaction of surface polaritons and plasmons with surface roughness, and to present some new results. It is not my intention, however, merely to present a compilation of results and references to the literature. These lectures have been written with the deliberate aim of describing the theoretical techniques that have been developed for dealing with such problems in sufficient detail that they can serve as a self-contained guide for anyone interested in carrying out such calculations for himself/herself.

At the same time, some experimental results bearing on the theoretical results derived here will be presented in the text at appropriate points. Much more extensive reviews of the experimental side of this subject can be found in two articles by RAETHER [16,17].

Finally, I note an earlier review of theoretical work on the interaction of surface plasmons and polaritons with rough surfaces that covers some aspects of the subject that are not covered here [18].

II. Characterization of a Rough Surface

When one speaks of rough surfaces one instinctively thinks of randomly rough surfaces. However, that is only one type of rough surface. In these lectures I will consider three types of rough surfaces, each of which would be planar in the absence of the roughness. The latter restriction is both a reflection of the paucity of work on surface polaritons on rough nonplanar surfaces [19,20] and a consequence of restrictions imposed by both time and space.

As the planar surface perturbed by the roughness we take the plane $x_3 = 0$. For all three types of roughness the equation defining the surface is given by

$$x_3 = \zeta(\vec{x}_\parallel), \tag{2.1}$$

where the real function $\zeta(\vec{x}_\parallel)$ is called the surface profile function. It is a function of the two-dimensional position vector

$$\vec{x}_\parallel = \hat{x}_1 x_2 + \hat{x}_2 x_2, \tag{2.2}$$

where $\hat{x}_1$ and $\hat{x}_2$ are two mutually perpendicular unit vectors in the plane $x_3 = 0$. The methods that will be used in these lectures require the assumption that $\zeta(\vec{x}_\parallel)$ is a single-valued function of $\vec{x}_\parallel$.

It should be noted that this definition of a rough surface is independent of whether the medium bounded by the surface occupies the upper half-space $(x_3 > \zeta(\vec{x}_\parallel))$ or the lower $x_3 < \zeta(\vec{x}_\parallel))$. In fact, the surface defined by Eq. (2.1) could be the interface between two different media.

We will assume in each case that the region $x_3 > \zeta(\vec{x}_\parallel)$ is vacuum, while the region $x_3 < \zeta(\vec{x}_\parallel)$ is filled by a dielectric medium characterized by an isotropic, frequency-dependent dielectric constant $\varepsilon(\omega)$. The region $\zeta(\vec{x}_\parallel)_{min} < x_3 < \zeta(\vec{x}_\parallel)_{max}$ is called the selvedge region.

In many problems the unit vector normal to the surface at each point is required. If we further specify this vector to be directed from the region $x_3 < \zeta(\vec{x}_\parallel)$ into the region $x_3 > \zeta(\vec{x}_\parallel)$, it is given by the normalized gradient of the function $f(\vec{x}) = x_3 - \zeta(\vec{x}_\parallel)$, and has the explicit form

$$\hat{n} = \left[1 + \left(\frac{\partial\zeta(\vec{x}_\parallel)}{\partial x_1}\right)^2 + \left(\frac{\partial\zeta(\vec{x}_\parallel)}{\partial x_2}\right)^2\right]^{-1/2} \left(-\frac{\partial\zeta(\vec{x}_\parallel)}{\partial x_1}, -\frac{\partial\zeta(\vec{x}_\parallel)}{\partial x_2}, 1\right) . \tag{2.3}$$

It is sometimes convenient to introduce the unnormalized normal vector $\vec{n}$ defined by

$$\vec{n} = \left(-\frac{\partial\zeta(x_\parallel)}{\partial x_1}, -\frac{\partial\zeta(\vec{x}_\parallel)}{\partial x_2}, 1\right) . \tag{2.4}$$

Also often required is the derivative along the normal to the surface at each point. In the case that the normal is directed from the region $x_3 < \zeta(\vec{x}_\parallel)$ into the region $x_3 > \zeta(\vec{x}_\parallel)$ the normal derivative is given by

$$\frac{\partial}{\partial n} = \hat{n}\cdot\nabla = \left[1 + \left(\frac{\partial\zeta(\vec{x}_\parallel)}{\partial x_1}\right)^2 + \left(\frac{\partial\zeta(\vec{x}_\parallel)}{\partial x_2}\right)^2\right]^{-1/2} \times$$

$$\times \left(-\frac{\partial\zeta(\vec{x}_\parallel)}{\partial x_1}\frac{\partial}{\partial x_1} - \frac{\partial\zeta(\vec{x}_\parallel)}{\partial x_2}\frac{\partial}{\partial x_2} + \frac{\partial}{\partial x_3}\right) . \tag{2.5}$$

Finally, the differential element of surface area, dS, on the surface defined by Eq. (2.1) can be expressed as

$$dS = \frac{d^2x_\parallel}{(\hat{n}\cdot\hat{x}_3)} = \left[1 + \left(\frac{d\zeta(\vec{x}_\parallel)}{\partial x_1}\right)^2 + \left(\frac{\partial\zeta(\vec{x}_\parallel)}{\partial x_2}\right)^2\right]^{1/2} d^2x_\parallel , \tag{2.6}$$

where $d^2x_\parallel$ is the differential element of area on the plane $x_3 = 0$ and $\hat{x}_3 = (0, 0, 1)$ is the unit vector normal to the plane $x_3 = 0$ in the $+x_3$-direction.

We now turn to a discussion of the three types of surface roughness that can be described by Eq. (2.1).

II.A. A Deterministic Nonperiodic Profile

For the first type of rough surface $\zeta(\vec{x}_\parallel)$ is deterministic and is essentially nonzero only for a finite range of values of x_1 and/or x_2. Examples of such profiles that depend on a single coordinate only are

$$\zeta(x_1) = A\exp(-x_1^2/R^2) \tag{2.7a}$$

$$\zeta(x_1) = \frac{AR^2}{x_1^2 + R^2} , \tag{2.7b}$$

and describe ridges (A > 0) or grooves (A < 0) on an otherwise planar surface. Examples of profiles of this type that depend on both x_1 and x_2 are

$$\zeta(\vec{x}_\parallel) = A\exp(-x_\parallel^2/R^2) \tag{2.8a}$$

$$\zeta(\vec{x}_\parallel) = A\exp(-x_\parallel/R) \tag{2.8b}$$

$$\zeta(\vec{x}_\parallel) = \frac{A}{e^{\beta(x_\parallel - R)} + 1} , \tag{2.8c}$$

and describe an isolated protuberance ($A > 0$) or indentation ($A < 0$) on an otherwise planar surface.

II.B. A Deterministic Periodic Profile

For the second type of rough surface $\zeta(\vec{x}_\parallel)$ is again deterministic, but is a periodic function of x_1 and/or of x_2. It describes a classical grating or a bigrating on a material surface.

II.B.1. A One-Dimensional Profile

The simplest example of a deterministic, periodic surface profile is a planar surface on which a one-dimensional grating has been ruled. The equation defining such a surface can be written in the form $x_3 = \zeta(x_1)$, where the surface profile function has the periodicity property

$$\zeta(x_1 + a) = \zeta(x_1) , \tag{2.9}$$

with a the period of the grating. The function $\zeta(x_1)$ can therefore be expanded in the Fourier series

$$\zeta(x_1) = \sum_{n=-\infty}^{\infty} \hat{\zeta}(n) e^{i\frac{2\pi n}{a} x_1} , \tag{2.10}$$

where the Fourier coefficients $\{\hat{\zeta}(n)\}$ are given by

$$\hat{\zeta}(n) = \frac{1}{a} \int_0^a dx_1 \, \zeta(x_1) \, e^{-i \frac{2\pi n}{a} x_1} = \hat{\zeta}^*(-n) . \tag{2.11}$$

It is convenient to assume that $\zeta(x_1)$ has been defined in such a way that

$$\hat{\zeta}(0) = \frac{1}{a} \int_0^a dx_1 \, \zeta(x_1) = 0 . \tag{2.12}$$

To define the degree of roughness of the grating surface $x_3 = \zeta(x_1)$ it is useful to introduce the notion of the <u>corrugation strength</u>. We will define this generally as the ratio of the root mean square deviation of the surface from flatness to the period of the grating,

$$\text{corrugation strength} \equiv \delta/a , \tag{2.13a}$$

where

$$\delta = \left[\frac{1}{a} \int_0^a dx_1 \, \zeta^2(x_1)\right]^{1/2} = \left[\sum_{n=-\infty}^{\infty} |\hat{\zeta}(n)|^2\right]^{1/2} . \tag{2.13b}$$

It should be noted, however, that in the case of such simple grating profiles as the sinusoidal,

$$\zeta(x_1) = \zeta_0 \cos \frac{2\pi x_1}{a} \tag{2.14}$$

or the symmetric sawtooth profile,

$$\begin{aligned}\zeta(x_1) &= h + 4\frac{h}{a}x_1 \qquad -\frac{a}{2} < x_1 < 0 \\ &= h - 4\frac{h}{a}x_1 \qquad 0 < x_1 < \frac{a}{2},\end{aligned} \tag{2.15}$$

it has become almost conventional to define the corrugation strengths as ζ_0/a and h/a, respectively, rather than by the values $(\sqrt{2}/2)\zeta_0/a$ and $(\sqrt{3}/3)\,h/a$ given by Eq. (2.13). That is, in the case of these simple profiles it is ζ_{max}/a rather than δ/a that is commonly used as the definition of the corrugation strength.

II.B.2. A Two-Dimensional Profile

A second type of surface profile that is not statistical but deterministic and periodic is one that is doubly periodic in the plane $x_3 = 0$. In this case we describe the periodicity of the function $\zeta(\vec{x}_\parallel)$ by the equation

$$\zeta(\vec{x}_\parallel + \vec{a}_1) = \zeta(\vec{x}_\parallel + \vec{a}_2) = \zeta(\vec{x}_\parallel), \tag{2.16}$$

where $\vec{a}_1$ and $\vec{a}_2$ are the two, non-parallel, primitive translation vectors of the surface profile function. The parallelogram two of whose adjacent edges are the vectors $\vec{a}_1$ and $\vec{a}_2$ drawn from a common origin is called the <u>surface unit cell</u>. It is the period of the function $\zeta(\vec{x}_\parallel)$. Its area will be denoted by a_c, where

$$a_c = |\vec{a}_1||\vec{a}_2|\sin\theta. \tag{2.17}$$

In this expression $\theta(<\pi)$ is the angle through which $\vec{a}_1$ must be turned in a counterclockwise sense to line up with $\vec{a}_2$.

It is convenient to introduce the set of vectors $\{\vec{x}_\parallel(\ell)\}$, whose typical element is defined by

$$\vec{x}_\parallel(\ell) = \ell_1\vec{a}_1 + \ell_2\vec{a}_2. \tag{2.18}$$

Here ℓ_1 and ℓ_2 are any two integers, positive, negative, or zero, which we denote collectively by ℓ. We can associate a surface unit cell with each lattice point $\vec{x}_\parallel(\ell)$ of the two-dimensional lattice defined by the vectors $\{\vec{x}_\parallel(\ell)\}$ by identifying the common origin from which $\vec{a}_1$ and $\vec{a}_2$ are drawn with the point $\vec{x}_\parallel(\ell)$. The array of surface unit cells generated in this way covers the entire plane $x_3 = 0$ with no cracks and with no overlaps. The periodicity of $\zeta(\vec{x}_\parallel)$ expressed by Eqs. (2.16) can be expressed in terms of the vectors $\{\vec{x}_\parallel(\ell)\}$ as

$$\zeta(\vec{x}_\parallel + \vec{x}_\parallel(\ell)) = \zeta(\vec{x}_\parallel) \ , \tag{2.19}$$

for all $\{\vec{x}_\parallel(\ell)\}$.

We can now introduce two vectors $\vec{b}_1$ and $\vec{b}_2$ by the conditions

$$\vec{a}_i \cdot \vec{b}_j = 2\pi\delta_{ij} \ . \tag{2.20}$$

Explicit expressions for these vectors are

$$\vec{b}_1 = \frac{2\pi}{a_c}(a_{22}, -a_{21}), \qquad \vec{b}_2 = \frac{2\pi}{a_c}(-a_{12}, a_{11}) \ , \tag{2.21}$$

where a_{ij} is the jth Cartesian component of $\vec{a}_i$. The set of vectors $\{\vec{G}_\parallel(h)\}$ defined by

$$\vec{G}_\parallel(h) = h_1\vec{b}_1 + h_2\vec{b}_2 \tag{2.22}$$

are the translation vectors of the lattice reciprocal to the one in real space defined by the vectors $\{\vec{x}_\parallel(\ell)\}$. In Eq. (2.22) h_1 and h_2 are any two integers, positive, negative, or zero, which we denote collectively by h. With the aid of the vectors $\{\vec{G}_\parallel(h)\}$ the surface profile function $\zeta(\vec{x}_\parallel)$ can be expanded in the Fourier series

$$\zeta(\vec{x}_\parallel) = \sum_{\vec{G}_\parallel} \hat{\zeta}(\vec{G}_\parallel) e^{i\vec{G}_\parallel \cdot \vec{x}_\parallel} \qquad \text{where} \tag{2.23}$$

$$\hat{\zeta}(\vec{G}_\parallel) = \frac{1}{a_c}\int_{a_c} d^2x_\parallel \ \zeta(\vec{x}_\parallel) e^{-i\vec{G}_\parallel \cdot \vec{x}_\parallel} \ . \tag{2.24}$$

The integral in Eq. (2.24) is taken over the surface unit cell. We will assume in this case that $\zeta(\vec{x}_\parallel)$ is defined in such a way that

$$\hat{\zeta}(\vec{0}) = \frac{1}{a_c}\int_{a_c} d^2x_\parallel \ \zeta(\vec{x}_\parallel) = 0 \ . \tag{2.25}$$

The corrugation strength of a bigrating can be defined to be $\delta/a_c^{1/2}$, where the root mean square deviation of the surface from flatness is now given by

$$\delta = \left[\frac{1}{a_c}\int_{a_c} d^2x_\parallel \zeta^2(\vec{x}_\parallel)\right]^{1/2} = \left[\sum_{\vec{G}_\parallel} |\hat{\zeta}(\vec{G}_\parallel)|^2\right]^{1/2} . \tag{2.26}$$

II.C. A Random Profile

For the third kind of rough surface $\zeta(\vec{x}_\parallel)$ is a stochastic function of $\vec{x}_\parallel$, and describes a randomly rough surface.

The fact that the surface profile function for a randomly rough surface is unknown in general forces us to characterize it by certain statistical properties. Underlying this characterization is the assumption that there is not a single function $\zeta(\vec{x}_\parallel)$. There is instead an ensemble of realizations of this function. Physical properties associated with a statistically rough surface are to be averaged over this ensemble, and it is assumed that this ensemble average does not differ appreciably from the spatial average over a single surface. The probability that $\zeta(\vec{x}_\parallel)$ has a certain value at the point $\vec{x}_\parallel$ is given by a probability distribution function. An explicit form for this distribution function is usually not required. What are specified are the first few of its moments. In common with most theoretical treatments of surface roughness, the one presented here is based on the assumption that $\zeta(\vec{x}_\parallel)$ is a stationary stochastic process, and that the first two moments of its probability distribution function are

$$\langle\zeta(\vec{x}_\parallel)\rangle = 0 \tag{2.27}$$

$$\langle\zeta(\vec{x}_\parallel)\zeta(\vec{x}_\parallel')\rangle = \delta^2 W(|\vec{x}_\parallel-\vec{x}_\parallel'|). \tag{2.28}$$

In Eqs. (2.27) - (2.28) the angular brackets $\langle\ldots\rangle$ denote an average over the ensemble of realizations of the function $\zeta(\vec{x}_\parallel)$. The quantity δ^2 appearing in Eq. (2.28) is the mean square departure of the surface from flatness,

$$\langle\zeta^2(\vec{x}_\parallel)\rangle = \delta^2 . \tag{2.29}$$

Equation (2.27) expresses the assumption that the nominal surface is the plane $x_3 = 0$. This assumption involves no loss of generality. The fact that the correlation function $W(|\vec{x}_\parallel-\vec{x}_\parallel'|)$ depends on $\vec{x}_\parallel$ and $\vec{x}_\parallel'$ only through their difference (the stationarity assumption) has the consequence that infinitesimal translational invariance is restored to our rough surface system when physical properties are averaged with respect to the ensemble of realizations of the function $\zeta(\vec{x}_\parallel)$. The fact that this correlation function depends on the magnitude of the vector $\vec{x}_\parallel-\vec{x}_\parallel'$ is an expression of the assumption that all directions are equivalent on the rough surface. If the medium bounded by the rough surface is isotropic in its properties in the plane $x_3 = 0$, a consequence of the assumption of the equivalence of all directions on the rough surface is that isotropy in the plane $x_3 = 0$ is restored to physical properties by the averaging process.

The correlation function $W(|\vec{x}_\parallel|)$ possesses some important general properties. It follows from Eqs. (2.28) and (2.29) that

$$W(0) = 1. \tag{2.30}$$

$W(|\vec{x}_\parallel|)$ is clearly an even function of $\vec{x}_\parallel$. In addition, it tends to zero as $|\vec{x}_\parallel| \to \infty$. This property follows from the fact that on a statistically rough surface the heights of the surface at two widely separated points are uncorrelated. We will define the <u>transverse correlation length</u> a as the distance over which the function $W(|\vec{x}_\parallel|)$ decreases appreciably. It can be shown [21] that the transverse correlation length is also a measure of the average distance between consecutive peaks and valleys on the surface (Fig. 1).

The corrugation strength of a randomly rough surface is then defined to be δ/a.

For calculational purposes it is convenient, even necessary, to introduce the Fourier transform of the surface roughness profile function:

$$\zeta(\vec{x}_\parallel) = \int \frac{d^2k_\parallel}{(2\pi)^2} e^{i\vec{k}_\parallel\cdot\vec{x}_\parallel}\hat{\zeta}(\vec{k}_\parallel). \tag{2.31}$$

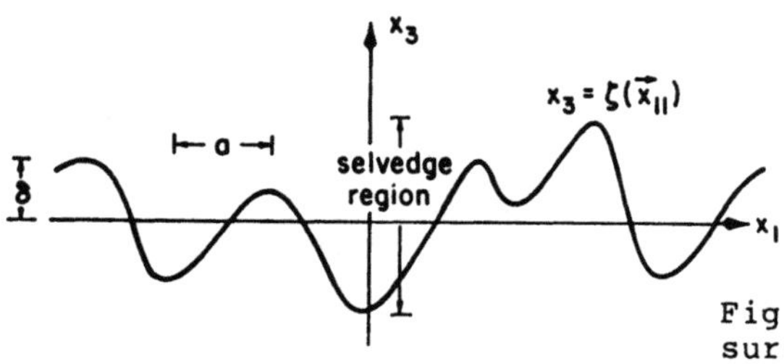

Fig. 1. A randomly rough surface profile.

Here $\vec{k}_\parallel$ is a two-dimensional wave vector,

$$\vec{k}_\parallel = \hat{x}_1k_1 + \hat{x}_2k_2 , \tag{2.32}$$

whose components are parallel to the plane $x_3 = 0$. The Fourier coefficients $\{\hat{\zeta}(\vec{k}_\parallel)\}$ are now random quantities. With the aid of the Fourier inversion theorem and the results given by Eqs. (2.27) and (2.28) we find that the statistical properties of these coefficients are characterized by

$$\langle\hat{\zeta}(\vec{k}_\parallel)\rangle = 0 \tag{2.33}$$

$$\langle\hat{\zeta}(\vec{k}_\parallel)\hat{\zeta}(\vec{k}'_\parallel)\rangle = \delta^2 g(k_\parallel)(2\pi)^2\delta(\vec{k}_\parallel + \vec{k}'_\parallel) . \tag{2.34}$$

The function $g(k_\parallel)$ appearing in Eq. (2.34) is called the <u>surface structure factor</u>, and is defined by

$$g(k_\parallel) = \int d^2x_\parallel\, e^{-i\vec{k}_\parallel\cdot\vec{x}_\parallel}W(|\vec{x}_\parallel|)$$

$$= 2\pi \int_0^\infty dx_\parallel\, x_\parallel J_0(k_\parallel x_\parallel)W(|x_\parallel|) , \tag{2.35}$$

where $J_0(k_\parallel x_\parallel)$ is a Bessel function. From the inversion formula

$$W(|\vec{x}_\parallel|) = \int \frac{d^2k_\parallel}{(2\pi)^2} e^{i\vec{k}_\parallel \cdot \vec{x}_\parallel} g(k_\parallel) \tag{2.36}$$

and Eq. (2.30) we see that $g(k_\parallel)$ is normalized according to

$$\int \frac{d^2k_\parallel}{(2\pi)^2} g(k_\parallel) = 1. \tag{2.37}$$

Up to now we have not had to specify an explicit analytic form for the correlation function $W(|\vec{x}_\parallel|)$. However, in order to obtain explicit, quantitative results such a form is required. In these lectures we will generally adopt a Gaussian form for $W(|\vec{x}_\parallel|)$ [22-28],

$$W(|\vec{x}_\parallel|) = \exp(-x_\parallel^2/a^2) . \tag{2.38a}$$

The surface structure factor $g(k_\parallel)$ obtained by combining Eqs. (2.35) and (2.38a) is

$$g(k_\parallel) = \pi a^2 e^{-\frac{1}{4} a^2 k_\parallel^2} . \tag{2.38b}$$

Although it possesses the general properties required of $W(|\vec{x}_\parallel|)$ that have been listed above, it should be emphasized that there is nothing unique about our choice of the Gaussian form. Other analytic forms for $W(|\vec{x}_\parallel|)$ have been used by various authors from time to time [29,30]. The experimental determinations of $g(k_\parallel)$ have produced mixed results. There are measurements [31-34] that show that $g(k_\parallel)$ is well represented by a Gaussian function. There are other data that show that $W(|\vec{x}_\parallel|)$ is not a Gaussian form in some cases [35,36]. Still other data [37-43] suggest that at least under some methods of surface preparation $g(\vec{k}_\parallel)$, and therefore $W(|\vec{x}_\parallel|)$, is not Gaussian at all. One such result is shown in Fig. 2, for a silver deposit. It was obtained from measurements of the surface profile along a single line (a one-dimensional surface structure factor) [43]. Since it is assumed that all directions are equivalent on the randomly rough surface of a medium that is isotropic in the plane of the surface in the absence of the roughness, there is no loss of generality in this procedure. We see in this case that $g(k)$ has its maximum away from $k = 0$, and decreases to zero with increasing k in a slightly nonmonotonic fashion. There is also some evidence that $W(|\vec{x}_\parallel|)$ is not a monotonically decreasing function of $|x_\parallel|$ [37,42,44]. Finally, there are experiments that show that in some cases more than one transverse length scale is required [44].

From a theoretical standpoint the chief advantage the Gaussian form for $W(|\vec{x}_\parallel|)$ possesses over others that could

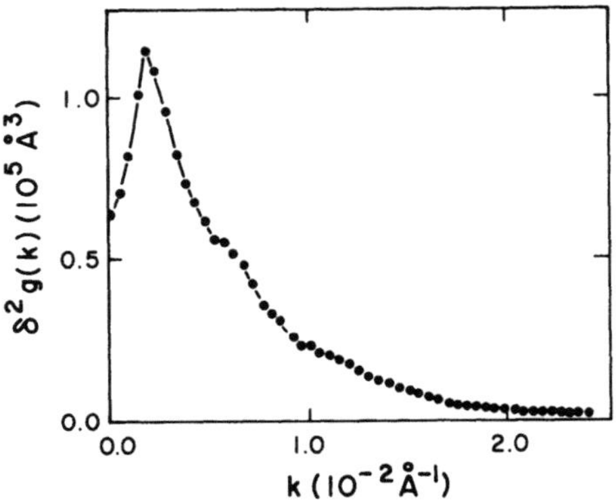

Fig. 2. The one-dimensional surface structure factor $\delta^2 g(k)$ for the rough surface of a silver deposit [Ref. 43].

be chosen is that it allows many of the integrals in which $W(|\vec{x}_\parallel|)$, or equivalently $g(k_\parallel)$, appears to be carried out analytically rather than numerically. This often makes it possible to obtain more explicit results than would be the case if some other form for $W(|\vec{x}_\parallel|)$ were assumed. It is the combination of physical reasonableness and computational ease present in the Gaussian form for $W(|\vec{x}_\parallel|)$ that prompts its use here.

However, to avoid tying our theoretical results completely to a particular form for $g(k_\parallel)$, and in the expectation that experimental results for $g(k_\parallel)$ will become more readily available for the samples on which studies of surface roughness are carried out, we will present them in terms of an arbitrary $g(k_\parallel)$ before proceeding to evaluate them for the particular choice given by Eq. (2.38b).

Although for many applications knowledge of only the first two moments of the probability distribution function of $\zeta(\vec{x}_\parallel)$ is sufficient, it is sometime convenient to be able to specify higher order moments as well. A commonly made assumption that allows this to be done is that the $\{\zeta(\vec{x}_\parallel)\}$ are Gaussianly distributed random variables. The consequences of this assumption are that (i) the average of a product of an odd number of $\zeta(\vec{x}_\parallel)$'s vanishes, e.g.

$$\langle\zeta(\vec{x}_\parallel)\zeta(\vec{x}_\parallel')\zeta(\vec{x}_\parallel'')\rangle = 0 \quad , \tag{2.39}$$

where $\vec{x}_\parallel$, $\vec{x}_\parallel'$, $\vec{x}_\parallel''$ do not need to be different points; and (ii) the average of the product of an even number of $\zeta(\vec{x}_\parallel)$'s is given by the sum of the products of the averages of the $\zeta(\vec{x}_\parallel)$'s taken two-by-two different in all possible ways, e.g.

$$\langle\zeta(\vec{x}_\parallel)\zeta(\vec{x}_\parallel')\zeta(\vec{x}_\parallel'')\zeta(\vec{x}_\parallel''')\rangle$$

$$= \langle\zeta(\vec{x}_\parallel)\zeta(\vec{x}_\parallel')\rangle\langle\zeta(\vec{x}_\parallel'')\zeta(\vec{x}_\parallel''')\rangle +$$

$$+ \langle\zeta(\vec{x}_\parallel)\zeta(\vec{x}''_\parallel)\rangle\langle\zeta(\vec{x}'_\parallel)\zeta(\vec{x}'''_\parallel)\rangle +$$

$$+ \langle\zeta(\vec{x}_\parallel)\zeta(\vec{x}'''_\parallel)\rangle\langle\zeta(\vec{x}'_\parallel)\zeta(\vec{x}''_\parallel)\rangle \quad . \tag{2.40}$$

Each average of a pair of $\zeta(\vec{x}_\parallel)$'s on the right-hand side of the equation, called a contraction, is given by Eq. (2.28).

The consequence for the Fourier coefficients $\{\hat{\zeta}(\vec{k}_\parallel)\}$ of the assumption that the $\{\zeta(\vec{x}_\parallel)\}$ are Gaussianly distributed random variables is that they possess the following properties: (i) the average of the product of an odd number of $\hat{\zeta}(\vec{k}_\parallel)$'s vanishes, e.g.

$$\langle\hat{\zeta}(\vec{k}_\parallel)\hat{\zeta}(\vec{k}'_\parallel)\hat{\zeta}(\vec{k}''_\parallel)\rangle = 0 \ , \tag{2.41}$$

and (ii) the average of the product of an even number of $\hat{\zeta}(\vec{k}_\parallel)$'s is given by the sum of the products of the averages of the $\hat{\zeta}(\vec{k}_\parallel)$'s taken two-by-two different in all possible ways, e.g.

$$\langle\hat{\zeta}(\vec{k}_\parallel)\hat{\zeta}(\vec{k}'_\parallel)\hat{\zeta}(\vec{k}''_\parallel)\hat{\zeta}(\vec{k}'''_\parallel)\rangle$$

$$= \langle\hat{\zeta}(\vec{k}_\parallel)\hat{\zeta}(\vec{k}'_\parallel)\rangle\langle\hat{\zeta}(\vec{k}''_\parallel)\hat{\zeta}(\vec{k}'''_\parallel)\rangle +$$

$$+ \langle(\hat{\zeta}(\vec{k}_\parallel)\hat{\zeta}(\vec{k}''_\parallel)\rangle\langle\hat{\zeta}(\vec{k}'_\parallel)\hat{\zeta}(\vec{k}'''_\parallel)\rangle +$$

$$+ \langle\hat{\zeta}(\vec{k}_\parallel)\hat{\zeta}(\vec{k}'''_\parallel)\rangle\langle\hat{\zeta}(\vec{k}'_\parallel)\hat{\zeta}(\vec{k}''_\parallel)\rangle \ . \tag{2.42}$$

Each average of a pair of $\hat{\zeta}(\vec{k}_\parallel)$'s on the right-hand side of this equation is given by Eq. (2.34).

At the present time there is only a single study aimed at testing the assumption that the $\{\zeta(\vec{x}_\parallel)\}$ are Gaussianly distributed random variables. Using an electron micrographic method RASIGNI et al. [45] have studied surface profiles $\zeta(x)$ along straight lines on the surfaces of several metals and ionic crystals. From their data they have constructed the quantities

$$\gamma_1 = \langle\zeta^3(x)\rangle/\delta^3 \tag{2.43}$$

$$\gamma_2 = [\langle\zeta^4(x)\rangle - 3\delta^4]/\delta^4 \ , \tag{2.44}$$

where $\delta^2 = \langle\zeta^2(x)\rangle$. If the $\{\zeta(x)\}$ are Gaussianly distributed random variables both γ_1 and γ_2 should vanish.

Rasigni *et al*. find that γ_1 is practically zero for several of the surfaces, and for the others seldom exceeds 0.3 in magnitude. They argue that if the errors in the determination of the surface profile function $\zeta(x)$, and in the determination of $\delta(\sim 10\%)$, are taken into account, it may be considered that the equation $\langle\zeta^3(x)\rangle = 0$ is reasonably checked.

At the same time the quantity γ_2 was found to depart appreciably from zero for most of the samples studied, although it remained smaller than unity for many of them. However, it must be kept in mind that the variance of a moment depends on the population moment of twice the order [46], so that it becomes very large for higher moments. In view of this and the present limitations of the experimental method used to obtain γ_2, Rasigni *et al*. conclude that the equation $\langle\zeta^4(x)\rangle = 3\delta^4$ is roughly checked.

In view of the simplifications that the assumption that the $\{\zeta(\vec{x}_\parallel)\}$ are Gaussianly distributed random variables introduces into theoretical studies of surface roughness, it is hoped that the pioneering study by Rasigni *et al*. of the higher moments of the probability distribution of the surface profile function will be followed by further work to test this assumption. It would be desirable to know $\langle\zeta^3(\vec{x}_\parallel)\rangle$ and $\langle\zeta^4(\vec{x}_\parallel)\rangle$ with greater accuracy, as well as two-, three-, and four-point correlation functions of the type appearing on the left-hand sides of Eqs.(2.40) and (2.41). For if the $\{\zeta(\vec{x}_\parallel)\}$ are not Gaussianly distributed, it will be necessary to include correlation functions such as $\langle\zeta(\vec{x}_\parallel)\zeta(\vec{x}_\parallel^{\,\prime})\zeta(\vec{x}_\parallel^{\,\prime\prime})\rangle$ into theoretical calculations, and it will be necessary to have some notions about their form.

III. Surface Plasmons and Polaritons on a Rough Surface

In pursuit of our aim to present a unified treatment of the propagation of surface plasmons and surface polaritons on rough surfaces in these lectures, in this section we derive the equations governing this propagation for a metal-vacuum interface defined by Eq. (2.1), without specifying the nature of the surface profile function $\zeta(\vec{x}_\parallel)$. In the following three sections the equations derived here will be used in discussions of electrostatic and electromagnetic surface shape resonances; of surface plasmons and surface polaritons on grating surfaces; and of surface plasmons and surface polaritons on randomly rough surfaces.

III.A. Surface Plasmons

The study of surface plasmons on a rough dielectric surface is somewhat easier than the study of their electromagnetic counterparts, as we shall see, and provides a good introduction to the latter. In this subsection we first derive the equation governing the propagation of surface plasmons across a rough surface defined by a surface profile function $\zeta(\vec{x}_\parallel)$ that is a function of both x_1 and x_2, and in the following subsection specialize this result to the case of a surface defined by a surface profile function $\zeta(x_1)$ that depends on only the single coordinate x_1.

III.A.1. A Two-Dimensional Profile

The problem to be addressed here is to obtain the solution of Laplace's equation,

$$\nabla^2\phi^>(\vec{x}|\omega) = 0 \qquad x_3 > \zeta(\vec{x}_\parallel) \tag{3.1a}$$

$$\nabla^2\phi^<(\vec{x}|\omega) = 0 \qquad x_3 < \zeta(\vec{x}_\parallel) \quad , \tag{3.1b}$$

for the electrostatic scalar potential in the vacuum and in the dielectric medium, respectively, that vanishes as $|x_3| \to \infty$, and satisfies the boundary conditions

$$\phi^>(\vec{x}|\omega)\Big|_{x_3=\zeta(\vec{x}_\parallel)} = \phi^<(\vec{x}|\omega)\Big|_{x_3=\zeta(\vec{x}_\parallel)} \tag{3.2a}$$

$$\frac{\partial}{\partial n}\phi^>(\vec{x}|\omega)\Big|_{x_3=\zeta(\vec{x}_\parallel)} = \varepsilon(\omega)\frac{\partial}{\partial n}\phi^<(\vec{x}|\omega)\Big|_{x_3=\zeta(\vec{x}_\parallel)} \tag{3.2b}$$

at the interface $x_3 = \zeta(\vec{x}_\parallel)$. In Eq. (3.2b) $\partial/\partial n$ is the derivative along the normal to the interface at each point, directed from the dielectric into the vacuum. It is given by Eq. (2.5a).

The solution of Eq. (3.1a) in the region $x_3 > \zeta(\vec{x}_\parallel)_{max}$ that satisfies the boundary condition at $x_3 = \infty$ can be written in the form of a Fourier integral,

$$\phi^>(\vec{x}|\omega) = \int \frac{d^2k_\parallel}{(2\pi)^2} A(\vec{k}_\parallel\omega)e^{i\vec{k}_\parallel\cdot\vec{x}_\parallel - k_\parallel x_3} \qquad x_3>\zeta(\vec{x}_\parallel)_{max}. \tag{3.3}$$

A similar integral representation can be written for $\phi^<(\vec{x}|\omega)$ in the region $x_3 < \zeta(\vec{x}_\parallel)$, and used together with Eq. (3.3) in satisfying the boundary conditions (3.2)

We will proceed in a somewhat different fashion here, however, by eliminating the potential in the region $x_3<\zeta(\vec{x}_\parallel)$ altogether, and working with the potential $\phi^>(\vec{x}|\omega)$ alone. To accomplish this we apply Green's theorem [47]

$$\int_V d^3x(u\nabla^2v-v\nabla^2u) = \int_\Sigma dS\left(u\,\frac{\partial v}{\partial n} - v\,\frac{\partial u}{\partial n}\right) \quad , \tag{3.4}$$

to the volume V defined by the region $x_3 < \zeta(\vec{x}_\parallel)$ occupied by the dielectric. The surface Σ bounding this volume consists of the interface $x_3 = \zeta(\vec{x}_\parallel)$, which we denote by S, and a hemispherical cap of infinite radius in the lower half-space, which we denote by $S^{(-\infty)}$. The derivative $\partial/\partial n$ on the right-hand side of Eq. (3.4) is along the normal to the surface $\sum$ directed outward from the volume V. We assume that u in Eq. (3.4) is $\phi^<(\vec{x}|\omega)$, while v is the Green's function $G(\vec{x};\vec{x}')$ that is the solution of the equation

$$\nabla^2 G(\vec{x};\vec{x}') = -4\pi\delta(\vec{x}-\vec{x}') \quad , \tag{3.5}$$

subject to vanishing boundary conditions at infinity. A convenient representation for this function is

$$G(\vec{x};\vec{x}') = \int \frac{d^2q_\parallel}{(2\pi)^2} \frac{2\pi}{q_\parallel} e^{i\vec{q}_\parallel\cdot(\vec{x}_\parallel-\vec{x}'_\parallel)-q_\parallel|x_3-x'_3|} \quad . \tag{3.6}$$

With these identifications, and Eqs. (3.1b) and (3.5), Eq. (3.4) becomes

$$\theta_V(\vec{x})\phi^<(\vec{x}|\omega) = -\frac{1}{4\pi}\int_S dS' \left[\frac{\partial G(\vec{x};\vec{x}')}{\partial n'}\phi^<(\vec{x}'|\omega) - \right.$$

$$\left. - G(\vec{x};\vec{x}')\frac{\partial\phi^<(\vec{x}'|\omega)}{\partial n'}\right] \quad , \tag{3.7}$$

where $\theta_V(\vec{x})$ is unity if $\vec{x}\epsilon V$ and vanishes if $\vec{x}\notin V$. In writing Eq. (3.7) we have used the result that since we are concerned with potentials $\phi^<(\vec{x}|\omega)$ that vanish as $x_3 \to -\infty$, the contribution to the right-hand side of Eq. (3.4) from the integral over $S^{(-\infty)}$ vanishes. We have also used the symmetry of the Green's function expressed by $G(\vec{x};\vec{x}') = G(\vec{x}';\vec{x})$.

We now use the boundary conditions (3.2) and the assumption that $x_3 > \zeta(\vec{x}_\parallel)_{max}$ to rewrite Eq. (3.7) as

$$0 = -\frac{1}{4\pi}\int_S dS'\left[\frac{\partial G(\vec{x};\vec{x}')}{\partial n'}\phi^>(\vec{x}'|\omega) - \right.$$

$$\left. - \frac{1}{\varepsilon(\omega)} G(\vec{x};\vec{x}')\frac{\partial\phi^>(\vec{x}'|\omega)}{\partial n'}\right] \qquad x_3 > \zeta(\vec{x}_\parallel)_{max} \ . \tag{3.8}$$

This is the boundary condition on $\phi^>(\vec{x}|\omega)$ that we will now use to determine the Fourier coefficient $A(\vec{k}_\parallel\omega)$ in Eq. (3.3). We first go from integration over dS' in Eq. (3.8) to integration over $d^2x'_\parallel$ with the aid of the relation (2.6). The boundary condition (3.8) thus becomes

$$0 = -\frac{1}{4\pi}\int d^2x'_\parallel\left[(\vec{n}'\cdot\nabla' G(\vec{x};\vec{x}'))\phi^>(\vec{x}'|\omega) - \right.$$

$$\left. - \frac{1}{\varepsilon(\omega)} G(\vec{x};\vec{x}')(\vec{n}'\cdot\nabla'\phi^>(\vec{x}'|\omega))\right]_{x'_3=\zeta(\vec{x}'_\parallel)} \qquad x_3>\zeta(\vec{x}_\parallel)_{max}, \tag{3.9}$$

where the vector $\vec{n}$ has been defined in Eq. (2.4).

When we substitute Eqs. (3.3) and (3.6) into Eq. (3.9) it is very convenient to introduce at the same time the representations

$$e^{\alpha\zeta(\vec{x}_\parallel)} = 1 + \alpha \int \frac{d^2Q_\parallel}{(2\pi)^2} J(\alpha|Q_\parallel) e^{i\vec{Q}_\parallel\cdot\vec{x}_\parallel} \tag{3.10}$$

and

$$\frac{\partial\zeta(\vec{x}_\parallel)}{\partial x_\beta} e^{\alpha\zeta(\vec{x}_\parallel)} = \int \frac{d^2Q_\parallel}{(2\pi)^2} iQ_\beta J(\alpha|\vec{Q}_\parallel) e^{i\vec{Q}_\parallel\cdot\vec{x}_\parallel}, \quad \beta=1,2, \tag{3.11}$$

where

$$J(\alpha|\vec{Q}_\parallel) = \int d^2x_\parallel\, e^{-i\vec{Q}_\parallel\cdot\vec{x}_\parallel} \frac{e^{\alpha\zeta(\vec{x}_\parallel)}-1}{\alpha} . \tag{3.12}$$

As a result, Eq. (3.9) is transformed into

$$\int \frac{d^2q_\parallel}{(2\pi)^2} e^{i\vec{q}_\parallel\cdot\vec{x}_\parallel} \frac{e^{-q_\parallel x_3}}{2q_\parallel} A(\vec{q}_\parallel\omega)q_\parallel[1 + \frac{1}{\varepsilon(\omega)}] +$$

$$+ \int \frac{d^2q_\parallel}{(2\pi)^2} e^{i\vec{q}_\parallel\cdot\vec{x}_\parallel} \frac{e^{-q_\parallel x_3}}{(2\pi)^2} \int \frac{d^2p_\parallel}{(2\pi)^2} A(\vec{p}_\parallel\omega)J(q_\parallel-p_\parallel|\vec{q}_\parallel-\vec{p}_\parallel) \times$$

$$\times [(-i\vec{q}_\parallel + \hat{x}_3q_\parallel) + i(\vec{p}_\parallel - \hat{x}_3p_\parallel)]\cdot[(-i\vec{q}_\parallel + \hat{x}_3q_\parallel) -$$

$$- \frac{1}{\varepsilon(\omega)}(i\vec{p}_\parallel - \hat{x}_3p_\parallel)] = 0, \qquad x_3 > \zeta(\vec{x}_\parallel)_{max} . \tag{3.13}$$

On equating to zero the $\vec{q}_\parallel$-th Fourier coefficient in this equation, we obtain finally the integral equation satisfied by $A(\vec{q}_\parallel\omega)$ [48]:

$$\frac{\varepsilon(\omega)+1}{\varepsilon(\omega)-1} A(\vec{q}_\parallel\omega) = \int \frac{d^2p_\parallel}{(2\pi)^2} J(q_\parallel-p_\parallel|\vec{q}_\parallel-\vec{p}_\parallel)(1-\hat{q}_\parallel\cdot\hat{p}_\parallel)p_\parallel A(\vec{p}_\parallel\omega). \tag{3.14}$$

III.A.2. A One-Dimensional Profile

Equation (3.14) was obtained on the assumption that the surface profile function is a function of both x_1 and x_2. In the case of a one-dimensional surface profile function $\zeta(x_1)$, which corresponds to a ridge or groove on an otherwise planar surface, Eq. (3.14) takes the form

$$\frac{\varepsilon(\omega)+1}{\varepsilon(\omega)-1} A(q\omega) = \int \frac{dp}{2\pi} J(|q|-|p||q-p)(1-\mathrm{sgn}q\,\mathrm{sgn}p)|p|A(p\omega), \tag{3.15a}$$

where

$$J(\alpha|Q) = \int dx_1 e^{-iQx_1} \frac{e^{\alpha\zeta(x_1)}-1}{\alpha} . \tag{3.15b}$$

The associated electrostatic scalar potential is given by

$$\phi^{>}(x_1 x_3|\omega) = \int \frac{dk}{2\pi} A(k\omega) e^{ikx_1-|k|x_3} \qquad x_3>\zeta(x_1)_{max}. \tag{3.16}$$

III.B. Surface Polaritons

Unlike the situation in the electrostatic approximation, electromagnetic theory allows for the radiation of electromagnetic energy. Therefore, for greater generality we approach the study of surface polaritons on rough surfaces by first considering the scattering of light, incident from the vacuum side, from a metal surface defined by the equation $x_3 = \zeta(\vec{x}_\parallel)$. We then specialize to the homogeneous case to obtain the equations describing the propagation of surface polaritons across rough surfaces.

We consider first the case of a two-dimensional surface profile function and then turn to the case of a one-dimensional surface profile function.

III.B.1. A Two-Dimensional Profile

The solution of Maxwell's equations for the electric field in the vacuum region $x_3 > \zeta(\vec{x}_\parallel)_{max}$,

$$\nabla\times\nabla\times\vec{E}(\vec{x},t) + \frac{1}{c^2}\frac{\partial^2}{\partial t^2}\vec{E}(\vec{x},t) = 0, \tag{3.17}$$

that satisfies the boundary conditions at $x_3 = \infty$ can be written as

$$\vec{E}(\vec{x},t) = \vec{E}(\vec{x}|\omega)e^{-i\omega t}, \tag{3.18a}$$

where

$$\begin{aligned}\vec{E}(\vec{x}|\omega) = \{&-\frac{c}{\omega}[\hat{k}_\parallel\alpha_0(k_\parallel\omega) + \hat{x}_3 k_\parallel]B_\parallel(\vec{k}_\parallel\omega) + \\ &+ (\hat{x}_3\times\hat{k}_\parallel)B_\perp(\vec{k}_\parallel\omega)\}e^{i\vec{k}_\parallel\cdot\vec{x}_\parallel - i\alpha_0(k_\parallel\omega)x_3} + \\ &+ \int\frac{d^2q_\parallel}{(2\pi)^2}\{\frac{c}{\omega}[\hat{q}_\parallel\alpha_0(q_\parallel\omega) - \hat{x}_3 q_\parallel]A_\parallel(\vec{q}_\parallel\omega) + \\ &+ (\hat{x}_3\times\hat{q}_\parallel)A_\perp(\vec{q}_\parallel\omega)\}e^{i\vec{q}_\parallel\cdot\vec{x}_\parallel + i\alpha_0(q_\parallel\omega)x_3} \qquad x_3 > \zeta(\vec{x}_\parallel)_{max}.\end{aligned} \tag{3.18b}$$

The first term on the right-hand side of this equation represents the incident wave; the second describes the scattered wave. The subscripts $\parallel$ and $\perp$ denote the p-polarized (TM) and s-polarized (TE) components of each of these fields. The function $\alpha_0(q_\parallel\omega)$ appearing in Eq. (3.18) is defined by

$$\alpha_0(q_\parallel\omega) = \left(\frac{\omega^2}{c^2} - q_\parallel^2\right)^{1/2} \qquad q_\parallel^2 < \frac{\omega^2}{c^2} \tag{3.19a}$$

$$= i\left(q_\parallel^2 - \frac{\omega^2}{c^2}\right)^{1/2} \qquad q_\parallel^2 > \frac{\omega^2}{c^2} \ . \tag{3.19b}$$

The function $\alpha_0(k_\parallel\omega)$ is always real and positive, since $k_\parallel^2 < (\omega^2/c^2)$.

We can write a similar expression for the refracted field in the dielectric medium, and use it together with Eq. (3.18) in satisfying the boundary conditions at the surface $x_3 = \zeta(\vec{x}_\parallel)$. However, in diffraction problems it is generally only the electromagnetic field in the vacuum region above the dielectric in which we are interested. Thus, in this section we will eliminate the electric field in the dielectric medium from the problem, and will work only with the field in the vacuum, given by Eq. (3.18), in obtaining the coefficients $A_{\parallel,\perp}(\vec{q}_\parallel\omega)$ in terms of $B_{\parallel,\perp}(\vec{k}_\parallel\omega)$.

We do this by substituting Eq. (3.18) into the boundary condition on the field in the vacuum derived in Appendix A with the aid of the vectorial equivalent of the Kirchhoff integral [2] and the extinction theorem [3]. When this is done we obtain the following equation for $A_{\parallel,\perp}(\vec{q}_\parallel\omega)$:

$$\int \frac{d^2q_\parallel}{(2\pi)^2} \frac{I(\alpha(Q_\parallel\omega)-\alpha_0(q_\parallel\omega)|\vec{Q}_\parallel-\vec{q}_\parallel)}{\alpha(Q_\parallel\omega) - \alpha_0(q_\parallel\omega)} \left[\frac{\omega^2}{c^2}\vec{\varepsilon}(\vec{q}\omega) - \frac{1}{\varepsilon(\omega)}(\vec{\gamma}\cdot\vec{\varepsilon}(\vec{q}\omega))\vec{\gamma}\right]$$

$$= -\frac{I(\alpha(Q_\parallel\omega) + \alpha_0(k_\parallel\omega)|\vec{Q}_\parallel-\vec{k}_\parallel)}{\alpha(Q_\parallel\omega) + \alpha_0(k_\parallel\omega)} \left[\frac{\omega^2}{c^2}\vec{E}(\vec{k}\omega) - \frac{1}{\varepsilon(\omega)}(\vec{\gamma}\cdot\vec{E}(\vec{k}\omega))\vec{\gamma}\right] \tag{3.20}$$

where, to simplify the notation, we have defined

$$\vec{E}(\vec{k}\omega) = -\frac{c}{\omega}[\hat{k}_\parallel\alpha_0(k_\parallel\omega)+\hat{x}_3k_\parallel]B_\parallel(\vec{k}_\parallel\omega)+(\hat{x}_3\times\hat{k}_\parallel)B_\perp(\vec{k}_\parallel\omega) \tag{3.21a}$$

$$\vec{\varepsilon}(\vec{q}\omega) = \frac{c}{\omega}[\hat{q}_\parallel\alpha_0(q_\parallel\omega)-\hat{x}_3q_\parallel]A_\parallel(\vec{q}_\parallel\omega)+(\hat{x}_3\times\hat{q}_\parallel)A_\perp(\vec{q}_\parallel\omega) \tag{3.21b}$$

$$\vec{k} = \vec{k}_\parallel - \hat{x}_3\alpha_0(k_\parallel\omega), \qquad \vec{q} = \vec{q}_\parallel + \hat{x}_3\alpha_0(q_\parallel\omega) \ , \tag{3.22}$$

$$\vec{\gamma} = \vec{Q}_\parallel + \hat{x}_3\alpha(Q_\parallel\omega) \ . \tag{3.23}$$

The integral $I(\alpha|\vec{Q}_\parallel)$ is given by

$$I(\alpha|\vec{Q}_\parallel) = \int d^2x_\parallel e^{-i\vec{Q}_\parallel\cdot\vec{x}_\parallel} e^{-i\alpha\zeta(\vec{x}_\parallel)} \ . \tag{3.24}$$

Equation (3.20) is a vector equation, i.e. it consists of three equations. However, there are only two unknown amplitudes $A_{\parallel}(\vec{q}_{\parallel}\omega)$ and $A_{\perp}(\vec{q}_{\parallel}\omega)$. Consequently, one of these three equations is redundant. To obtain the two independent equations we first introduce three mutually perpendicular unit vectors by

$$\hat{\gamma}(\vec{Q}_{\parallel}\omega) = \frac{c}{\omega(\varepsilon(\omega))^{1/2}}(\vec{Q}_{\parallel} + \hat{x}_3\alpha(Q_{\parallel}\omega)) \tag{3.25a}$$

$$\hat{a}_{\parallel}(\vec{Q}_{\parallel}\omega) = \frac{c}{\omega(\varepsilon(\omega))^{1/2}}(\vec{Q}_{\parallel}\alpha(Q_{\parallel}\omega) - \hat{x}_3 Q_{\parallel}) \tag{3.25b}$$

$$\hat{a}_{\perp}(\vec{Q}_{\parallel}\omega) = \hat{x}_3 x \hat{Q}_{\parallel}. \tag{3.25c}$$

We then take the scalar product of Eq. (3.20) with each of these unit vectors in turn. The scalar product of Eq. (3.20) with $\hat{\gamma}(\vec{Q}_{\parallel}\omega)$ yields the identity $0 = 0$. The vectors $\hat{a}_{\parallel}(\vec{Q}_{\parallel}\omega)$ and $\hat{a}_{\perp}(\vec{Q}_{\parallel}\omega)$ project out of Eq. (3.20) a pair of coupled integral equations for $A_{\parallel}(\vec{q}_{\parallel}\omega)$ and $A_{\perp}(\vec{q}_{\parallel}\omega)$ that can be written in the form [49]

$$\int\frac{d^2q_{\parallel}}{(2\pi)^2}\,\frac{I(\alpha(p_{\parallel}\omega)-\alpha_o(q_{\parallel}\omega)|\vec{p}_{\parallel}-\vec{q}_{\parallel})}{\alpha(p_{\parallel}\omega)-\alpha_o(q_{\parallel}\omega)}\times$$

$$\begin{pmatrix} [p_{\parallel}q_{\parallel}+\hat{p}_{\parallel}\cdot\hat{q}_{\parallel}\alpha(p_{\parallel}\omega)\alpha_o(q_{\parallel}\omega)] & -\frac{\omega}{c}\alpha(p_{\parallel}\omega)(\hat{p}_{\parallel}\times\hat{q}_{\parallel})_3 \\ \frac{\omega}{c}(\hat{p}_{\parallel}\times\hat{q}_{\parallel})_3\ \alpha_o(q_{\parallel}\omega) & \frac{\omega^2}{c^2}\hat{p}_{\parallel}\cdot\hat{q}_{\parallel} \end{pmatrix}\begin{pmatrix} A_{\parallel}(\vec{q}_{\parallel}\omega) \\ A_{\perp}(\vec{q}_{\parallel}\omega) \end{pmatrix}$$

$$= -\frac{I(\alpha(p_{\parallel}\omega)+\alpha_o(k_{\parallel}\omega)|\vec{p}_{\parallel}-\vec{k}_{\parallel})}{\alpha(p_{\parallel}\omega)+\alpha_o(k_{\parallel}\omega)}\times$$

$$\begin{pmatrix} [p_{\parallel}k_{\parallel}-\hat{p}_{\parallel}\cdot\hat{k}_{\parallel}\alpha(p_{\parallel}\omega)\alpha_o(k_{\parallel}\omega)] & -\frac{\omega}{c}\alpha(p_{\parallel}\omega)(\hat{p}_{\parallel}\times\hat{k}_{\parallel})_3 \\ -\frac{\omega}{c}(\hat{p}_{\parallel}\times\hat{k}_{\parallel})_3\ \alpha_o(k_{\parallel}\omega) & \frac{\omega^2}{c^2}\hat{p}_{\parallel}\cdot\hat{k}_{\parallel} \end{pmatrix}\begin{pmatrix} B_{\parallel}(\vec{k}_{\parallel}\omega) \\ B_{\perp}(\vec{k}_{\parallel}\omega) \end{pmatrix}. \tag{3.26}$$

Since in these lectures we are primarily concerned with the free oscillations of our vacuum-dielectric system, we equate the incident field to zero: $B_{\parallel}(\vec{k}_{\parallel}\omega) = B_{\perp}(\vec{k}_{\parallel}\omega) = 0$. In addition, because we seek modes that are localized in the vicinity of the surface, it is convenient to make the replacements

$$\alpha_0(q_\parallel\omega) = i\beta_0(q_\parallel\omega), \quad \alpha(q_\parallel\omega) = i\beta(q_\parallel\omega) \tag{3.27}$$

where

$$\beta_0(q_\parallel\omega) = \left(q_\parallel^2 - \frac{\omega^2}{c^2}\right)^{1/2} \qquad q_\parallel^2 > \frac{\omega^2}{c^2} \tag{3.28a}$$

$$= -i\left(\frac{\omega^2}{c^2} - q_\parallel^2\right)^{1/2} \qquad q_\parallel^2 < \frac{\omega^2}{c^2} \tag{3.28b}$$

$$\beta(q_\parallel\omega) = \left(q_\parallel^2 - \varepsilon(\omega)\frac{\omega^2}{c^2}\right)^{1/2} \quad \mathrm{Re}\,\beta(q_\parallel\omega) > 0,\ \mathrm{Im}\,\beta(q_\parallel\omega) < 0. \tag{3.29}$$

With these changes Eqs. (3.26) become

$$\begin{pmatrix} \dfrac{\varepsilon(\omega)\beta_0(p_\parallel\omega)+\beta(p_\parallel\omega)}{\varepsilon(\omega)-1} & 0 \\ 0 & \dfrac{\beta_0(p_\parallel\omega)+\beta(p_\parallel\omega)}{\varepsilon(\omega)-1} \end{pmatrix} \begin{pmatrix} A_\parallel(\vec{p}_\parallel\omega) \\ A_\perp(\vec{p}_\parallel\omega) \end{pmatrix}$$

$$= \int \frac{d^2q_\parallel}{(2\pi)^2} J(\beta(p_\parallel\omega) - \beta_0(q_\parallel\omega)|\vec{p}_\parallel - \vec{q}_\parallel) \times$$

$$\begin{pmatrix} p_\parallel q_\parallel - \beta(p_\parallel\omega)(\hat{p}_\parallel\cdot\hat{q}_\parallel)\beta_0(q_\parallel\omega) & -i\,\frac{\omega}{c}\,\beta(p_\parallel\omega)(\hat{p}_\parallel\times\hat{q}_\parallel)_3 \\ i\,\frac{\omega}{c}\,(\hat{p}_\parallel\times\hat{q}_\parallel)_3\beta_0(q_\parallel\omega) & \frac{\omega^2}{c^2}\,(\hat{p}_\parallel\cdot\hat{q}_\parallel) \end{pmatrix} \begin{pmatrix} A_\parallel(\vec{q}_\parallel\omega) \\ A_\perp(\vec{q}_\parallel\omega) \end{pmatrix}, \tag{3.30}$$

where the function $J(\alpha|\vec{Q}_\parallel)$ has been defined in Eq. (3.12).

III.B.2. A One-Dimensional Profile

To obtain the equations describing the diffraction of light from a rough surface defined by a one-dimensional surface profile function $\zeta(x_1)$ we could proceed from Eqs. (3.30). However, the result would be a pair of coupled integral equations for the amplitudes of the p- and s-polarized components of the scattered electric field. We can simplify the analysis if we consider the particular case of p-polarized light incident on the surface from the vacuum side, with the x_1x_3-plane as the plane of polarization. This case is of particular interest because surface polaritons can be resonantly excited in this polarization. The electromagnetic field in this case has the form

$$\vec{H}(\vec{x},t) = (0, H_2(x_1x_3|\omega), 0)e^{-i\omega t} \tag{3.31a}$$

$$\vec{E}(\vec{x},t) = (E_1(x_1x_3|\omega), 0, E_3(x_1x_3|\omega)e^{-i\omega t}, \tag{3.31b}$$

where

$$E_1(x_1x_3|\omega) = -\frac{ic}{\omega\varepsilon}\frac{\partial H_2(x_1x_3|\omega)}{\partial x_3} \tag{3.32a}$$

$$E_3(x_1x_3|\omega) = \frac{ic}{\omega\varepsilon}\frac{\partial H_2(x_1x_3|\omega)}{\partial x_1}, \tag{3.32b}$$

and ε denotes the dielectric constant of the medium in which the electric field is being calculated.

In view of Eqs. (3.32) we regard $H_2(x_1x_3|\omega)$ as the primary field component, and $E_{1,3}(x_1x_3|\omega)$ as derived field components. The equations satisfied by $H_2(x_1x_3|\omega)$ in the regions $x_3 \gtrless \zeta(x_1)$ are

$$\left(\frac{\partial^2}{\partial x_1^2} + \frac{\partial^2}{\partial x_3^2} + \frac{\omega^2}{c^2}\right) H_2^>(x_1x_3|\omega) = 0 \tag{3.33a}$$

$$\left(\frac{\partial^2}{\partial x_1^2} + \frac{\partial^2}{\partial x_3^2} + \varepsilon(\omega)\frac{\omega^2}{c^2}\right) H_2^<(x_1x_3|\omega) = 0 \quad . \tag{3.33b}$$

The boundary conditions at the interface $x_3 = \zeta(x_1)$ satisfied by $H_2(x_1x_3|\omega)$ are

$$H_2^>(x_1x_3|\omega)\Big|_{x_3=\zeta(x_1)} = H_2^<(x_1x_3|\omega)\Big|_{x_3=\zeta(x_1)} \tag{3.34a}$$

$$\frac{\partial}{\partial n} H_2^>(x_1x_3|\omega)\Big|_{x_3=\zeta(x_1)} = \frac{1}{\varepsilon(\omega)}\frac{\partial}{\partial n} H_2^<(x_1x_3|\omega)\Big|_{x_3=\zeta(x_1)}, \tag{3.34b}$$

where $(\partial/\partial n)$ is the derivative along the normal to the interface $x_3 = \zeta(x_1)$ at each point, directed from the dielectric into the vacuum:

$$\frac{\partial}{\partial n} = [1+(\zeta'(x_1))^2]^{-1/2}\left(-\zeta'(x_1)\frac{\partial}{\partial x_1} + \frac{\partial}{\partial x_3}\right) \quad . \tag{3.35}$$

In addition, $H_2(x_1x_3|\omega)$ satisfies vanishing or radiation boundary condtions at $x_3 = \pm\infty$.

The solution of Eq. (3.33a) that satisfies the boundary conditions at infinity can be written in the form

$$H_2^>(x_1x_3|\omega) = e^{ikx_1-i\alpha_0(k\omega)x_3} + \int \frac{dq}{2\pi} A(q|k)e^{iqx_1+i\alpha_0(q\omega)x_3} \tag{3.36}$$

where

$$\alpha_0(q\omega) = \left(\frac{\omega^2}{c^2} - q^2\right)^{1/2} \tag{3.37a}$$

$$= i\left(q^2 - \frac{\omega^2}{c^2}\right)^{1/2} , \tag{3.37b}$$

while $\alpha_o(k\omega)$ is always real $(k^2 < \omega^2/c^2)$.

The solution to Eq. (3.33b) can be written in a form similar to that of Eq. (3.36). However, as in the preceding subsection we will simplify the problem by eliminating the field in the medium and by working only with the field in the vacuum.

For this purpose we denote by V the region $x_3 < \zeta(x_1)$. Then let $\sum$ be the surface consisting of the union of the surface S- the surface $x_3 = \zeta(x_1)$ - and the surface $S^{(-\infty)}$ - a hemicylinder of infinite radius in the lower half-space. We then use Green's second integral theorem in the following form:

$$\begin{aligned}
&\int_V d^3x\{H_2^<(x_1x_3|\omega)[\nabla^2+\varepsilon(\omega)\frac{\omega^2}{c^2}]G_\varepsilon(x_1x_3;x_1'x_3'|\omega) - \\
&\quad - G_\varepsilon(x_1x_3;x_1'x_3'|\omega)[\nabla^2+\varepsilon(\omega)\frac{\omega^2}{c^2}]H_2^<(x_1x_3|\omega)\} \\
&= \int_V d^3x\{H_2^<(x_1x_3|\omega)\nabla^2 G_\varepsilon(x_1x_3;x_1'x_3'|\omega) - \\
&\quad - G_\varepsilon(x_1x_3;x_1'x_3'|\omega)\nabla^2 H_2^<(x_1x_3|\omega)\} \\
&- \int_\Sigma dS\{H_2^<(x_1x_3|\omega)\frac{\partial}{\partial n}G_\varepsilon(x_1x_3;x_1'x_3'|\omega) - \\
&\quad - G_\varepsilon(x_1x_3;x_1'x_3'|\omega)\frac{\partial}{\partial n}H_2^<(x_1x_3|\omega)\} ,
\end{aligned} \tag{3.38}$$

where $(\partial/\partial n)$ is the derivative along the normal to the surface $\sum$, directed outward from the volume V. The Green's function $G_\varepsilon(x_1x_3;x_1'x_3'|\omega)$ is the solution of

$$\left(\frac{\partial^2}{\partial x_1^2} + \frac{\partial^2}{\partial x_3^2} + \varepsilon(\omega)\frac{\omega^2}{c^2}\right) G_\varepsilon(x_1x_3;x_1'x_3'|\omega) = -4\pi\delta(x_1-x_1')\delta(x_3-x_3') \tag{3.39}$$

subject to radiation or vanishing boundary conditions at $x_3 = \pm\infty$. A convenient representation for $G_\varepsilon(x_1x_3;x_2'x_3'|\omega)$ is

$$G_\varepsilon(x_1x_3;x_1'x_3'|\omega) = i\int_{-\infty}^{\infty}\frac{dp}{\alpha(p\omega)}e^{ip(x_1-x_1')+i\alpha(p\omega)|x_3-x_3'|} \tag{3.40a}$$

where

$$\alpha(p\omega) = \left(\varepsilon(\omega)\frac{\omega^2}{c^2} - p^2\right)^{1/2} \qquad \mathrm{Re}\,\alpha(p\omega) > 0,\ \mathrm{Im}\,\alpha(p\omega) > 0. \tag{3.40b}$$

The surface $S^{(-\infty)}$ does not contribute to the integral over $\sum$ in Eq. (3.38) because of the boundary conditions at infinity satisfied by $H_2^<(x_1x_3|\omega)$ and $G_\varepsilon(x_1x_3;x_1'x_3'|\omega)$. Thus, we can use this fact, (Eq. (3.33b), and Eq. (3.39) to rewrite Eq. (3.38) as

$$\begin{aligned} - 4\pi\theta_V(x_3')H_2^<(x_1'x_3'|\omega) \\ = \int_S ds_1\Big\{H_2^<(x_1x_3|\omega)\,\frac{\partial}{\partial n}\,G_\varepsilon(x_1x_3;x_1'x_3'|\omega) - \\ - G_\varepsilon(x_1x_3;x_1'x_3'|\omega)\,\frac{\partial}{\partial n}\,H_2^<(x_1x_3|\omega)\Big\}\ , \end{aligned} \tag{3.41}$$

where

$$\begin{aligned} \theta_V(x_3) &= 1 \qquad x_3 < \zeta(x_1) &(3.42a)\\ &= 0 \qquad x_3 > \zeta(x_1)\ , &(3.42b)\end{aligned}$$

and S now denotes the curve $x_3 = \zeta(x_1)$ while ds_1 denotes the element of arc length along it.

We now replace s_1 as the integration variable by x_1 with the use of the relation

$$ds_1 = [1 + (\zeta'(x_1))^2]^{1/2}dx_1\ , \tag{3.43}$$

which is the one-dimensional form of Eq. (2.6). We also assume that $x_3' > \zeta(x_1')$, and use the boundary conditions (3.34) and the definition (3.35) to transform Eq. (3.41) into

$$\begin{aligned} \int dx_1\Big\{H_2^>(x_1x_3|\omega)\Big(-\zeta'(x_1)\,\frac{\partial}{\partial x_1} + \frac{\partial}{\partial x_3}\Big)\,G_\varepsilon(x_1x_3;x_1'x_3'|\omega) - \\ - G_\varepsilon(x_1x_3;x_1'x_3'|\omega)\,\varepsilon(\omega)\Big(-\zeta'(x_1)\,\frac{\partial}{\partial x_1} + \frac{\partial}{\partial x_3}\Big)\times \\ H_2^>(x_1x_3|\omega)\Big\}_{x_3=\zeta(x_1)} = 0 \qquad x_3' > \zeta(x_1'). \end{aligned} \tag{3.44}$$

This is the boundary condition satisfied by $H_2^>(x_1x_3)|\omega)$.

We finally substitute the expression (3.36) for $H_2^>(x_1x_3|\omega)$ and the representation (3.40a) for $G_\varepsilon(x_1x_3;x_1'x_3'|\omega)$ into Eq. (3.44), and equate to zero the pth Fourier coefficient. The result is the following integral equation for the amplitude $A(p\omega)$ of the scattered wave [50]:

$$\int \frac{dq}{2\pi} \frac{I(\alpha(p\omega)-\alpha_o(q\omega)|p-q)}{\alpha(p\omega)-\alpha_o(q\omega)} [\alpha(p\omega)\alpha_o(q\omega)+pq]A(q|k)$$

$$= \frac{I(\alpha(p\omega)+\alpha_o(k\omega)|p-k)}{\alpha(p\omega)+\alpha_o(k\omega)} [\alpha(p\omega)\alpha_o(k\omega)-pk] \quad , \tag{3.45}$$

where

$$I(\alpha|Q) = \int dx_1 e^{-iQx_1-i\alpha\zeta(x_1)} \quad . \tag{3.46}$$

For the study of the propagation of a surface polariton in the x_1-direction across the surface $x_3 = \zeta(x_1)$ we need the homogeneous form of Eq. (3.45). Since we are seeking solutions that are localized to the surface, it is convenient to make the replacements

$$\alpha_o(q\omega) = i\beta_o(q\omega) \quad , \qquad \alpha(q\omega) = i\beta(q\omega) \quad , \tag{3.47}$$

so that

$$\beta_o(q\omega) = \left(q^2 - \frac{\omega^2}{c^2}\right)^{1/2} \qquad q^2 > \frac{\omega^2}{c^2} \tag{3.48a}$$

$$= -\, i\left(\frac{\omega^2}{c^2} - q^2\right)^{1/2} \qquad q^2 < \frac{\omega^2}{c^2} \quad , \tag{3.48b}$$

and

$$\beta(q\omega) = \left(q^2 - \varepsilon(\omega)\frac{\omega^2}{c^2}\right)^{1/2} \quad , \qquad \mathrm{Re}\beta(q\omega) > 0, \ \mathrm{Im}\beta(q\omega) < 0 \quad . \tag{3.49}$$

With these changes, Eq. (3.45) becomes

$$\frac{\varepsilon(\omega)\beta_o(p\omega)+\beta(p\omega)}{\varepsilon(\omega)-1} A(p\omega) = \int \frac{dq}{2\pi} J(\beta(p\omega) - \beta_o(q\omega)|p-q) \times$$

$$\times [pq-\beta(p\omega)\beta_o(q\omega)] A(q\omega) \quad , \tag{3.50}$$

where $J(\alpha|Q)$ has been defined in Eq. (3.15b). In writing Eq. (3.50) we have suppressed reference to the wave vector k, since there is no incident wave in the homogeneous problem, and have indicated explicitly that the amplitude $A(q\omega)$ depends on the frequency ω.

Equation (3.50) is the equation governing the propagation of a surface polariton in the x_1-direction across the surface $x_3 = \zeta(x_1)$.

IV. Surface Shape Resonances

We begin a discussion of surface electromagnetic waves on rough surfaces with a consideration of electrostatic and electromagnetic surface shape resonances. These are solutions of Laplace's and Maxwell's equations, respectively, that are spatially localized in the vicinity of a protuberance or indentation on an otherwise planar surface of a dielectric medium in contact with vacuum. The frequencies of these modes, and the field distributions associated with them, depend on the shape of the protuberance or indentation [51-58].

Although surface shape resonances are not propagating surface waves, they are of interest in the present context because certain kinds of rough surfaces can be viewed as consisting of a random array of identical protuberances on a planar dielectric surface. Consequently, understanding the resonances associated with an isolated protuberance should help us to understand the surface waves on such rough surfaces.

The physical structure we will be considering in this section is depicted in Fig. 3. It consists of vacuum in the region $x_3 > \zeta(\vec{x}_\parallel)$ and a dielectric medium, characterized by an isotropic, real, frequency-dependent dielectric constant $\varepsilon(\omega)$, in the region $x_3<\zeta(\vec{x}_\parallel)$. We will be interested in frequencies in the range where $\varepsilon(\omega)$ is negative. The surface profile function $\zeta(\vec{x}_\parallel)$ is assumed to be localized in the sense that it is sensibly zero for values of $|\vec{x}_\parallel|$ greater than some characteristic radius R in two dimensions, or for values of $|x_1|$ greater than R in the one-dimensional case.

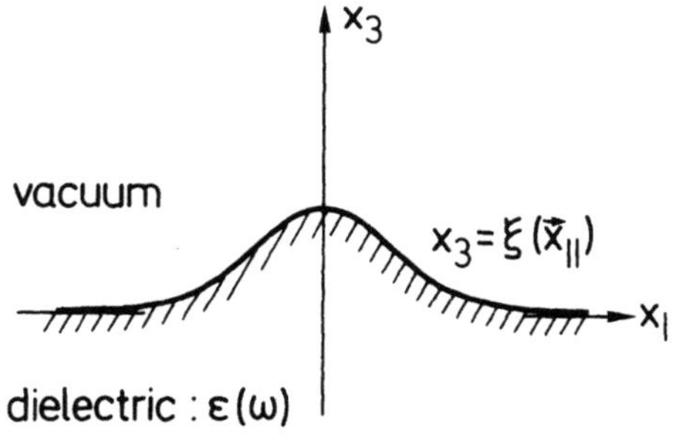

Fig. 3. An isolated protuberance on an otherwise planar dielectric surface.

IV.A. Electrostatic Surface Shape Resonances

The theory of electrostatic surface shape resonances is simpler than that of their electromagnetic counterparts, and provides a good introduction to the latter. At the same time, they have an intrinsic interest because of the possibility that they can be excited by the passage of a slow charged particle in their vicinity. In this case the electrostatic approximation is a good one, and becomes better the longer the wavelength of the charged particles relative to the linear dimensions of the surface perturbation supporting the electrostatic surface shape resonances.

In this section we present a theory of such surface shape resonances for surface perturbations described by both one-dimensional surface profile functions $\zeta(x_1)$ and by two-dimensional surface profile functions $\zeta(\vec{x}_\parallel)$. In the latter case the analysis is simplified by assuming that $\zeta(\vec{x}_\parallel)$ depends on $\vec{x}_\parallel$ only through its magnitude, i.e. it possesses cylindrical symmetry about the x_3-axis.

IV.A.1. A One-Dimensional Surface Profile

We first consider electrostatic surface shape resonances associated with a one-dimensional surface profile. We present the discussion in some detail for this simple case, because the same methods will be used in the succeeding parts of this section and it will then not be necessary to repeat the details of the computational approach there.

To simplify matters a bit, let us assume that the surface profile function $\zeta(x_1)$ is an even function of x_1. The electrostatic potential in this case is either even or odd in x_1:

$$\phi^{>}(-x_1, x_3|\omega) = \pm\, \phi^{>}(x_1 x_3|\omega) \; .$$

From Eq. (3.16) this relation translates into the conditions

$$A(-q\omega) = \pm\, A(q\omega) \; . \tag{4.1}$$

Thus we can restrict our attention to positive values of q. In this case only negative values of p give a nonvanishing contribution to the integral in Eq. (3.15). With a suitable change of variable, the latter becomes

$$\frac{\varepsilon(\omega)+1}{\varepsilon(\omega)-1} A(q\omega) = \pm \frac{1}{\pi} \int_0^\infty dp\; J(q-p|q+p)\, pA(p\omega) \; . \tag{4.2}$$

From Eq. (4.2) we see that if we solve the eigenvalue problem

$$\lambda_s a_s(q) = \frac{1}{\pi} \int_0^\infty dp J(q-p|q+p)\, p a_s(p) \; , \tag{4.3}$$

where s = 1, 2, 3, ... labels the solutions, the frequencies of the surface shape resonances are obtained from the equation

$$\frac{\varepsilon(\omega)+1}{\varepsilon(\omega)-1} = \pm\, \lambda_s \; . \tag{4.4}$$

In the case that $\varepsilon(\omega)$ has the simple free electron form

$$\varepsilon(\omega) = 1 - \frac{\omega_p^2}{\omega^2} \; , \tag{4.5}$$

where ω_p is the plasma frequency of the conduction electrons in the bulk, the solutions of Eq. (4.4) are

$$\omega_s = \frac{\omega_p}{\sqrt{2}} (1 \mp \lambda_s)^{1/2}, \tag{4.6}$$

where the upper (lower) sign corresponds to solutions that are even (odd) in x_1.

To illustrate the preceding results, we consider the case of a ridge or groove described by a Lorentzian profile

$$\zeta(x_1) = \frac{AR^2}{x_1^2 + R^2} . \tag{4.7}$$

When A is positive we are dealing with a ridge; when A is negative we have a groove. The function $J(q-p|q+p)$ defined by Eq. (3.15b) can be written in the form

$$J(q-p|q+p) = 2 \sum_{n=1}^{\infty} (AR^2)^n \frac{(q-p)^{n-1}}{n!} \int_0^{\infty} dx_1 \frac{\cos(q+p)x_1}{(x_1^2+R^2)^n}$$

$$= 2\sqrt{\pi}\, R \sum_{n=1}^{\infty} \frac{A^n}{2^{n-1/2}} \frac{(q-p)^{n-1}}{n!(n-1)!} [R(q+p)]^{n-1/2} K_{n-1/2}(R(q+p)) , \tag{4.8}$$

where $K_\nu(z)$ is a modified Bessel function. We use the result that

$$z^{n-1/2} K_{n-1/2}(z) = (\tfrac{\pi}{2})^{1/2} e^{-z} g_{n-1}(z), \tag{4.9}$$

where

$$g_o(z) = 1 \tag{4.10a}$$

$$g_1(z) = z + 1, \tag{4.10b}$$

and in general

$$g_{n+1}(z) = (2n+1) g_n(z) + z^2 g_{n-1}(z), \tag{4.11}$$

to rewrite $J(q-p|q+p)$ in the form

$$J(q-p|q+p) = \pi A R e^{-R(q+p)} \sum_{n=0}^{\infty} [\frac{A(q-p)}{2}]^n \frac{g_n(R(q+p))}{n!(n+1)!} . \tag{4.12}$$

The changes of variables

$$Rq = x, \quad Rp = y, \quad a_s(x/R) = b_s(x)/x^{1/2} , \tag{4.13}$$

convert Eq. (4.3) into

$$\lambda_s(A/R)b_s(x) = \int_0^\infty dy\ x^{1/2}e^{-x}K(x;y|A/R)e^{-y}y^{1/2}b_s(y), \qquad (4.14)$$

where

$$K(x;y)|A/R) = \frac{A}{R}\sum_{n=o}^{\infty}\left(\frac{A}{2R}(x-y)\right)^n \frac{g_n(x+y)}{n!(n+1)!}$$

$$= - K(y;x|-A/R) , \qquad (4.15)$$

and we have indicated explicitly that the eigenvalue λ_s depends on the value of A/R. In view of the second of Eqs. (4.15) and the fact that the eigenvalues of the transpose of a kernel equal those of the kernel itself, we can conclude that

$$\lambda_s(-A/R) = -\lambda_s(A/R) \quad . \qquad (4.16)$$

Consequently only positive values of A/R need to be considered explicitly.

Equation (4.14) was solved by converting it into a matrix equation by the use of an N-point Gauss-Laguerre numerical quadrature scheme. The latter is defined by the equation

$$\int_0^\infty dy\ e^{-y}f(y) \cong \sum_{j=1}^{N} w_j f(y_j) , \qquad (4.17)$$

where the weights $\{w_j\}$ and abscissas $\{y_j\}$ can be found in standard references [59]. In this way Eq. (4.14) is converted into the matrix equation

$$\lambda_s(A/R)b_s(x_i) = \sum_{j=1}^{N} M_{ij}(A/R)b_s(x_j) \quad \text{where} \qquad (4.18)$$

$$M_{ij}(A/R) = x_i^{1/2}e^{-x_i}\ K(x_i;x_j|A/R)w_j x_j^{1/2} . \qquad (4.19)$$

The frequencies $\omega_s(A/R)$ corresponding to the two largest eigenvalues $\lambda_1(A/R)>\lambda_2(A/R)>\ldots$ of Eqs. (4.18)-(4.19) are plotted as functions of A/R in Fig. 4. [60]

We see that the frequencies of the surface shape resonances corresponding to an electrostatic potential that is even in x_1 are lower than the frequency $\omega_p/\sqrt{2}$ of a surface plasmon on a planar surface in the case of a ridge (A/R > 0), and are higher in the case of a groove (A/R < 0). The reverse

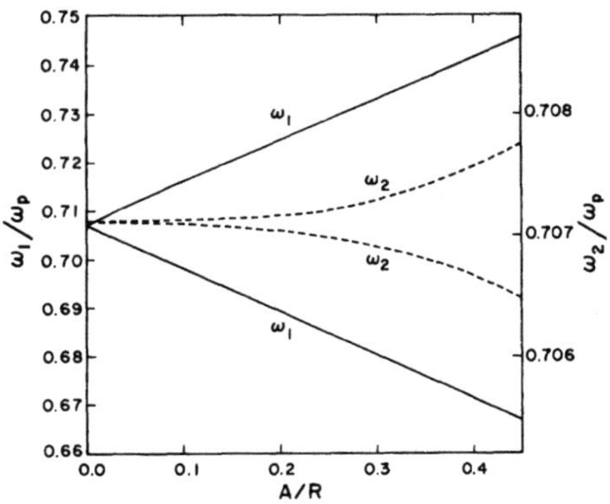

Fig. 4. The frequencies ω_1 and ω_2 corresponding to the two largest eigenvalues of the matrix $M_{ij}(A/R)$ defined by Eq. (4.19). The upper (lower) branches of these curves correspond to solutions for the electrostatic scalar potential that are odd (even) in x_1.

is the case for the modes corresponding to an electrostatic potential that is odd in x_1.

IV.A.2. A Two-Dimensional Surface Profile

In the case of a surface profile function $\zeta(\vec{x}_\parallel)$ that depends on both x_1 and x_2 it is convenient to assume that it possesses cylindrical symmetry about the x_3-axis. In that case, with the representations

$$\vec{q}_\parallel = q_\parallel(\cos\theta,\sin\theta),\ \vec{p}_\parallel = p_\parallel(\cos\theta',\sin\theta') \tag{4.20}$$

and

$$A(\vec{q}_\parallel\omega) = \sum_{\ell=-\infty}^{\infty} A_\ell(q_\parallel\omega)e^{i\ell\theta}\ , \tag{4.21}$$

Eq. (3.14) is converted into a set of uncoupled equations for the $\{A_\ell(q_\parallel\omega)\}$:

$$\frac{\varepsilon(\omega)+1}{\varepsilon(\omega)-1}A_\ell(q_\parallel\omega) = \int_0^\infty dp_\parallel K_\ell(q_\parallel|p_\parallel)p_\parallel^2\ A_\ell p_\parallel\omega)\ . \tag{4.22}$$

The kernel $K_\ell(q_\parallel|p_\parallel)$ is given by

$$K_\ell(q_\parallel|p_\parallel) = J_\ell(q_\parallel|p_\parallel) - \tfrac{1}{2}[J_{\ell-1}(q_\parallel|p_\parallel) + J_{\ell+1}(q_\parallel|p_\parallel)] \tag{4.23}$$

where

$$J_\ell(q_\parallel|p_\parallel) = \int_0^\infty dx_\parallel x_\parallel \frac{e^{(q_\parallel-p_\parallel)\zeta(x_\parallel)}-1}{(q_\parallel - p_\parallel)} J_\ell(q_\parallel x_\parallel)J_\ell(p_\parallel x_\parallel)\ , \tag{4.24}$$

and $J_\ell(x)$ is a Bessel function.

Equation (4.22) has been solved numerically for the modes with $\ell = 0$ in the case of a Gaussian surface profile function

$$\zeta(x_\parallel) = A\exp(-x_\parallel^2/R^2) \tag{4.25}$$

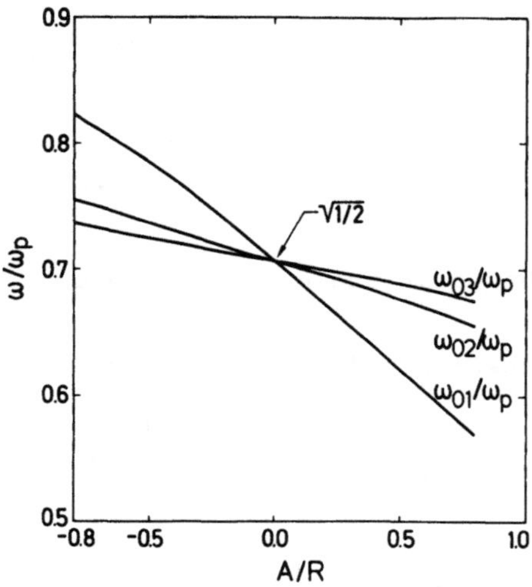

Fig. 5. The frequencies of the first three electrostatic surface shape resonances, as functions of A/R, for the Gaussian profile (2.8a).

by methods similar to those described in the preceding subsection [58]. The free electron metal form for $\varepsilon(\omega)$, Eq. (4.5), was used in these calculations. The results for the frequencies ω_{os}, s = 1, 2, 3 corresponding to the three largest eigenvalues of the kernel $K_o(g_\parallel|p_\parallel)p_\parallel^2$ are plotted as functions of A/R in Fig. 5. We see that surface shape resonances associated with a protuberance on the surface (A/R > 0) have frequencies that are lower than the frequency of surface plasmons on a planar surface. The opposite is the case for surface shape resonances associated with an indentation (A/R < 0).

IV.B. Electromagnetic Surface Shape Resonances

When the effects of retardation are included in the theory of surface shape resonances, a feature is introduced into the results that is absent in the electrostatic approximation, viz. the damping of these localized modes by radiation damping. The consequence of this is that the frequencies of surface shape resonances become complex, with the imaginary part of the frequency a measure of the finite lifetime of the mode caused by the radiation of energy into the vacuum above the surface perturbation.

The starting point for our discussion of electromagnetic surface shape resonances is the pair of coupled, homogeneous integral equations, Eqs. (3.30), for the Fourier coefficients of the electric field in the vacuum above the metal surface. In the derivation of these equations no assumptions were made about the nature of the surface profile function $\zeta(\vec{x}_\parallel)$. To simplify the subsequent analysis, we now assume that it is a function of $\vec{x}_\parallel$ only through its magnitude: $\zeta(\vec{x}_\parallel) \to \zeta(x_\parallel)$. At the same time, we introduce the expansions

$$A_{\parallel,\perp}(\vec{p}_\parallel\omega) = \sum_{\ell=-\infty}^{\infty} a_\ell^{(\parallel,\perp)}(p_\parallel\omega)e^{i\ell\theta} \tag{4.26a}$$

$$A_{\parallel,\perp}(\vec{q}_\parallel\omega) = \sum_{\ell=-\infty}^{\infty} a_\ell^{(\parallel,\perp)}(q_\parallel\omega)e^{i\ell\theta'} , \tag{4.26b}$$

where the angles θ and θ' are the polar angles of $\vec{p}_\parallel$ and $\vec{q}_\parallel$, respectively,

$$\vec{p}_\parallel = p_\parallel(\cos\theta,\sin\theta), \quad \vec{q}_\parallel = q_\parallel(\cos\theta',\sin\theta'). \tag{4.27}$$

When these expansions are substituted into Eqs. (3.30) and the angular integrations carried out.we obtain the following pair of coupled integral equations for the amplitudes

$a_\ell^{(\parallel)}(p_\parallel\omega)$ and $a_\ell^{(\perp)}(p_\parallel\omega)$:

$$\frac{\varepsilon(\omega)\beta_o(p_\parallel\omega)+\beta(p_\parallel\omega)}{\varepsilon(\omega)-1} a_\ell^{(\parallel)}(p_\parallel\omega)$$
$$= \frac{1}{2\pi}\int_0^\infty dq_\parallel q_\parallel \{m_\ell^{(\parallel,\parallel)}(p_\parallel|q_\parallel)a_\ell^{(\parallel)}(q_\parallel\omega)+m_\ell^{(\parallel,\perp)}(p_\parallel|q_\parallel)\ a_\ell^{(\perp)}(q_\parallel\omega)\} \tag{4.28a}$$

$$\frac{\beta_o(p_\parallel\omega)+\beta(p_\parallel\omega)}{\varepsilon(\omega)-1} a_\ell^{(\perp)}(p_\parallel\omega)$$
$$= \frac{1}{2\pi}\int_0^\infty dq_\parallel q_\parallel \{m_\ell^{(\perp,\parallel)}(p_\parallel|q_\parallel)a_\ell^{(\parallel)}(q_\parallel\omega)+m_\ell^{(\perp,\perp)}(p_\parallel|q_\parallel)a_\ell^{(\perp)}(q_\parallel\omega)\} , \tag{4.28b}$$

where

$$m_\ell^{(\parallel,\parallel)}(p_\parallel|q_\parallel) = p_\parallel J_\ell(p_\parallel|q_\parallel)q_\parallel - \frac{1}{2}\beta(p_\parallel\omega) \times$$
$$\times [J_{\ell-1}(p_\parallel|q_\parallel) + J_{\ell+1}(p_\parallel|q_\parallel)]\beta_o(q_\parallel\omega) \tag{4.29a}$$

$$m_\ell^{(\parallel,\perp)}(p_\parallel|q_\parallel) = \frac{1}{2}\frac{\omega}{c}\beta(p_\parallel\omega)[J_{\ell-1}(p_\parallel|q_\parallel) - J_{\ell+1}(p_\parallel|q_\parallel)] \tag{4.29b}$$

$$m_\ell^{(\perp,\parallel)}(p_\parallel|q_\parallel) = -\frac{1}{2}\frac{\omega}{c}[J_{\ell-1}(p_\parallel|q_\parallel) - J_{\ell+1}(p_\parallel|q_\parallel)]\beta_o(q_\parallel\omega) \tag{4.29c}$$

$$m_\ell^{(\perp,\perp)}(p_\parallel|q_\parallel) = \frac{1}{2}\frac{\omega^2}{c^2}[J_{\ell-1}(p_\parallel|q_\parallel) + J_{\ell+1}(p_\parallel|q_\parallel)] \tag{4.29d}$$

with

$$J_\ell(p_\parallel|q_\parallel) = \frac{2\pi}{\beta(p_\parallel\omega)-\beta_o(q_\parallel\omega)}\int_0^\infty dx_\parallel x_\parallel \times$$

$$\times \left(e^{(\beta(p_\parallel\omega)-\beta_o(q_\parallel\omega))\zeta(x_\parallel)} - 1\right)J_\ell(p_\parallel x_\parallel)J_\ell(q_\parallel x_\parallel). \qquad (4.30)$$

In what follows, we focus our attention on the case $\ell = 0$. In this case the equations for $a_o^{(\parallel)}(p_\parallel\omega)$ and $a_o^{(\perp)}(p_\parallel\omega)$ decouple, and we obtain the pair of homogeneous equations

$$\frac{\varepsilon(\omega)\beta_o(p_\parallel\omega)+\beta(p_\parallel\omega)}{\varepsilon(\omega)-1}\, a_o^{(\parallel)}(p_\parallel\omega)$$

$$= \frac{1}{2\pi}\int_0^\infty dq_\parallel [p_\parallel J_o(p_\parallel|q_\parallel)q_\parallel - \beta(p_\parallel\omega)J_1(p_\parallel|q_\parallel)\beta_o(q_\parallel\omega)]q_\parallel a_o^{(\parallel)}(q_\parallel\omega) \qquad (4.31a)$$

$$\frac{\beta_o(p_\parallel\omega)+\beta(p_\parallel\omega)}{\varepsilon(\omega)-1}\, a_o^{(\perp)}(p_\parallel\omega) = \frac{1}{2\pi}\int_0^\infty dq_\parallel\, \frac{\omega^2}{c^2}\, J_1(p_\parallel|q_\parallel)q_\parallel a_o^{(\perp)}(q_\parallel\omega). \qquad (4.31b)$$

The question of whether a surface perturbation can induce a surface shape resonance of s-polarization is an interesting one. We will not pursue it here, however, and instead concentrate on the solution of Eq. (4.31a) to obtain the frequencies of surface shape resonances of p-polarization. To this end we assume the Gaussian profile (2.8a).

To obtain the values of ω for which solutions of Eq. (4.31a) exist we proceed as follows. We first note that its kernel has a square root singularity at $q_\parallel = \omega/c$, so that if we discretize the equation by using some sort of quadrature scheme on the integral, the quadrature scheme should be chosen for accuracy in the neighborhood of this value of $q_\parallel$. (Of course, in the actual calculations dimensionless values of all the variables are used.) One way to do this is to split the range of integration into three parts: (1) $0 < q_\parallel < \omega/c$; (2) $\omega/c < q_\parallel < 2\omega/c$; and (3) $2\omega/c < q_\parallel < \infty$. Gauss-Laguerre quadrature can be used in (3), and Gauss-Legendre quadrature should be used in (1) and (2). But there is no particular need for the high density of quadrature abscissae at $q_\parallel \gtrsim 0$ in (1) and at $q_\parallel \lesssim 2\omega/c$ in (2) that would be afforded by the latter choice of quadrature scheme. Thus we use only half of the Gauss-Legendre points in each of the regions (1) and (2), chosen to give a high density of abscissae near $q_\parallel = \omega/c$. The number of abscissae in each region is N_1, $N_2(=N_1)$, N_3, respectively.

When Eq. (4.31a) is discretized in this way, we obtain a set of $N_1 + N_2 + N_3$ homogeneous, linear equations for an equal number of unknowns which are $a_o^{(\parallel)}(p_\parallel^{(i)}\omega)$ for $i = 1, 2, \ldots, N_1 + N_2 + N_3$. The condition for a solution to exist is that the determinant $D(\omega)$ of the matrix of coefficients in this set of equations should vanish.

In the preceding two paragraphs we have acted as if the frequency ω is real. In fact, the values of ω for which Eq. (4.31a) has solutions should be complex due to radiation damping. Since we have assumed a time dependence of $\exp(-i\omega t)$ for our fields, we should have that $\mathrm{Im}\,\omega < 0$, even if $\varepsilon(\omega)$ is real, as it is assumed to be in the present work. We therefore write

$$\omega = \omega_R - i\omega_I, \quad \omega_R > 0, \quad \omega_I > 0. \tag{4.32}$$

Nevertheless, it is still possible to solve for ω_R and ω_I by keeping ω real in Eq. (4.31a), as we now show.

Even for real values of ω the determinant $D(\omega)$ is complex because, according to Eq. (3.28), $\beta_o(p_\parallel\omega)$ is purely real or purely imaginary depending on the value of $p_\parallel$. Thus, for real ω we can write

$$D(\omega) = D_R(\omega) + iD_I(\omega) \quad , \tag{4.33}$$

where $D_R(\omega)$ and $D_I(\omega)$ are real functions of ω. If we assume that each of these functions is analytic in the vicinity of the real axis, we obtain

$$\begin{aligned} D(\omega_R - i\omega_I) &= D_R(\omega_R - i\omega_I) + iD_I(\omega_R - i\omega_I) \\ &\cong D_R(\omega_R) + \omega_I D_I'(\omega_R) + i[D_I(\omega_R) - \omega_I D_R'(\omega_R)], \end{aligned} \tag{4.34}$$

where we have assumed that $\omega_I \ll \omega_R$. We will see below that this assumption is amply justified. The primes in Eq. (4.34) denote differentiation with respect to argument. When we equate to zero the real and imaginary terms on the right-hand side of Eq. (4.34) we obtain the pair of equations

$$\omega_I = -\frac{D_R(\omega_R)}{D_I'(\omega_R)} \quad , \qquad \omega_I = \frac{D_I(\omega_R)}{D_R'(\omega_R)} \quad . \tag{4.35}$$

If we eliminate ω_I from this pair of equations, we obtain an equation for ω_R in the form

$$D_R(\omega_R)D_R'(\omega_R) + D_I(\omega_R)D_I'(\omega_R) = 0 \quad , \tag{4.36}$$

or

$$\left.\frac{\partial|D(\omega)|^2}{\partial\omega}\right|_{\omega=\omega_R} = 0 \quad . \tag{4.37}$$

Thus, ω_R is the value of (real) ω for which $|D(\omega)|^2$, and hence $|D(\omega)|$, is a minimum. Once ω_R has been obtained in this way, ω_I can be obtained from either of Eqs. (4.35).

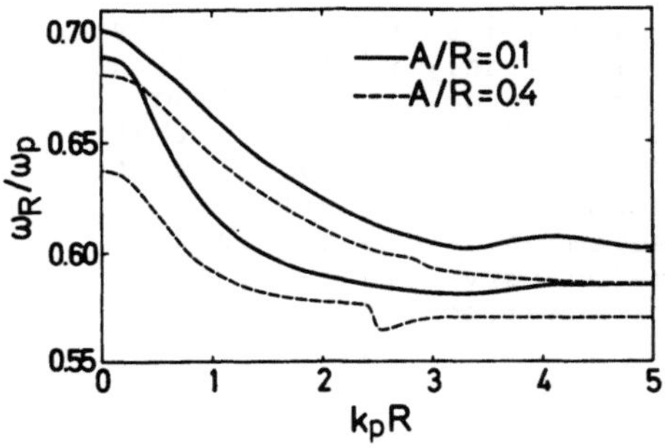

Fig. 6. The real parts of the frequencies of the first two electromagnetic surface shape resonances, as functions of k_pR, for the Gaussian profile (2.8a) in the structure depicted in Fig. 3. The frequencies have been calculated for A/R = 0.1 and 0.4.

Figure 6 shows some results, computed in this way [58], for the real part of the frequencies of the modes that correspond to the two lowest modes in the electrostatic case (Fig. 5), for two protuberances, with A/R = 0.1 and A/R = 0.4, as functions of k_pR, where $k_p = \omega_p/c$. For silver, $\hbar\omega_p$=3.78 eV, and $k_pR = 2$ if R = 104nm. The resonant frequencies are substantially lowered as the scale of the protruberance increases relative to the radiation wavelength.

We have not plotted the imaginary parts of ω on Fig. 6, because they are insignificantly small. ω_I/ω_p is of the order of 10^{-4} or smaller, and is generally larger in magnitude for taller protuberances with a given R. This is the radiation damping. It is small compared to the ohmic damping, which can be phenomenologically described with a complex dielectric constant $\varepsilon(\omega) = 1 - \omega_p^2/\omega(\omega+i/\tau)$, with $\tau = 31 \times 10^{-15}$sec for silver [61]. If this form of $\varepsilon(\omega)$ is substituted into Eq. (4.31a) the real part of ω is little affected, but the imaginary part is increased (in absolute value) by about two orders of magnitude. Thus, for example, the mode with $\omega_R/\omega_p = 0.58$ at $k_pR = 2$ (A/R = 0.4) has $\omega_I/\omega_p = 1.9 \times 10^{-3}$, whereas it would have $\omega_I/\omega_p = 4.52 \times 10^{-5}$ if $\tau = \infty$ (radiation damping only). The corresponding mode for the protuberance with A/R = 0.1 at $k_pR = 2$ with $\omega_R/\omega_p = 0.59$ has a radiative width $\omega_I/\omega_p = 1.36 \times 10^{-5}$.

V. Grating Surfaces

An important category of rough surfaces is that in which the surface profile is described by a deterministic function of the coordinates in the plane of the surface, that is also a periodic function of those coordinates. An example of such rough surfaces is provided by diffraction gratings. Such surfaces are of interest for several reasons. They are rough surfaces whose profiles are known. They therefore provide an excellent testing ground for the predictions of theories of rough surface phenomena. This use for gratings has been exploited by several groups, most notably by RAETHER and his co-workers [62-63] in studies of the propagation of surface polaritons along such periodically corrugated surfaces. They are also widely used in optical systems and provide an experimental method for coupling volume electromagnetic waves into surface polaritons that can then be used in devices. This same coupling makes it possible to measure the dispersion curves of surface polaritons by reflectivity measurements. It is then of interest to know how the dispersion curve of a surface polariton is affected by the very grating that makes it observable.

In this section we consider, in turn, the propagation of surface plasmons and polaritons across classical gratings and bigratings.

V.A. Surface Plasmons

When a surface plasmon propagates across a grating, either a classical grating or a bigrating, its dispersion curve is changed in interesting ways in comparison with what it is for a flat surface. It becomes dispersive, and acquires an infinity of branches. In this section we present the theory of surface plasmons on classical gratings and bigratings in turn.

V.A.1. A Classical Grating

The starting point for deriving the dispersion relation for surface plasmons propagating in the x_1-direction across a grating surface defined by the equation $x_3 = \zeta(x_1)$ is Eq. (3.15a). We make the following *Ansatz* for the amplitude $A(q)$:

$$A(q) = 2\pi \sum_{m=-\infty}^{\infty} A_m(k)\,\delta(q-k_m), \tag{5.1}$$

where

$$k_m \equiv k + \frac{2\pi m}{a}, \tag{5.2}$$

in order that the corresponding electrostatic scalar potential $\phi^>(x_1x_3|\omega)$ satisfy the Bloch condition $\phi^>(x_1 + a, x_3|\omega) = \exp(ika)\phi^>(x_1x_3|\omega)$. In Eq. (5.1) k is the wave vector of the surface plasmon. When we substitute Eq. (5.1) into Eq. (3.15a) the latter becomes

$$\frac{\varepsilon(\omega)+1}{\varepsilon(\omega)-1} \sum_m A_m(k)\,\delta(q-k_m)$$
$$= \sum_n \int dx_1 e^{-i(q-k_n)x_1} \frac{e^{(|q|-|k_n|)\zeta(x_1)}-1}{(|q|-|k_n|)} \times$$
$$\times \left(|k_n| - \frac{q}{|q|} k_n\right) A_n(k) \tag{5.3}$$

where we have used Eq. (3.15b). We next rewrite the integral using the identity

$$\int_{-\infty}^{\infty} dx_1 f(x_1) = \sum_{p=-\infty}^{\infty} \int_0^a dx_1 f(x_1+pa) , \tag{5.4}$$

and the result that

$$\sum_{n=-\infty}^{\infty} e^{-iqna} = \frac{2\pi}{a} \sum_{n=-\infty}^{\infty} \delta\left(q - \frac{2\pi n}{a}\right) , \tag{5.5}$$

to obtain the equation

$$\frac{\varepsilon(\omega)+1}{\varepsilon(\omega)-1} \sum_m A_m(k)\delta(q-k_m)$$

$$= \sum_m 2\pi\delta(q-k_m) \sum_n \frac{1}{a}\int_0^a dx_1 e^{-i\frac{2\pi}{a}(m-n)x_1} \times$$

$$\times \frac{e^{(\alpha_m-\alpha_n)\zeta(x_1)}-1}{\alpha_m - \alpha_n} \frac{\alpha_m\alpha_n - k_m k_n}{\alpha_m} A_n(k) \tag{5.6}$$

where, to simplify the notation, we have defined

$$\alpha_m = |k_m| . \tag{5.7}$$

This equation has a solution only if the coefficients of $\delta(q-k_m)$ on both sides are the same. This yields the equation satisfied by the $\{A_m(k)\}$:

$$\frac{\varepsilon(\omega)+1}{\varepsilon(\omega)-1} A_m(k) = \sum_n M_{mn}(k)A_n(k) , \tag{5.8}$$

where

$$M_{mn}(k) = \frac{\alpha_m\alpha_n - k_m k_n}{\alpha_m(\alpha_m - \alpha_n)} \frac{1}{a} \int_0^a dx_1 e^{-i\frac{2\pi}{a}(m-n)x_1} e^{(\alpha_m-\alpha_n)\zeta(x_1)} \qquad m\neq n \tag{5.9a}$$

$$= 0 \qquad m = n . \tag{5.9b}$$

Thus, if we denote the eigenvalues of the matrix $\overleftrightarrow{M}(k)$ by $\{\lambda_s(k)\}$, where s labels the distinct eigenvalues, the dispersion relation for a surface plasmon on a grating can be written as

$$\frac{\varepsilon(\omega)+1}{\varepsilon(\omega)-1} = \lambda_s(k) , \tag{5.10}$$

and leads to a separate branch of the dispersion curve for each distinct eigenvalue.

It can be shown from the definition (5.9) of $M_{mn}(k)$ that these eigenvalues occur in pairs, $\pm\lambda_s(k)$, and are periodic in k with a period $2\pi/a$. The latter property allows us to restrict k to the first Brillouin zone of the grating. In the small roughness limit the largest pair of eigenvalues is proportional to δ/a, where δ/a is the corrugation strength of the grating; the next largest pair is proportional to $(\delta/a)^3$; the next largest to $(\delta/a)^5$; and so on.

The eigenvalues $\{\lambda_s(k)\}$ of $\overset{\leftrightarrow}{M}(k)$ were obtained numerically for the sinusoidal profile function $\zeta(x_1) = \zeta_0 \cos(2\pi x_1/a)$, and the corresponding surface plasmon frequencies obtained from Eq. (5.10) for the free electron dielectric constant (4.5) [10]. Good convergence could be obtained for values of ζ_0/a up to about 0.1, with matrices of up to 38 x 38 in size. It turned out to be difficult to calculate the smaller eigenvalues with high accuracy without going to large matrices: the rate of convergence of the eigenvalues decreases as their magnitudes decrease.

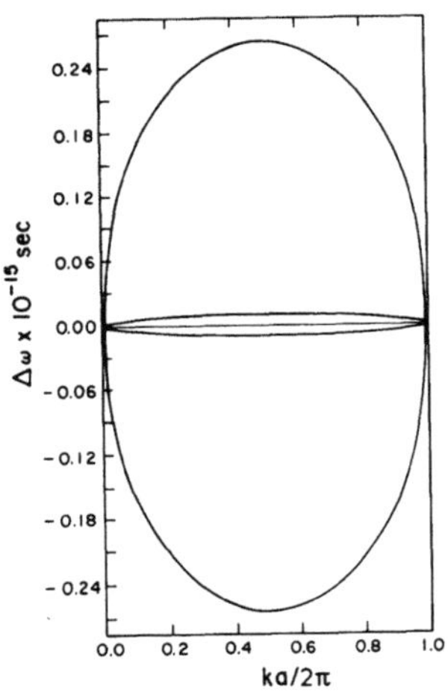

Fig. 7. The dispersion curve for surface plasmons on a grating characterized by a sinusoidal profile $\zeta(x_1) = \zeta_0 \cos(2\pi x_1/a)$ and a free electron form for the dielectric constant, as a function of $\xi = ka/2\pi$. In the present case $\zeta_0/a = 0.07$ and $\omega_p = 3.699 \times 10^{15} \text{sec}^{-1}$ [Ref. 18].

In Fig. 7 is plotted a typical result for $\Delta\omega(k) = \omega(k) - \omega_0$, where $\omega(k)$ is the frequency of a surface plasmon on a grating and $\omega_0 = 2^{-1/2}\omega_p$ is the frequency of a surface plasmon on a planar surface. In these calculations a value of $\zeta_0/a = 0.07$ was assumed. The curves shown were obtained from the diagonalization of a 38 x 38 matrix approximation to $\overset{\leftrightarrow}{M}(k)$. Only the branches corresponding to the four largest eigenvalues $\{\lambda_s(k)\}$ are plotted in this figure. The remaining branches are too close to the line $\Delta\omega(k) = 0$ to be depicted graphically.

It did not prove to be possible to obtain convergent results for the $\{\lambda_s(k)\}$ for values of ζ_0/a larger than about 0.1 by the present method. It was also not possible to obtain convergent results for a grating with a nonanalytic profile, viz. a symmetric sawtooth grating. However, convergent results can be obtained for larger values of ζ_o/a and for gratings with nonanalytic profile, if instead of eliminating the electrostatic scalar potential in the dielectric medium as we did here, one keeps it and determines the Fourier coefficients of the potential in the vacuum and in the dielectric from the usual Maxwell boundary conditions at the interface $x_3 = \zeta(x_1)$. The calculations by the latter approach involve matrices that are typically twice as large as those used in the approach described in these lectures. The reader is referred to Ref. 10 for a discussion of the alternate approach.

The sinusoidal grating studied here can be thought of as the result of adding a periodic array of ridges onto an otherwise planar surface at $x_3 = 0$. The surface profile function from this point of view is given by

$$\zeta(x_1) = \zeta_o(1 + \cos\frac{2\pi x_1}{a}).$$

If we expand this about the position of one of the maxima, say the one at $x_1 = 0$,

$$\zeta(x_1) = 2\zeta_o\left(1 - \frac{\pi^2}{a^2} x_1^2 + O(x_1^4)\right)$$

$$\cong \frac{2\zeta_o(a^2/\pi^2)}{x_1^2 + (a^2/\pi^2)},$$

we see that the grating can be viewed as the superposition of an infinite number of parallel, Lorentzian ridges whose centers are at $x_1 = 0, \pm a, \pm 2a, \ldots$ Since each such ridge in isolation gives rise to an infinite number of electrostatic surface shape resonances, as we have seen in subsection IV.A.1, with frequencies both above and below $\omega_p/\sqrt{2}$, it is tempting to associate the branches of the dispersion curve shown in Fig. 7 with the frequencies of these surface shape resonances that have been broadened into bands by being brought together to form a periodic array. A quantitative study of this point has not been carried out so far, however.

V.A.2. A Bigrating

Bigratings are becoming interesting at the present time as substrates for enhancing the intensity of Raman scattering from molecules adsorbed on them. LIAO *et al.* [64] have reported enhancements of $\sim 10^7$ for Raman scattering from CN^- adsorbed on an Ag surface covered with a lithographically fabricated ordered two-dimensional array of silver particles. Bigratings are also a closer periodic approximation to randomly rough surfaces than are classical gratings, because the surface profile function for a bigrating is a function of both x_1 and x_2, as in the case of a randomly rough surface.

In studying surface plasmons on a bigrating in this subsection we have two goals. The first is to explore the limits of convergence of the method used in these lectures as the corrugation strength of the grating is increased. The second is to obtain properties of the surface plasmons themselves on such a corrugated surface that can serve as a model for a periodic array of the kinds of surface protrusions that have been studied in isolation in subsection IV.A.2.

Our starting point is Eq. (3.14), into which we substitute the expansion

$$A(\vec{q}_\parallel) = (2\pi)^2 \sum_{\vec{G}_\parallel} A_{\vec{k}_\parallel}(\vec{G}_\parallel)\delta(\vec{q}_\parallel - \vec{k}_\parallel - \vec{G}_\parallel), \tag{5.11}$$

which ensures that $\phi^>(\vec{x}|\omega)$, Eq. (3.3) have the Bloch form. With the aid of the identity

$$\int d^2x_\parallel f(\vec{x}_\parallel) = \sum_\ell \int_{a_c} d^2x_\parallel f(\vec{x}_\parallel + \vec{x}_\parallel(\ell)), \tag{5.12}$$

the periodicity of $\zeta(\vec{x}_\parallel)$, and the result that

$$\sum_\ell e^{-i(\vec{q}_\parallel - \vec{k}_\parallel)\cdot\vec{x}_\parallel(\ell)} = \frac{(2\pi)^2}{a_c} \sum_{\vec{G}_\parallel} \delta(\vec{q}_\parallel - \vec{k}_\parallel - \vec{G}_\parallel), \tag{5.13}$$

we transform Eq. (3.14) into the following matrix equation for the $\{A_{\vec{k}_\parallel}(\vec{G}_\parallel)\}$:

$$\frac{\varepsilon(\omega)+1}{\varepsilon(\omega)-1} A_{\vec{k}_\parallel}(\vec{G}_\parallel) = \sum_{\vec{G}'_\parallel} M_{\vec{k}_\parallel}(\vec{G}_\parallel, \vec{G}'_\parallel) A_{\vec{k}_\parallel}(\vec{G}'_\parallel), \tag{5.14a}$$

where

$$M_{\vec{k}_\parallel}(\vec{G}_\parallel, \vec{G}'_\parallel) = \frac{|\vec{k}_\parallel + \vec{G}_\parallel||\vec{k}_\parallel + \vec{G}'_\parallel| - (\vec{k}_\parallel + \vec{G}_\parallel)\cdot(\vec{k}_\parallel + \vec{G}'_\parallel)}{|\vec{k}_\parallel + \vec{G}_\parallel|[|\vec{k}_\parallel + \vec{G}_\parallel| - |\vec{k}_\parallel + \vec{G}'_\parallel|]} \times$$

$$\times \frac{1}{a_c} \int_{a_c} d^2x_\parallel e^{-i(\vec{G}_\parallel - \vec{G}'_\parallel)\cdot\vec{x}_\parallel} e^{(|\vec{k}_\parallel + \vec{G}_\parallel| - |\vec{k}_\parallel + \vec{G}'_\parallel|)\zeta(\vec{x}_\parallel)} \qquad \vec{G}_\parallel \neq \vec{G}'_\parallel$$

$$= 0 \qquad \vec{G}_\parallel = \vec{G}'_\parallel \quad . \tag{5.14b}$$

If we denote the sth distinct eigenvalue of $\overleftrightarrow{M}_{\vec{k}_\parallel}$ by $\lambda_s(\vec{k}_\parallel)$, then the dispersion relation for surface plasmons on a bigrating becomes

$$\frac{\varepsilon(\omega)+1}{\varepsilon(\omega)-1} = \lambda_s(\vec{k}_\parallel), \tag{5.15}$$

and leads to a separate branch of the dispersion curve for each distinct eigenvalue.

It is straightforward to show that $\lambda_s(\vec{k}_\parallel)$ is a periodic function of $\vec{k}_\parallel$ with a period given by the unit cell of the reciprocal lattice generated by the $\{\vec{G}_\parallel\}$:

$$\lambda_s(\vec{k}_\parallel + \vec{G}_\parallel) = \lambda_s(\vec{k}_\parallel), \tag{5.16}$$

so that

$$\omega_s(\vec{k}_\parallel+\vec{G}_\parallel) = \omega_s(\vec{k}_\parallel). \tag{5.17}$$

Consequently, in calculating the dispersion curves we can restrict $\vec{k}_\parallel$ to lie within a unit cell of the reciprocal lattice or, equivalently, within the two-dimensional first Brillouin zone defined by the periodicity of the grating.

The eigenvalues $\{\lambda_s(\vec{k}_\parallel)\}$ have been obtained [65] for the case where the dielectric constant is given by Eq. (4.5) and the surface profile function $\zeta(\vec{x}_\parallel)$ is

$$\zeta(\vec{x}_\parallel) = \zeta_0\left(\cos\frac{2\pi x_1}{a} + \cos\frac{2\pi x_2}{a}\right), \tag{5.18}$$

so that

$$\vec{G}_\parallel = \frac{2\pi}{a}(h_1,h_2). \tag{5.19}$$

They were calculated by diagonalizing an $N^2\times N^2$ matrix (N even) obtained by letting h_1 and h_2 run from $-(N/2) + 1$ to $(N/2)$ in integer steps. Convergence was sought by increasing N. The eigenvalues were found to occur in pairs, $\pm\lambda_1(\vec{k}_\parallel)$, $\pm\lambda_2(\vec{k}_\parallel),\ldots$, and decreased rapidly in magnitude with increasing order: $|\lambda_1| > |\lambda_2| > |\lambda_3| > \ldots$, so that the resulting infinity of branches of the dispersion curve merges rapidly with the dispersion curve for a flat surface, $\omega(\vec{k}_\parallel) = 2^{-1/2}\omega_p$. A set of dispersion curves for $\zeta_0/a = 0.04$ is shown in Fig. 8, where the first four branches are drawn for values of the wave vector $\vec{k}_\parallel$ lying along the symmetry directions in the first Brillouin zone for this grating depicted in Fig. 9. Values of N of the order of ten are sufficient to produce these frequencies with at least five figure accuracy. However, it did not seem possible to use the present approach for calculating surface plasmon dispersion curves for values of $\zeta_0/a \gtrsim 0.1$. For such values of ζ_0/a the eigenvalues $\{\lambda_s(\vec{k}_\parallel)\}$ either acquired imaginary parts or became larger than unity in magnitude.

Methods for obtaining convergent results for larger values of ζ_0/a are described in Ref. 65.

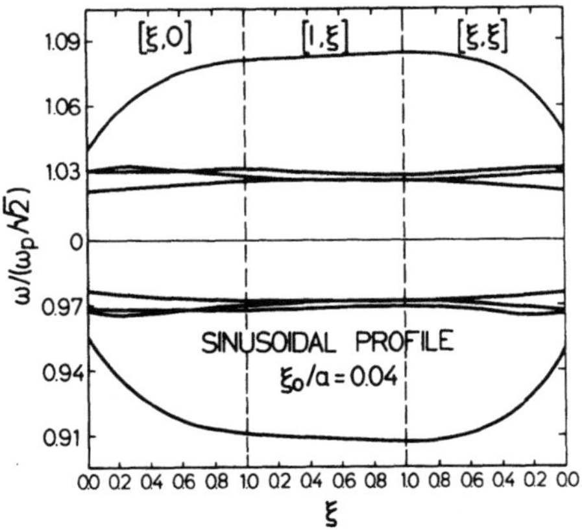

Fig. 8. First four pairs of branches (highest and lowest frequency) of the dispersion curve for surface plasmons on a sinusoidal bigrating of corrugation strength $\zeta_0/a = 0.04$. The normalized frequency $(\omega/(\omega_p/\sqrt{2})$ is plotted as a function of the reduced wave vector $\vec{\xi} = (a/\pi)(k_1,k_2)$ along the three symmetry directions in the surface first Brillouin zone, where ξ stands for ξ_1 and/or ξ_2 as appropriate [Ref. 65].

One of the interesting consequences of the results obtained by these methods is that the grating-induced downward shift of the frequency of a surface plasmon can be large enough, for sufficiently large values of ζ_0/a, to bring into the visible range a surface plasmon that on a flat surface would have a frequency in the ultraviolet portion of the optical spectrum.

A study of the dispersion curves for surface plasmons on a bigrating as bands formed from the frequencies of the surface shape resonances associated with the protuberances out of which the bigrating can be said to be constructed has not yet been carried out. It would be of considerable interest for the theory of surface excitations on such surfaces.

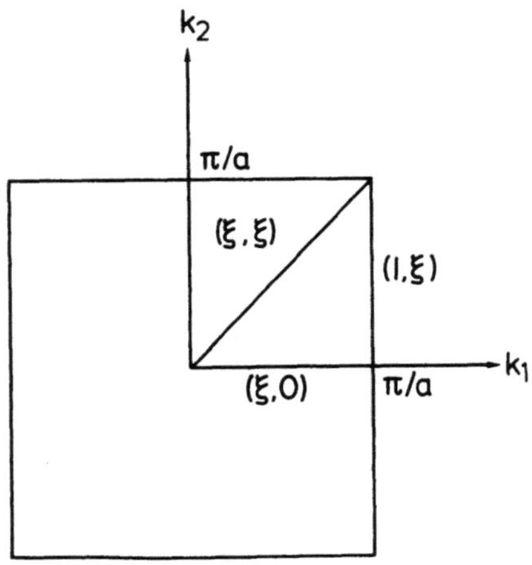

Fig. 9. The surface first Brillouin zone for a bigrating possessing the symmetry of a square with edge a. The three symmetry directions used in the calculations of the surface plasmon dispersion curves are labeled by the reduced wave vector

$$\vec{\xi} = (a/\pi)(k_1,k_2),$$

where ξ stands for ξ_1 and/or ξ_2 as appropriate.

V.B. Surface Polaritons

The dispersion curve for surface polaritons on a planar dielectric surface is dispersive, in contrast with the dispersion curve for surface plasmons on such a surface. At the same time, radiative effects are possible in electromagnetic theory that are absent in the electrostatic approximation. These two circumstances combine to rob the dispersion curves of surface polaritons on a periodically corrugated surface of some of the richness possessed by their electrostatic counterparts.

In this subsection we apply results obtained in Section III to the determination of the dispersion curves of surface polaritons on classical gratings and bigratings.

V.B.1. A Classical Grating

In obtaining the dispersion curve for surface polaritons propagating across a classical grating, it is convenient to start with the homogeneous version of Eq. (3.45), and to make the replacements given by Eqs. (3.47) in it. The result is

$$\int\frac{dq}{2\pi}\,\frac{K(\beta(p\omega)-\beta_o(q\omega)\,|\,p-q)}{\beta(p\omega)-\beta_o(q\omega)}\,[pq-\beta(p\omega)\beta_o(q\omega)]A(q\omega) = 0 \tag{5.20}$$

where

$$K(\beta|Q) = \int dx_1 e^{-iQx_1} e^{\beta\zeta(x_1)}. \tag{5.21}$$

We now use the expansion

$$A(q\omega) = 2\pi\sum_{n=-\infty}^{\infty} A_n(k)\,\delta(q-k_n) \tag{5.22}$$

together with Eqs. (5.4) and (5.5) to transform Eq. (5.20) into

$$\sum_m 2\pi\delta(p-k_m)\sum_n \frac{k_mk_n-\beta(k_m\omega)\beta_o(k_n\omega)}{\beta(k_m\omega)-\beta_o(k_n\omega)}\,\times$$

$$\times\,\frac{1}{a}\int_0^a dx_1 e^{-i\frac{2\pi}{a}(m-n)x_1} e^{(\beta(k_m\omega)-\beta_o(k_n\omega))\zeta(x_1)} A_n(k) = 0. \tag{5.23}$$

This equation can be satisfied only if the coefficient of $\delta(p-k_m)$ vanishes for each m. In this way we obtain the equations satisfied by the $\{A_n(k)\}$ [50]:

$$\sum_n \frac{k_mk_n-\beta(k_m\omega)\beta_o(k_n\omega)}{\beta(k_m\omega)-\beta_o(k_n\omega)}\,\frac{1}{a}\int_0^a dx_1\; e^{-i\frac{2\pi}{a}(m-n)x_1}\,\times$$

$$\times e^{(\beta(k_m\omega)-\beta_o(k_n\omega))\zeta(x_1)} A_n(k) = 0 \qquad (5.24)$$

$$m = 0, \pm 1, \pm 2, \pm 3, \ldots$$

Since the frequency we seek appears in all of the elements of the matrix multiplying the column vector $\vec{A}(k)$ in this equation, we cannot reduce Eq. (5.24) to an eigenvalue problem. Instead, the dispersion relation for surface polaritons on a grating is obtained by equating to zero the determinant of the matrix of coefficients in Eq. (5.24). This is computationally heavier than solving for the eigenvalues of a matrix.

One can show on the basis of Eq. (5.24) that the frequencies $\omega(k)$ that are the solutions of this equation are even functions of k and are periodic functions of k with a period given by $2\pi/a$. We can therefore confine k to the interval $(0,\pi/a)$ to obtain the distinct solutions.

What should we expect these solutions to look like? In answering this question it is useful to use two concepts from the theory of the electronic band structure of solids, viz. the empty lattice construction and the reduced zone scheme. In the upper half of Fig. 10 is plotted the dis-

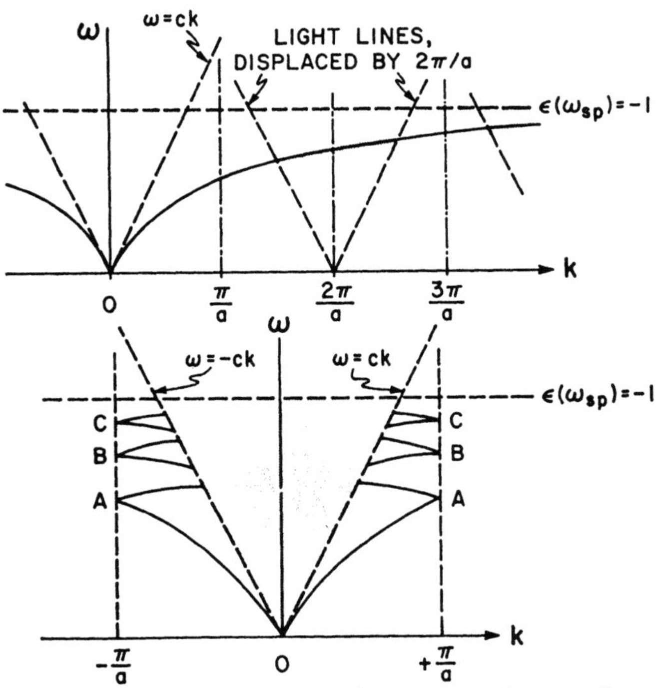

Fig. 10(a). The dispersion curve for surface polaritons on a planar metal surface in the extended-zone scheme. (b) The same curve in the reduced zone scheme

persion curve on the planar surface of a metal described by the free electron dielectric constant (4.5). Also indicated on this figure are the boundaries of the first, second, third, ... Brillouin zones of the grating structure. In addition, we have drawn the light lines $\omega = \pm ck$, and the light lines displaced by integer multiples of $(2\pi/a)$. The significance of these light lines is that in the region of the (ω,k)-plane above them, $\beta_0(k_m\omega)$ is pure imaginary for some m, and the solutions of Eq. (5.24) correspond to radiative modes rather than to the nonradiative surface polaritons. Two portions of the dispersion curve that fall into the radiative region are indicated by being hatched. It is only the portions of the dispersion curve that fall below the light lines and the displaced light lines that can belong to the dispersion curve of a surface polariton on a grating.

In the lower part of Fig. 10 we have shown the result of bringing the portions of the dispersion curve for surface polaritons on a planar surface that lie outside the first Brillouin zone into the first Brillouin zone by displacing them to the right or left by suitable integer multiples of $(2\pi/a)$. The radiative region is shown hatched here. The resulting dispersion curve now consists of several branches that extend from the light lines $\omega = \pm ck$ to the boundaries of the first Brillouin zone, $k = \pm\pi/a$, where the upper and lower signs go together. The points marked A, B, C,... on this figure are of particular interest. At these points two points on the flat surface dispersion curve, separated by a grating reciprocal lattice vector, are degenerate. When the grating profile is "turned on," so to speak, if the surface profile function $\zeta(x_1)$ has a nonzero Fourier coefficient for this grating reciprocal lattice vector, we know from degenerate perturbation theory that the degeneracies at the points A, B, C,..., will be lifted and gaps will open up in the dispersion curve for surface polaritons at these points.

This is indeed what is observed. In Fig. 11 [9] we have plotted the dispersion curves for surface polaritons on a sinusoidal grating, ruled on the surface of a free electron metal. The value of $\hbar\omega_p$ was chosen to be 2eV. The period a was assumed to be 5000Å, and dispersion curves for $\zeta_0 = 0$,

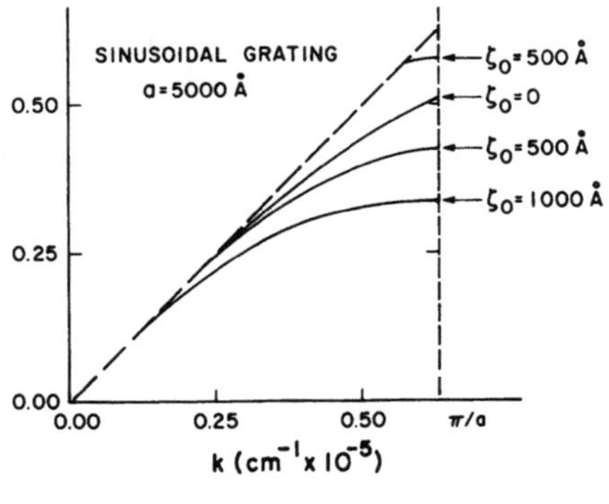

Fig. 11. The dispersion curves for surface polaritons on a grating defined by the profile function $\zeta(x_1) = \zeta_0\cos(2\pi x_1/a)$ for $a = 5000Å$ and the three values of ζ_0 [Ref. 9].

500Å, and 1000Å are presented. It is seen that for ζ_o = 500Å the dispersion curve consists of two branches with a gap at $k = \pi/a$. Any other branches in this case have higher frequencies and lie in the radiative region. When ζ_o = 1000Å the gap in the dispersion curve at the zone boundary has become so large that the upper branch is in the radiative region. The phenomenon of wave slowing is well displayed by the dispersion curves depicted in Fig. 11. Both the group velocity and the phase velocity of surface polaritons on the lowest branch are smaller in the presence of the grating than on the planar surface, for the same value of k. Related to this is the result that the frequency of the surface polariton on the lowest branch of the dispersion curve is lower in the presence of the grating than on the planar surface, for the same value of k.

The dispersion curve for a surface polariton on an InSb grating consisting of a periodic array of parallel triangular grooves was obtained by VAN DEN BERG and BORBURGH [66] by a somewhat different approach. These authors calculated the reflectivity of p-polarized light normally incident on such a grating as a function of the frequency of the incident light. From the positions of the minima in the frequency dependence of the reflectivity, for different grating periods, which are associated with the resonant excitation of surface polaritons, they constructed the dispersion curve. Their results are in qualitative agreement with those shown in Fig. 11.

V.B.2. A Bigrating

Although to our knowledge there has been no theoretical study of the propagation of surface polaritons across a bigrating, for completeness we derive in this subsection the equations from which the dispersion curves can be obtained in this case.

We start with the homogeneous form of Eq. (3.26) after the replacements (3.27) have been made in it:

$$\int \frac{d^2 q_\parallel}{(2\pi)^2} \frac{K(\beta(p_\parallel\omega)-\beta_o(q_\parallel\omega)|\vec{p}_\parallel-\vec{q}_\parallel)}{\beta(p_\parallel\omega) - \beta_o(q_\parallel\omega)} \times$$

$$\times \begin{pmatrix} p_\parallel q_\parallel - \beta(p_\parallel\omega)(\hat{p}_\parallel\cdot\hat{q}_\parallel)\beta_o(q_\parallel\omega) & -i\frac{\omega}{c}\beta(p_\parallel\omega)(\hat{p}_\parallel\times\hat{q}_\parallel)_3 \\ i\frac{\omega}{c}(\hat{p}_\parallel\times\hat{q}_\parallel)_3\,\beta_o(q_\parallel\omega) & \frac{\omega^2}{c^2}\hat{p}_\parallel\cdot\hat{q}_\parallel \end{pmatrix} \begin{pmatrix} A_\parallel(\vec{q}_\parallel\omega) \\ A_\perp(\vec{q}_\parallel\omega) \end{pmatrix} = 0 \quad (5.25)$$

where

$$K(\beta|\vec{Q}_\parallel) = \int d^2x_\parallel e^{-i\vec{Q}_\parallel\cdot\vec{x}_\parallel} e^{\beta\zeta(\vec{x}_\parallel)}$$

$$= (2\pi)^2 \sum_{\vec{G}_\parallel} \delta(\vec{Q}_\parallel-\vec{G}_\parallel)\frac{1}{a_c}\int_{a_c} d^2x_\parallel e^{-i\vec{G}_\parallel\cdot\vec{x}_\parallel} e^{\beta\zeta(\vec{x}_\parallel)}, \quad (5.26)$$

where the second form obtains in the case that $\zeta(\vec{x}_\parallel)$ is a periodic function of $\vec{x}_\parallel$. When we substitute Eq. (5.26) into Eq. (5.25), together with the expansions

$$A_{\parallel,\perp}(\vec{q}_\parallel\omega) = (2\pi)^2 \sum_{\vec{G}_\parallel} A^{(\parallel,\perp)}_{\vec{k}_\parallel}(\vec{G}_\parallel)\delta(\vec{q}_\parallel-\vec{k}_\parallel-\vec{G}_\parallel), \tag{5.27}$$

where $\vec{k}_\parallel$ is the wave vector of the surface polariton, we are led to the following pair of coupled matrix equations for the coefficients $\{A^{(\parallel,\perp)}_{\vec{k}_\parallel}(\vec{G}_\parallel)\}$:

$$\sum_{\vec{G}'_\parallel} \frac{1}{a_c}\int_{a_c} d^2x_\parallel \, e^{-i(\vec{G}_\parallel-\vec{G}'_\parallel)\cdot\vec{x}_\parallel} \frac{e^{(\beta(K_\parallel\omega)-\beta_0(K'_\parallel\omega))\zeta(\vec{x}_\parallel)}}{\beta(K_\parallel\omega)-\beta_0(K'_\parallel\omega)} \times$$

$$\times \begin{pmatrix} K_\parallel K'_\parallel-\beta(K_\parallel\omega)(\hat{K}_\parallel\cdot\hat{K}'_\parallel)\beta_0(K'_\parallel\omega) & -i\frac{\omega}{c}\beta(K_\parallel\omega)(\hat{K}_\parallel\times\hat{K}'_\parallel)_3 \\ i\frac{\omega}{c}(\hat{K}_\parallel\times\hat{K}'_\parallel)_3\beta_0(K'_\parallel\omega) & \frac{\omega^2}{c^2}\hat{K}_\parallel\cdot\hat{K}'_\parallel \end{pmatrix} \times$$

$$\times \begin{pmatrix} A^{(\parallel)}_{\vec{k}_\parallel}(\vec{G}'_\parallel) \\ A^{(\perp)}_{\vec{k}_\parallel}(\vec{G}'_\parallel) \end{pmatrix} = 0. \tag{5.28}$$

To simplify the notation we have introduced the vectors $\vec{K}_\parallel \equiv \vec{k}_\parallel + \vec{G}_\parallel$ and $\vec{K}'_\parallel \equiv \vec{k}_\parallel + \vec{G}'_\parallel$ in writing Eq. (5.28). The dispersion relation for surface polaritons on a bigrating is then obtained by equating to zero the determinant of the matrix of the coefficients in Eq. (5.28).

The solutions of this equation are even functions of $\vec{k}_\parallel$, $\omega_s(-\vec{k}_\parallel) = \omega_s(\vec{k}_\parallel)$, and are periodic functions of $\vec{k}_\parallel$ with a period given by the two-dimensional first Brillouin zone of the grating, $\omega_s(\vec{k}_\parallel+\vec{G}_\parallel) = \omega_s(\vec{k}_\parallel)$. The solutions can then be sought for values of $\vec{k}_\parallel$ inside the two-dimensional first Brillouin zone, and inside the nonradiative region defined by $k_\parallel > (\omega/c)$.

V.C. Convergence

We conclude this section devoted to the determination of the dispersion curves of surface plasmons and polaritons on periodically corrugated surfaces with a general remark about the convergence of the method used to obtain them. This should be done because our approach is based on the use of the Rayleigh hypothesis which, in other contexts, is known to be rigorous only for small surface corrugation strengths [5-8]. Thus, for example, it is known that in the case of the scattering of a scalar plane wave from a sinusoidal hard wall defined by $x_3 = \zeta_0\cos(2\pi x_1/a)$ the Rayleigh hypothesis is valid provided $\zeta_0/a < 0.071$ [5-7]. It is valid for the scattering of a scalar plane wave from a doubly periodic hard wall defined by $x_3 = \zeta_0(\cos(2\pi x_1/a) + \cos(2\pi x_2/a))$ provided that $\zeta_0/a < 0.047$ [7]. Comparable limits are not known for the problems studied here, yet if they are of comparable size the methods employed here are capable of yielding accurate dispersion curves for corrugation strengths that are comparable to, or even larger than, those for which the Rayleigh hypothesis is valid. It has been suggested [7,14] that the kind of convergence we find has an asymptotic nature. Namely, as more terms are included in a plane wave expansion of the electrostatic scalar potential or the components of the electric field in the vacuum above the surface of the dielectric medium, the frequencies of the corresponding surface plasmons or surface polaritons approach limiting values, only to diverge away from the latter as the number of plane waves increases beyond some critical value. This kind of behavior has been observed, e.g. in Ref. 12 in connection with the determination of the dispersion curves of shear horizontal surface acoustic waves propagating across a grating. The manner in which such calculations should be done, therefore, is to continue to increase the number of plane waves used in the expansions of the field components in the problem being studied until the differences between the frequencies obtained in going from N to N+2 (say) plane waves are no longer smaller than in going from N-2 to N plane waves. The result obtained with N plane waves are the ones to use. Three to six figure accuracy for the frequencies can be obtained in this way, with the accuracy decreasing as the corrugation strength increases.

The convergence of the calculations of the frequencies of surface shape resonances described in the preceding section may also have an asymptotic nature. The number of abscissae used in the Gaussian quadrature schemes there may play a role analogous to that of the number of plane waves used in the expansions of the present section.

VI. Randomly Rough Surfaces

In studying the propagation of surface plasmons and surface polaritons across a randomly rough surface our starting equations are Eqs. (3.14) and (3.30), respectively. However, the surface profile function $\zeta(\vec{x}_\parallel)$ appearing in these equations must now be regarded as a stationary stochastic process of the kind discussed in subsection II.C. Consequently, the solution $A(\vec{q}_\parallel\omega)$ of Eq. (3.14), and the solutions $A_{\parallel,\perp}(\vec{q}_\parallel\omega)$ of Eq. (3.30), are random functions. Just as $\zeta(\vec{x}_\parallel)$ is defined by the moments of its probability distribution function, so can these solutions be defined by the moments of their probability distribution functions. Of these, a particularly important one is the first moment, $\langle A(\vec{q}_\parallel\omega)\rangle$ or $\langle A_{\parallel,\perp}(\vec{q}_\parallel\omega)\rangle$, which describes the propagation of the mean wave across the randomly rough surface.

In this section, with the aid of the smoothing method described in Appendix B, we obtain the equations satisfied by $\langle A(\vec{q}_\parallel\omega)\rangle$ and $\langle A_{\parallel,\perp}(\vec{q}_\parallel\omega)\rangle$, present their solutions, and discuss their properties.

VI.A. Surface Plasmons

It is useful to preface a detailed study of Eq. (3.14) by an examination of it in the small roughness limit. The latter is obtained by retaining only the lowest nonvanishing term in the expansion of the function $J(\alpha|\vec{Q}_\parallel)$ defined by Eq. (3.12):

$$J(\alpha|\vec{Q}_\parallel) = \zeta^{(1)}(\vec{Q}_\parallel) + \frac{1}{2}\alpha\zeta^{(2)}(\vec{Q}_\parallel) + \frac{1}{6}\alpha^2\zeta^{(3)}(\vec{Q}_\parallel) + \dots \tag{6.1}$$

where

$$\zeta^{(n)}(\vec{Q}_\parallel) = \int d^2x_\parallel e^{-i\vec{Q}_\parallel\cdot\vec{x}_\parallel}\zeta^n(\vec{x}_\parallel). \tag{6.2}$$

In what follows we will denote $\zeta^{(1)}(\vec{Q}_\parallel)$ by $\zeta(\vec{Q}_\parallel)$. In the small roughness limit Eq. (3.14) becomes

$$\int\frac{d^2p_\parallel}{(2\pi)^2}[H(\vec{q}_\parallel|\vec{p}_\parallel) - V(\vec{q}_\parallel|\vec{p}_\parallel)]A(\vec{p}_\parallel\omega) = 0, \tag{6.3}$$

where

$$H(\vec{q}_\parallel|\vec{p}_\parallel) = (2\pi)^2\delta(\vec{q}_\parallel - \vec{p}_\parallel)\frac{\varepsilon(\omega)+1}{\varepsilon(\omega)-1} \tag{6.4a}$$

$$V(\vec{q}_\parallel|\vec{p}_\parallel) = \zeta(\vec{q}_\parallel-\vec{p}_\parallel)(1 - \hat{q}_\parallel\cdot\hat{p}_\parallel)p_\parallel. \tag{6.4b}$$

Equation (6.3) is just the homogeneous form of Eq. (B.1). We can therefore use the results given by Eqs. (B.4) and (B.5) to assert that the equation satisfied by $\langle A(\vec{q}_\parallel\omega)\rangle$ is

$$\int\frac{d^2p_\parallel}{(2\pi)^2}[H(\vec{q}_\parallel|\vec{p}_\parallel) - \langle M(\vec{q}_\parallel|\vec{p}_\parallel)\rangle]\langle A(\vec{p}_\parallel\omega)\rangle = 0, \tag{6.5}$$

where the averaged proper self-energy $\langle M(\vec{q}_\parallel|\vec{p}_\parallel)\rangle$ is given to the lowest nonzero order in $\zeta(\vec{x}_\parallel)$ by

$$\langle M(\vec{q}_\parallel|\vec{p}_\parallel)\rangle = \int\frac{d^2r_\parallel}{(2\pi)^2}\int\frac{d^2s_\parallel}{(2\pi)^2}\langle V(\vec{q}_\parallel|\vec{r}_\parallel)G_0(\vec{r}_\parallel|\vec{s}_\parallel)V(\vec{s}_\parallel|\vec{p}_\parallel)\rangle. \tag{6.6}$$

The operator $G_o(\vec{q}_\parallel|\vec{p}_\parallel)$ appearing in Eq. (6.6) is the inverse of $H(\vec{q}_\parallel|\vec{p}_\parallel)$, and is given by

$$G_o(\vec{q}_\parallel|\vec{p}_\parallel) = (2\pi)^2\delta(\vec{q}_\parallel-\vec{p}_\parallel)\ \frac{\varepsilon(\omega)-1}{\varepsilon(\omega)+1}\ . \tag{6.7}$$

The use of Eqs. (2.34), (6.4b), and (6.7) in Eq. (6.6) yields the result that

$$\langle M(\vec{q}_\parallel|\vec{p}_\parallel)\rangle = (2\pi)^2\delta(\vec{q}_\parallel-\vec{p}_\parallel)M(q_\parallel\omega), \tag{6.8a}$$

where

$$M(q_\parallel\omega) = \frac{\varepsilon(\omega)-1}{\varepsilon(\omega)+1}\int\frac{d^2p_\parallel}{(2\pi)^2}\ \delta^2 g(|\vec{q}_\parallel-\vec{p}_\parallel|)\ \times$$
$$\times\ q_\parallel(1-\hat{q}_\parallel\cdot\hat{p}_\parallel)^2 p_\parallel . \tag{6.8b}$$

If we introduce the representations

$$\vec{q}_\parallel = q_\parallel(\cos\theta,\sin\theta),\ \vec{p}_\parallel = p_\parallel(\cos\theta',\sin\theta'), \tag{6.9}$$

we find that

$$g(|\vec{q}_\parallel - \vec{p}_\parallel|) = \sum_{\ell=-\infty}^{\infty} g_\ell(q_\parallel|p_\parallel)e^{i\ell(\theta-\theta')}, \tag{6.10a}$$

where

$$g_\ell(q_\parallel|p_\parallel) = 2\pi\int_0^\infty dx_\parallel x_\parallel W(|\vec{x}_\parallel|)J_\ell(q_\parallel x_\parallel)J_\ell(p_\parallel x_\parallel). \tag{6.10b}$$

Then we can carry out the angular integration in Eq. (6.8b) and obtain

$$M(q_\parallel\omega) = \delta^2\ \frac{\varepsilon(\omega)-1}{\varepsilon(\omega)+1}\ q_\parallel\int_0^\infty \frac{dp_\parallel}{2\pi}\ p_\parallel^2\ G(q_\parallel|p_\parallel) \tag{6.11a}$$

with

$$G(q_\parallel|p_\parallel) = \frac{3}{2}\ g_o(q_\parallel|p_\parallel) - 2g_1(q_\parallel|p_\parallel) +$$
$$+\ \frac{1}{2}\ g_2(q_\parallel|p_\parallel). \tag{6.11b}$$

When Eqs. (6.4a) and (6.8a) are used in Eq. (6.5) the equation for $\langle A(\vec{q}_\parallel\omega)\rangle$ becomes an algebraic one,

$$\left[\frac{\varepsilon(\omega)+1}{\varepsilon(\omega)-1} - M(q_\parallel\omega)\right]\ \langle A(\vec{q}_\parallel\omega)\rangle = 0, \tag{6.12}$$

as a consequence of the restoration of infinitesimal translational invariance to our system by the averaging process. The dispersion relation for surface plasmons on a randomly rough surface in the small roughness limit therefore becomes

$$\left(\frac{\varepsilon(\omega)+1}{\varepsilon(\omega)-1}\right)^2 = \delta^2 q_\parallel \int_0^\infty \frac{dp_\parallel}{2\pi} p_\parallel^2 G(q_\parallel|p_\parallel) \tag{6.13}$$

or

$$\frac{\varepsilon(\omega)+1}{\varepsilon(\omega)-1} = \pm\,\delta\left[q_\parallel \int_0^\infty \frac{dp_\parallel}{2\pi} p_\parallel^2 G(q_\parallel|p_\parallel)\right]^{1/2}. \tag{6.14}$$

Thus, we see that to every solution of the equation

$$\varepsilon(\omega) + 1 = 0, \tag{6.15}$$

which is the dispersion relation for surface plasmons on a planar surface, correspond two solutions of the dispersion relation for surface plasmons on a randomly rough surface. This roughness-induced splitting of the surface plasmon dispersion curve has been studied theoretically by several authors [18,30,48,67,68] and has been observed experimentally by several groups [69-72]. The results of a recent experimental study of this effect are shown in Fig. 12, in which differential reflectivity spectra at normal incidence, $\Delta R/R = 2(R'-R)/(R'+R)$, are plotted for quenched silver deposits of four different thicknesses on silver surfaces. Here R' and R are the reflectivities with and without the quenched films. In Fig. 12 the minima assigned to surface plasmons are pointed out by arrows. Their separation in energy gives the magnitude of the surface plasmon splitting, which is seen to increase with increasing film thickness, perhaps because the roughness increases with increasing film thickness.

The physical reason for the splitting of the surface plasmon dispersion curve is the following [30]. A rough

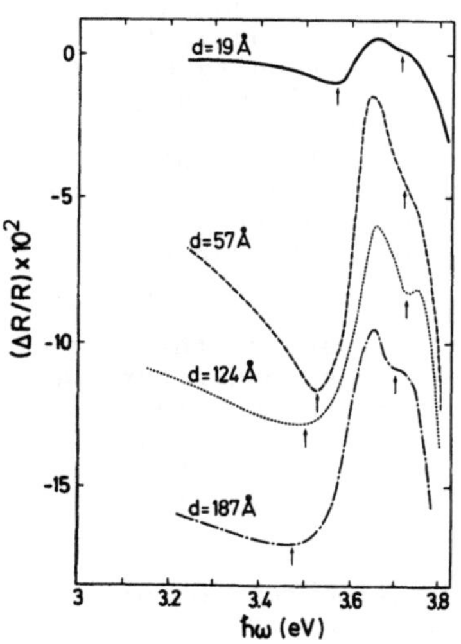

Fig. 12. Differential reflectivity spectra at normal incidence for Ag deposits of different thickness on an Ag substrate at 140K [Ref. 72).

surface can be regarded as a superposition of diffraction gratings, each with its own spacing, amplitude, and orientation in the x_1x_2-plane, which vary continuously from one grating to the next. As we have seen in subsection V.A.1, each grating can split the surface plasmon dispersion relation. This splitting occurs at a given frequency on the dispersion curve if two degenerate surface plasmons with different wave vectors can couple through the wave vector of the grating. Since the dispersion curve for a surface plasmon given by Eq. (6.15) is flat, i.e. it depends neither on the magnitude nor on the direction of the wave vector $\vec{k}_\parallel$, all wave vectors entering the Fourier decomposition of $\zeta(\vec{x}_\parallel)$ couple two degenerate surface plasmons with different wave vectors, and split the dispersion curve thereby. The fact that we obtain only two branches to the dispersion curve in the present case, instead of the infinite number found in Section V for surface plasmons on classical gratings or bigratings, is due to our retaining only the term linear in $\zeta(\vec{x}_\parallel)$ in the integral for $J(\alpha|\vec{Q}_\parallel)$ in going from Eq. (3.14) to Eq. (6.3). If higher order terms were kept, more branches would be obtained.

When we assume for $W(|\vec{x}_\parallel|)$ the Gaussian expression (2.38a), the dispersion relation (6.14) takes the form

$$\frac{\varepsilon(\omega)+1}{\varepsilon(\omega)-1} = \pm\sqrt{2}\,\frac{\delta}{a}\,f(\xi), \tag{6.16a}$$

where

$$f(\xi) = \frac{1}{\xi}\left\{e^{-\frac{1}{4}\xi^2}\int_0^\infty du\, u^2\, e^{-\frac{u^2}{\xi^2}}\left[3I_0(u) - 4I_1(u) + {}\right.\right.$$
$$\left.\left. + I_2(u)\right]\right\}^{1/2}, \tag{6.16b}$$

where $\xi = q_\parallel a$, and $I_n(x)$ is a modified Bessel function of the first kind. The function $f(\xi)$ has been evaluated, and Eq. (3.16a) solved, in Ref. 67. More interesting is the result obtained there that $\sqrt{2}f(\xi)$ is of order unity over a broad range of ξ (from $\xi \sim 0.1$ to $\xi \sim 4$). This means, from Eq. (6.16a), that when the latter is satisfied $(\varepsilon(\omega)+1)/(\varepsilon(\omega)-1)$ is of $O(\delta/a)$ or equivalently, $(\delta/a)\times(\varepsilon(\omega)-1)/(\varepsilon(\omega)+1)$ is of $O(1)$. This has the consequence that for frequencies ω in the vicinity of the solution of Eq. (6.16a) all of the terms in the expansion of Eq. (B.4) for $\langle M(\vec{q}_\parallel|\vec{p}_\parallel)\rangle$ in powers of $V(\vec{q}_\parallel|\vec{p}_\parallel)$, of which the nth is proportional to $[(\delta/a)(\varepsilon(\omega)-1)/(\varepsilon(\omega)+1)]^n$, are of the same order of magnitude at resonance as the lowest order term we have considered here so far. It is therefore incorrect to retain only the lowest order term in $(\delta/a)(\varepsilon(\omega)-1)/(\varepsilon(\omega)+1)$ from this infinite set of terms: since all the rest are of the same order of magnitude, they must all be taken into account in obtaining the dispersion relation.

We can do this by employing the expansion for $\langle M\rangle$ given by Eqs. (B.24)-(B.28) and using the fact that $\langle V\rangle = 0$ in the present case. The first two terms in this expansion are given formally by

$$\langle M\rangle = \langle V\langle G\rangle V\rangle + \\ + [\langle V\langle G\rangle V\langle G\rangle V\langle G\rangle V\rangle - \\ - \langle V\langle G\rangle\langle V\langle G\rangle V\rangle\langle G\rangle V\rangle - \\ - \langle V\langle G\rangle V\rangle\langle G\rangle\langle V\langle G\rangle V\rangle] + 0(V^6), \tag{6.17}$$

where

$$\langle G\rangle = (G_o^{-1} - \langle M\rangle)^{-1}. \tag{6.18}$$

An estimate [48] shows that the fourth order terms are numerically smaller than the second order terms. Consequently, as our approximation to $\langle M(\vec{q}_\parallel|\vec{p}_\parallel)\rangle$ we take

$$\langle M(\vec{q}_\parallel|\vec{p}_\parallel)\rangle = \int\frac{d^2r_\parallel}{(2\pi)^2}\int\frac{d^2s_\parallel}{(2\pi)^2}\langle V(\vec{q}_\parallel|\vec{r}_\parallel)\langle G(\vec{r}_\parallel|\vec{s}_\parallel)\rangle V(\vec{s}_\parallel|\vec{p}_\parallel)\rangle. \tag{6.19}$$

Since the averaging process restores infinitesimal translational invariance and isotropy in the plane $x_3 = 0$ to our system, Eq. (6.8a) is still valid, and from Eqs. (6.7) and (6.18) we find that

$$\langle G(\vec{q}_\parallel|\vec{p}_\parallel)\rangle = (2\pi)^2\delta(\vec{q}_\parallel-\vec{p}_\parallel)G(q_\parallel\omega), \tag{6.20a}$$

where

$$G(q_\parallel\omega) = \frac{1}{\frac{\varepsilon(\omega)+1}{\varepsilon(\omega)-1} - M(q_\parallel\omega)}. \tag{6.20b}$$

The equation for $M(q_\parallel\omega)$ obtained from Eqs. (6.8a), (6.19), and (6.20) is

$$M(q_\parallel\omega) = \int\frac{d^2p_\parallel}{(2\pi)^2}\delta^2 g(|\vec{q}_\parallel-\vec{p}_\parallel|)\frac{q_\parallel(1-\hat{q}_\parallel\cdot\hat{p}_\parallel)^2 p_\parallel}{\frac{\varepsilon(\omega)+1}{\varepsilon(\omega)-1} - M(p_\parallel\omega)}$$

$$= \delta^2 q_\parallel\int_0^\infty\frac{dp_\parallel}{2\pi}\frac{p_\parallel^2 G(q_\parallel|p_\parallel)}{\frac{\varepsilon(\omega)+1}{\varepsilon(\omega)-1} - M(p_\parallel\omega)}, \tag{6.21}$$

where $G(q_\parallel|p_\parallel)$ has been defined in Eq. (6.11b).

Equation (6.21) has been solved numerically by iteration [48], and the result used with Eq. (6.20b) to calculate the surface plasmon spectral density

$$\rho(q_\parallel\omega) = \frac{1}{2\pi i}\left[G(q_\parallel\omega+io) - G(q_\parallel\omega-io)\right]. \tag{6.22}$$

For a given value of $q_\parallel$ the spectral density has peaks at the values of ω for which the surface plasmon dispersion relation is satisfied.

Two results for the dimensionless spectral density $\rho(\xi\omega) = a\rho(q_\parallel\omega)$, where $\xi = aq_\parallel$, are plotted in Fig. 13. The Gaussian form (2.38a) for $W(|\vec{x}_\parallel|)$ was assumed in these calculations, together with the free electron form for $\varepsilon(\omega)$ given by Eq. (4.5). Shown as well are the spectral densities obtained from the approximation to $M(q_\parallel\omega)$ given by Eq. (6.11). The value of ξ chosen in these calculations, viz. $\xi = 1$, is the value for which the splitting of the surface plasmon dispersion curve in the latter approximation is largest. It is seen that the spectral densities calculated in both approximations display two peaks, signifying the roughness-induced splitting of the surface plasmon dispersion curve into two branches. However, the magnitude of the splitting, as measured by the distance between the centers of the two peaks, is larger in the results based on the self-consistent calculation of $M(q_\parallel\omega)$. To obtain a splitting of 0.15-0.2 eV in the surface plasmon dispersion curve for Ag, as has been observed experimentally [70-72], a value of $\delta/a \cong 0.05$ is required for $\xi = 1$ on the basis of the results shown in Fig. 13.

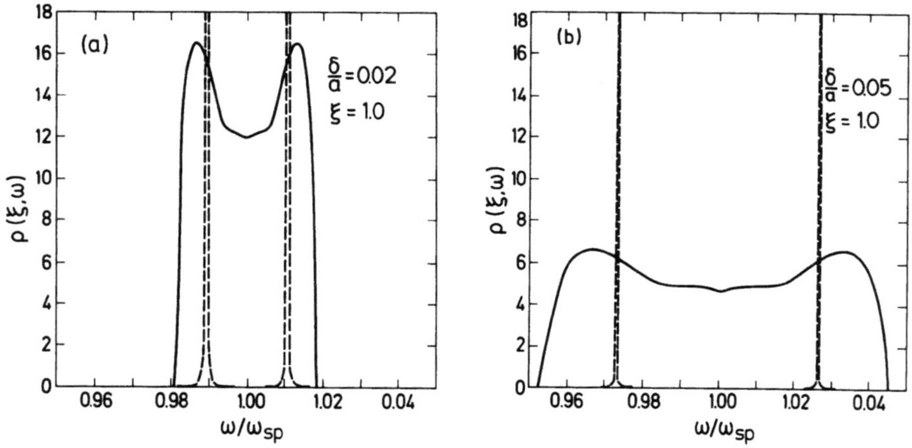

Fig. 13. Spectral density of surface plasmons on a randomly rough surface calculated self-consistently (solid line) and in the lowest order approximation in the surface profile function (dashed line) REf.[48].

At the same time, each of the two peaks in the self-consistent result for the spectral density is significantly broadened in comparison with the peaks obtained in the lowest order approximation. Indeed, in the latter approximation the peaks are rigorously δ-functions. The widths they display in Fig. 13, and their finite amplitudes, are due to the replacement of $\omega + io$ by $\omega + i\eta$ in the calculation of the spectral densities, where η is small but finite.

The widths of the peaks in the spectral densities calculated self-consistently are due to the attenuation of the surface plasmon as it propagates along a randomly rough surface by its roughness-induced multiple scattering into other surface plasmons with the same frequency but with different wave vectors, which removes energy from the incident beam. We see that the spectral density acquires a significant nonzero value in the frequency range between the two peaks, where it is zero in the lowest order approximation. This, too, is in qualitative agreement with experimental results [71,73].

VI.B. Surface Polaritons

In obtaining the equations satisfied by $\langle A_{\parallel,\perp}(\vec{q}_\parallel\omega)\rangle$ from Eqs. (3.30) we will also make the small roughness approximation. In this approximation Eqs. (3.30) become

$$\sum_\beta \int \frac{d^2q_\parallel}{(2\pi)^2} \left[H_{\alpha\beta}(\vec{p}_\parallel|\vec{q}_\parallel) - V_{\alpha\beta}(\vec{p}_\parallel|\vec{q}_\parallel)\right] A_\beta(\vec{q}_\parallel\omega) = 0, \tag{6.23}$$

where the subscripts α and β take the two values $\parallel$ and $\perp$, and where

$$H_{\parallel,\parallel}(\vec{p}_\parallel|\vec{q}_\parallel) = (2\pi)^2\delta(\vec{p}_\parallel-\vec{q}_\parallel)\,\frac{\varepsilon(\omega)\beta_0(p_\parallel\omega)+\beta(p_\parallel\omega)}{\varepsilon(\omega)-1} \tag{6.24a}$$

$$H_{\perp,\perp}(\vec{p}_\parallel|\vec{q}_\parallel) = (2\pi)^2\delta(\vec{p}_\parallel-\vec{q}_\parallel)\,\frac{\beta_0(p_\parallel\omega)+\beta(p_\parallel\omega)}{\varepsilon(\omega)-1} \tag{6.24b}$$

$$V_{\parallel,\parallel}(\vec{p}_\parallel|\vec{q}_\parallel) = \zeta(\vec{p}_\parallel-\vec{q}_\parallel)\left[p_\parallel q_\parallel - \beta(p_\parallel\omega)(\hat{p}_\parallel\cdot\hat{q}_\parallel)\beta_0(q_\parallel\omega)\right] \tag{6.25a}$$

$$V_{\parallel,\perp}(\vec{p}_\parallel|\vec{q}_\parallel) = \zeta(\vec{p}_\parallel-\vec{q}_\parallel)\left[-i\,\frac{\omega}{c}\,\beta(p_\parallel\omega)(\hat{p}_\parallel\times\hat{q}_\parallel)_3\right] \tag{6.25b}$$

$$V_{\perp,\parallel}(\vec{p}_\parallel|\vec{q}_\parallel) = \zeta(\vec{p}_\parallel-\vec{q}_\parallel)\left[i\,\frac{\omega}{c}\,(\hat{p}_\parallel\times\hat{q}_\parallel)_3\beta_0(q_\parallel\omega)\right] \tag{6.25c}$$

$$V_{\perp,\perp}(\vec{p}_\parallel|\vec{q}_\parallel) = \zeta(\vec{p}_\parallel-\vec{q}_\parallel)\,\frac{\omega^2}{c^2}\,(\hat{p}_\parallel\cdot\hat{q}_\parallel). \tag{6.25d}$$

According to the results given by Eqs. (B.4) and (B.5) the equation for $\langle A_{\parallel,\perp}(\vec{q}_\parallel\omega)\rangle$ is

$$\sum_\beta \int \frac{d^2q_\parallel}{(2\pi)^2} \left[H_{\alpha\beta}(\vec{p}_\parallel|q_\parallel) - \langle M_{\alpha\beta}(\vec{p}_\parallel|\vec{q}_\parallel)\rangle\right]\langle A_\beta(\vec{q}_\parallel\omega)\rangle = 0. \tag{6.26}$$

The averaged proper self energy in the present case is given to lowest nonzero order in $\zeta(\vec{x}_\parallel)$ by

$$\langle M_{\alpha\beta}(\vec{p}_\parallel|\vec{q}_\parallel)\rangle = \sum_{\mu\nu}\int\frac{d^2r_\parallel}{(2\pi)^2}\int\frac{d^2s_\parallel}{(2\pi)^2}\langle V_{\alpha\mu}(\vec{p}_\parallel|\vec{r}_\parallel)\times$$

$$\times G^{(o)}_{\mu\nu}(\vec{r}_\parallel|\vec{s}_\parallel)V_{\nu\beta}(\vec{s}_\parallel|\vec{q}_\parallel)\rangle, \tag{6.27}$$

where the nonzero elements of $G^{(o)}_{\alpha\beta}(\vec{p}_\parallel|\vec{q}_\parallel)$ are

$$G^{(o)}_{\parallel,\parallel}(\vec{p}_\parallel|\vec{q}_\parallel) = (2\pi)^2\delta(\vec{p}_\parallel-\vec{q}_\parallel)\frac{\varepsilon(\omega)-1}{\varepsilon(\omega)\beta_o(p_\parallel\omega)+\beta(p_\parallel\omega)} \tag{6.28a}$$

$$G^{(o)}_{\perp,\perp}(\vec{p}_\parallel|\vec{q}_\parallel) = (2\pi)^2\delta(\vec{p}_\parallel-\vec{q}_\parallel)\frac{\varepsilon(\omega)-1}{\beta_o(p_\parallel\omega)+\beta(p_\parallel\omega)}\ . \tag{6.28b}$$

When the results given by Eqs. (6.25) and (6.28) are substituted into Eq. (6.27), the result is found to have the form

$$\langle M_{\alpha\beta}(\vec{p}_\parallel|\vec{q}_\parallel)\rangle = \delta_{\alpha\beta}(2\pi)^2\delta(\vec{p}_\parallel-\vec{q}_\parallel)M_{\alpha\alpha}(p_\parallel\omega), \tag{6.29}$$

where

$$M_{\parallel,\parallel}(p_\parallel\omega) = \delta^2(\varepsilon(\omega)-1)\int\frac{d^2q_\parallel}{(2\pi)^2}g(|\vec{p}_\parallel-\vec{q}_\parallel|)\times$$

$$\times\left\{\frac{[p_\parallel q_\parallel-\beta(p_\parallel\omega)\beta_o(q_\parallel\omega)\cos(\theta-\theta')][q_\parallel p_\parallel-\beta(q_\parallel\omega)\beta_o(p_\parallel\omega)\cos(\theta'-\theta)]}{\varepsilon(\omega)\beta_o(q_\parallel\omega)+\beta(q_\parallel\omega)}\right.$$

$$\left.-\frac{\omega^2}{c^2}\frac{\beta(p_\parallel\omega)\beta_o(p_\parallel\omega)\sin^2(\theta-\theta')}{\beta_o(q_\parallel\omega)+\beta(q_\parallel\omega)}\right\} \tag{6.30a}$$

$$M_{\perp,\perp}(p_\parallel\omega) = \delta^2(\varepsilon(\omega)-1)\int\frac{d^2q_\parallel}{(2\pi)^2}g(|\vec{p}_\parallel-\vec{q}_\parallel|)\times$$

$$\times\left\{-\frac{\omega^2}{c^2}\frac{\beta_o(q_\parallel\omega)\beta(q_\parallel\omega)\sin^2(\theta-\theta')}{\varepsilon(\omega)\beta_o(q_\parallel\omega)+\beta(q_\parallel\omega)}+\frac{\omega^4}{c^4}\frac{\cos^2(\theta-\theta')}{\beta_o(q_\parallel\omega)+\beta(q_\parallel\omega)}\right\}, \tag{6.30b}$$

where θ and θ' are the polar angles of $\vec{p}_\parallel$ and $\vec{q}_\parallel$, respectively. With the use of the relation

$$[q_\parallel^2-\beta_o(q_\parallel\omega)\beta(q_\parallel\omega)][\beta_o(q_\parallel\omega)+\beta(q_\parallel\omega)]$$

$$=\frac{\omega^2}{c^2}[\varepsilon(\omega)\beta_o(q_\parallel\omega)+\beta(q_\parallel\omega)] \tag{6.31}$$

we can simplify Eqs. (6.30) somewhat:

$$M_{\parallel,\parallel}(p_\parallel\omega) = \delta^2(\varepsilon(\omega)-1)\int\frac{d^2q_\parallel}{(2\pi)^2}\frac{g(|\vec{p}_\parallel-\vec{q}_\parallel|)}{\varepsilon(\omega)\beta_o(q_\parallel\omega)+\beta(q_\parallel\omega)}\times$$

$$\times\ \{[p_\parallel q_\parallel-\beta(p_\parallel\omega)\beta_o(q_\parallel\omega)\cos(\theta-\theta')][q_\parallel p_\parallel-\beta(q_\parallel\omega)\beta_o(p_\parallel\omega)\ \times$$

$$\times\ \cos(\theta-\theta')]\ -$$

$$-\ \beta(p_\parallel\omega)\beta_o(p_\parallel\omega)[q_\parallel^2-\beta_o(q_\parallel\omega)\beta(q_\parallel\omega)]\sin^2(\theta-\theta')\} \tag{6.32a}$$

$$M_{\perp,\perp}(p_\parallel\omega) = \delta^2(\varepsilon(\omega)-1)\ \frac{\omega^2}{c^2}\int\frac{d^2q_\parallel}{(2\pi)^2}\frac{g(|\vec{p}_\parallel-\vec{q}_\parallel|)}{\varepsilon(\omega)\beta_o(q_\parallel\omega)+\beta(q_\parallel\omega)}\times$$

$$\times\ [q_\parallel^2\cos^2(\theta-\theta')\ -\ \beta_o(q_\parallel\omega)\beta(q_\parallel\omega)]. \tag{6.32b}$$

The fact that the averaged self-energy is not only diagonal in $\vec{p}_\parallel$ and $\vec{q}_\parallel$ but is also diagonal in the polarization indices is a consequence of the restoration of isotropy in the plane $x_3 = 0$ by the averaging process.

From Eqs. (6.24), (6.26), and (6.29), we see that the equations for $\langle A_{\parallel,\perp}(\vec{q}_\parallel\omega)\rangle$ become

$$\left[\frac{\varepsilon(\omega)\beta_o(p_\parallel\omega)+\beta(p_\parallel\omega)}{\varepsilon(\omega)-1}\ -\ M_{\parallel,\parallel}(p_\parallel\omega)\right]\langle A_\parallel(p_\parallel\omega)\rangle\ =\ 0 \tag{6.33a}$$

$$\left[\frac{\beta_o(p_\parallel\omega)+\beta(p_\parallel\omega)}{\varepsilon(\omega)\ -\ 1}\ -\ M_{\perp,\perp}(p_\parallel\omega)\right]\langle A_\perp(\vec{p}_\parallel\omega)\rangle\ =\ 0. \tag{6.33b}$$

Thus the dispersion relations for surface polaritons of p- and s-polarization propagating across a randomly rough surface are

$$\varepsilon(\omega)\beta_o(p_\parallel\omega)+\beta(p_\parallel\omega)\ =\ \delta^2(1-\varepsilon(\omega))^2\int\frac{d^2q_\parallel}{(2\pi)^2}\frac{g(|\vec{p}_\parallel-\vec{q}_\parallel|)}{\varepsilon(\omega)\beta_o(q_\parallel\omega)+\beta(q_\parallel\omega)}\times$$

$$\times\{[p_\parallel q_\parallel-\beta(p_\parallel\omega)\beta_o(q_\parallel\omega)\cos(\theta-\theta')][q_\parallel p_\parallel-\beta(q_\parallel\omega)\beta_o(p_\parallel\omega)\cos(\theta-\theta')]$$

$$-\ \beta_o(p_\parallel\omega)\beta(p_\parallel\omega)[q_\parallel^2-\beta_o(q_\parallel\omega)\beta(q_\parallel\omega)]\sin^2(\theta-\theta')\} \tag{6.34a}$$

and

$$\beta_o(p_\parallel\omega)+\beta(p_\parallel\omega)\ =\ \delta^2(1-\varepsilon(\omega))^2\ \frac{\omega^2}{c^2}\int\frac{d^2q_\parallel}{(2\pi)^2}\frac{g(|\vec{p}_\parallel-\vec{q}_\parallel|)}{\varepsilon(\omega)\beta_o(q_\parallel\omega)+\beta(q_\parallel\omega)}\times$$

$$\times\ [q_\parallel^2\cos^2(\theta-\theta')-\beta_o(q_\parallel\omega)\beta(q_\parallel\omega)]. \tag{6.34b}$$

respectively. No solutions of Eq. (6.34b) have been found to date, and we will not consider it farther here.

We can simplify Eq. (6.34a) still more. We rewrite the expression in braces in it in the following way:

$$[p_\parallel q_\parallel - \beta(p_\parallel\omega)\beta_0(q_\parallel\omega)\cos(\theta-\theta')][q_\parallel p_\parallel - \beta(q_\parallel\omega)\beta_0(p_\parallel\omega)\cos(\theta-\theta')] -$$

$$-\beta_0(p_\parallel\omega)\beta(p_\parallel\omega)[q_\parallel^2 - \beta_0(q_\parallel\omega)\beta(q_\parallel\omega)]\sin^2(\theta-\theta')$$

$$= [p_\parallel q_\parallel\cos(\theta-\theta') - \beta_0(p_\parallel\omega)\beta(q_\parallel\omega)] \times$$

$$\times [q_\parallel p_\parallel\cos(\theta-\theta') - \beta_0(q_\parallel\omega)\beta(p_\parallel\omega)] +$$

$$+ (p_\parallel^2 - \beta_0(p_\parallel\omega)\beta(p_\parallel\omega))q_\parallel^2\sin^2(\theta-\theta'). \qquad (6.35)$$

The last term, in view of Eq. (6.31) is $(\omega^2/c^2) \times ([\varepsilon(\omega)\beta_0(p_\parallel\omega)+\beta(p_\parallel\omega)]/[\beta_0(p_\parallel\omega)+\beta(p_\parallel\omega)])q_\parallel^2\sin^2(\theta-\theta')$ and, according to Eq. (6.34), is therefore of $O(\delta^2)$. It therefore does not contribute to the dispersion relation for p-polarized surface polaritons on a randomly rough surface to $O(\delta^2)$. The latter dispersion relation then becomes

$$\varepsilon(\omega)\beta_0(p_\parallel\omega)+\beta(p_\parallel\omega) = \delta^2(1-\varepsilon(\omega))^2 \int\frac{d^2q_\parallel}{(2\pi)^2}\,\frac{g(|\vec{p}_\parallel-\vec{q}_\parallel|)}{\varepsilon(\omega)\beta_0(q_\parallel\omega)+\beta(q_\parallel\omega)}\times$$

$$\times[p_\parallel q_\parallel\cos(\theta-\theta')-\beta_0(p_\parallel\omega)\beta(q_\parallel\omega)][q_\parallel p_\parallel\cos(\theta'-\theta)-\beta_0(q_\parallel\omega)\beta(p_\parallel\omega)]. \qquad (6.36)$$

A surface polariton propagating along the perfectly planar surface of a pure crystal is attenuated by processes that can be called intrinsic. By this we mean the dissipative process present in the bulk of the solid, i.e. the processes that give rise to the imaginary part of the dielectric constant. An expression for the attenuation length of the surface polariton in this case can be obtained by inserting the complex dielectric constant $\varepsilon(\omega) = \varepsilon_1(\omega) + i\varepsilon_2(\omega)$ of the material into the dispersion relation for surface polaritons on a planar surface,

$$\varepsilon(\omega)\beta_0(p_\parallel\omega) + \beta(p_\parallel\omega) = 0, \qquad (6.37)$$

and obtaining from it the imaginary part of the wave vector of the surface polariton, $k_\parallel^{(I)}(\omega)$. The attenuation length for energy flow is then $(2k_\parallel^{(I)}(\omega))^{-1}$.

However, in the presence of surface roughness a surface polariton can be attenuated even if the dielectric constant

$\varepsilon(\omega)$ is real. The physical reason for this is the roughness-induced scattering of the surface polaritons into radiative modes in the vacuum, and into other surface polariton modes. Both scattering processes remove energy from the incident beam, and the surface polariton is attenuated thereby. This is the situation we would like to focus our attention on. Therefore, to separate the attenuation of surface polaritons that has its origin in surface roughness from that which is due to the dissipative processes present in the bulk of the material, we will assume $\varepsilon(\omega)$ to be real everywhere in Eq. (6.36), except in the denominator $\varepsilon(\omega)\beta_o(q_\parallel\omega)+\beta(q_\parallel\omega)$ of the integrand on the right-hand side. The retention of the small, positive imaginary part of the dielectric constant here serves only to define the manner in which the pole in the integrand that occurs when $\varepsilon(\omega)\beta_o(q_\parallel\omega)+\beta(q_\parallel\omega) = 0$ is to be treated in the evaluation of the integral over $q_\parallel$.

If we introduce the definitions

$$F(p_\parallel\omega) = \varepsilon(\omega)\beta_o(p_\parallel\omega) + \beta(p_\parallel\omega) \tag{6.38a}$$

$$\begin{aligned} G(p_\parallel\omega) &= \int\frac{d^2q_\parallel}{(2\pi)^2}\,\frac{g(|\vec{p}_\parallel-\vec{q}_\parallel|)}{\varepsilon(\omega)\beta_o(q_\parallel\omega)+\beta(q_\parallel\omega)}\times \\ &\times\left[p_\parallel q_\parallel\cos(\theta-\theta')-\beta_o(p_\parallel\omega)\beta(q_\parallel\omega)\right]\times \\ &\times\left[q_\parallel p_\parallel\cos(\theta-\theta')-\beta_o(q_\parallel\omega)\beta(p_\parallel\omega)\right] \\ &= G^{(1)}(p_\parallel\omega) - iG^{(2)}(p_\parallel\omega), \end{aligned} \tag{6.38b}$$

and denote the solution of the equation $F(p_\parallel\omega) = 0$ by $\omega_o(p_\parallel)$, the frequency of a surface polariton on a planar surface, then the solution of Eq. (6.36) for the frequency of a surface polariton on a randomly rough surface can be written in the form

$$\omega(p_\parallel) = \omega_o(p_\parallel) + \Delta(p_\parallel) - i\Gamma(p_\parallel) , \tag{6.39}$$

where

$$\Delta(p_\parallel) = \delta^2\left(1-\varepsilon(\omega_o(p_\parallel))\right)^2\,\frac{G^{(1)}(p_\parallel,\omega_o(p_\parallel))}{[(d/d\omega)F(p_\parallel\omega)]_{\omega=\omega_o(p_\parallel)}} \tag{6.40a}$$

$$\Gamma(p_\parallel) = \delta^2\left(1-\varepsilon(\omega_o(p_\parallel))\right)^2\,\frac{G^{(2)}(p_\parallel,\omega_o(p_\parallel))}{[d/d\omega)F(p_\parallel\omega)]_{\omega=\omega_o(p_\parallel)}} . \tag{6.40b}$$

The attenuation length of the surface polariton, $\ell(p_\parallel)$, is the distance over which the energy of the polariton decays to 1/e of its initial value. It is given by

$$\ell(p_\parallel) = \frac{V_E(p_\parallel)}{2\Gamma(p_\parallel)} \ . \tag{6.41}$$

Here $V_E(p_\parallel)$ is the energy transport velocity of the surface polariton and, in the absence of damping, is equal to the group velocity of the surface polariton,

$$V_E(p_\parallel) = c^2 \frac{p_\parallel}{\omega_0(p_\parallel)} \left. \frac{[\varepsilon(\omega)+1]^2}{\varepsilon(\omega)[\varepsilon(\omega)+1] + 1/2\ \omega\varepsilon'(\omega)} \right|_{\omega=\omega_0(p_\parallel)} . \tag{6.42}$$

Numerical results for $\Delta(p_\parallel)$, $\Gamma(p_\parallel)$, and $\ell^{-1}(p_\parallel)$ for aluminum are shown in Fig. 14 [73].

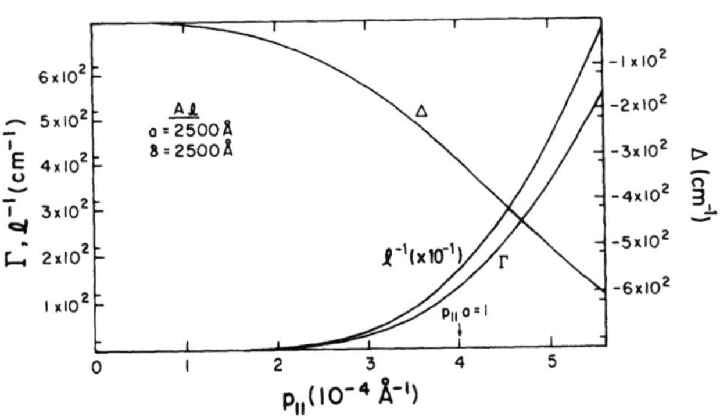

Fig. 14. The functions $\Delta(p_\parallel)$, $\Gamma(p_\parallel)$, $\ell^{-1}(p_\parallel)$ as functions of $p_\parallel$ for surface polaritons on a randomly rough Aℓ surface. Note that the graphical value of $\ell^{-1}(p_\parallel)$ is to be multiplied by 10 [Ref. 73].

The dielectric constant was chosen to have the free electron form with $h\omega_p = 1.2018 \times 10^5 \text{cm}^{-1}$. The values of δ and a used were $\delta = a = 2500$Å, so that the surface is a very rough one. From the figure we see that for a surface polariton wavelength of 2.1μm $(p_\parallel = 3 \times 10^{-4} \text{Å}^{-1})$, $\ell^{-1}(p_\parallel) \cong 4 \times 10^2\ \text{cm}^{-1}$, so that $\ell(p_\parallel) \cong 2.5 \times 10^{-3}$ cm.

In all cases studied $\Delta(p_\parallel)$ was found to be negative, i.e. surface roughness depresses the frequency of a surface polariton below its value on a planar surface. This is reminiscent of the depression of the frequency of (the lowest branch of) a surface polariton propagating across a grating by the periodic corrugations of the latter surface.

The numerical calculations also show that the dominant

contributions to $G^{(1)}(p_\parallel\omega)$ and $G^{(2)}(p_\parallel\omega)$ come from the part of the integral that comes from the region $q_\parallel > (\omega/c)$, and corresponds to the roughness-induced scattering into other surface polariton states. In a separate calculation Mills [74] has shown that a surface polariton is damped predominantly by scattering into other surface polariton states when $cp_\parallel >> \omega$ so that $|\varepsilon(\omega)| \approx 1$, and $p_\parallel a << 1$, which are reasonably well satisfied for the conditions depicted in Fig. 14. When $\omega << \omega_p$, so that $\omega \approx cp_\parallel$ the most effective mechanism for attenuating the surface polariton is the roughness-induced radiation into the vacuum above the material.

VII. Some Directions for Future Research

To conclude these lectures I should like to offer some suggestions of areas of research concerning the propagation of surface plasmons and surface polaritons across a rough surface, that seem to me to warrant consideration.

In the area of electrostatic and electromagnetic surface shape resonances, several interesting problems remain to be explored. In particular, the modes associated with protuberances whose profiles are more nearly nonanalytic than those considered here bear investigation. These can be regarded as the limit of a sequence of analytic profiles for which the approach described in these lectures should be applicable. For example, a flat-topped cylindrical post of radius R and height A can be described by the large β limit of the surface profile function given by Eq. (2.8c). For sufficiently large values of β the method described here may break down. However, it would be of interest to see if it still works for large enough values of β that Eq. (2.8c) provides a good approximation to a flat-topped cylindrical post.

We have also not discussed methods for the experimental study of surface shape resonances. One possibility is through the radiation emitted when a slow electron moves along a path in the near vicinity of a surface protuberance or indentation. In this case it is the electrostatic approximation that is applicable. Another is the diffraction of light from surface perturbations.

While on the subject of electrostatic and electromagnetic surface shape resonances, it should be mentioned that their study in the case that the protuberance or indentation that gives rise to them is on the surface of a nonlinear dielectric medium should yield some interesting results. A particularly interesting system for the study of such excitations would be a thin metal film on a nonlinear dielectric substrate, with a protuberance or indentation on the metal-vacuum interface. This is because optical nonlinearities are generally much larger in dielectric media than in metals, but the metal provides the negative dielectric constant needed for the existence of surface shape resonances that is usually not possessed by dielectric media. The frequencies of the surface shape resonances in this system have been studied in the case of a linear dielectric substrate [58]. In the presence of a nonlinear substrate these frequencies will depend on the intensity of the electric field at some point on the metal-substrate boundary.

In the theory of surface acoustic waves it is known that only waves whose displacement vector lies in the sagittal plane (the plane defined by the direction of propagation and the normal to the surface) can be trapped at the planar, stress-free surface of a semi-infinite isotropic elastic medium. Surface acoustic waves whose displacement vector is oriented perpendicular to the sagittal plane (shear horizontal polarization) cannot exist on such a surface [75]. However, surface acoustic waves of the latter polarization can exist when a grating is ruled on the surface and the displacement vector is parallel to the grooves of the grating. A similar situation exists for surface polaritons. They exist only in p-polarization on a planar interface between vacuum and an isotropic dielectric medium, not in s-polarization [18]. There is now evidence that s-polarized surface polaritons can exist when a grating with deep grooves is ruled on the surface of a dielectric medium. The evidence is indirect and is based on the recently discovered enhancement of the electric field near the bottom of each groove when the grating is illuminated by s-polarized light whose electric vector is parallel to the grooves of the grating [76]. This enhancement is due to the resonance excitation of s-polarized surface excitations trapped by the grating surface. The dispersion curves of these modes have not been determined yet, and it would be of considerable interest to have them.

Up to now the propagation of surface polaritons across periodically corrugated surfaces has been studied only for gratings described by one-dimensional surface profile functions, i.e. for classical gratings. Their propagation across doubly periodic gratings (bigratings) has not been studied yet. Such a study should yield dispersion curves with a rich variety of detail.

The propagation of surface polaritons across gratings ruled on a metal layer deposited on a nonlinear dielectric medium, in which the dielectric constant depends on the intensity of the electric field in it, should be of considerable interest. The gaps that open up in the dispersion curves of these surface waves at the boundaries of the first Brillouin zone for the grating will now be functions of the intensity of the electric field in the medium, so that the widths and locations of the pass bands and stop bands for surface polariton propagation can be modulated by the wave itself.

In the case of surface polaritons propagating across a rough surface consisting of a random array of identical protuberances on an otherwise planar dielectric surface, it would be desirable to have theoretical results for cylindrically symmetric protuberances, since experiments on such systems are now beginning to be done. The results would help us understand how surface polaritons interact with electromagnetic surface shape resonances.

All of the discussion of the propagation of surface plasmons and surface polaritons across either periodically corrugated or randomly rough surfaces in these lectures has been based on the assumption of a local relation between the electric displacement vector $\vec{D}(\vec{x};t)$ and the macroscopic electric field $\vec{E}(\vec{x};t)$. The interpretation of certain experimental data on the optical properties of bounded media, however, require the assumption of a nonlocal, integral relation between these two fields,

$$D_\mu(\vec{x};t) = \sum_\nu \int d^3x' \int dt' \varepsilon_{\mu\nu}(\vec{x},\vec{x}';t-t') E_\nu(\vec{x}';t'),$$

where $\varepsilon_{\mu\nu}(\vec{x},\vec{x}';t-t')$ is the dielectric tensor of the bounded medium [77]. The existence of a relation between $\vec{D}$ and $\vec{E}$ that is nonlocal in the spatial coordinates is usually referred to as a consequence of <u>spatial dispersion</u> in the medium. It is to be expected that the surfaces of crystals in which spatial dispersion manifests itself in their optical properties are rough to some extent. It is known that the dispersion curve for surface plasmons [78,79] and surface polaritons [80,81] on the planar surface of a spatially dispersive medium differ from those for a nonspatially dispersive medium. In particular, surface polaritons are damped in the presence of spatial dispersion, due to Landau damping, when in the absence of spatial dispersion they are not. Thus the interplay of surface roughness and spatial dispersion can be expected to lead to some interesting results for surface plasmons and polaritons, since each alone does.

Finally, we note that although there have been studies of surface plasmons and polaritons associated with nonplanar surfaces, e.g. with cylindrical and spherical surfaces, the surfaces considered have been perfectly smooth in the majority of these studies [82,83]. Recently, MAZUR and MILLS [19] have studied the propagation of guided waves through a cylindrical optical fiber whose surface was randomly corrugated in the direction parallel to the axis of the fiber, but whose cross-section at every point along the fiber remained circular. A generalization of this work to surface profiles that are functions of the azimuthal angle, in a cylindrical coordinate system whose z-axis is along the axis of the fiber, as well as of the coordinate z along the fiber would yield results for a more realistic system, and would be welcome. It has recently been shown [20] that frequencies of plasmons in a dielectric sphere separated from vacuum by a randomly rough surface are split into two levels each by the roughness, in much the same way that the dispersion curve of a surface plasmon on a planar dielectric surface is split into two branches by surface roughness.

Such a splitting might well be observable in absorption spectra of systems consisting of small metal particles embedded in dielectric matrices [84].

Acknowledgements

The work described in these notes was carried out over the past several years in collaboration with G. C. Brown, V. Celli, G. A. Farias, N. E. Glass, M. E. Haller, B. Laks, A. M. Marvin, A. R. Mcgurn, D. L. Mills, T. P. Shen, and W. M. Visscher. I am grateful to all of them for their help. I am also grateful to V. M. Agranovich and T. A. Leskova for discussions of several of the topics treated here.

This research was supported in part by NATO Research Grant No. 426/84. I would finally like to thank Professor Dr. H. Bilz and Dr. W. Hanke for the hospitality of the Max Planck Institut für Festkörperforschung, Stuttgart, during the preparation of a major portion of these lectures.

Appendix A. The Vectorial Form of Kirchhoff's Integral and the Extinction Theorem

We have seen in the text that the solution of the boundary value problem electromagnetic theory presents us in studying the propagation of surface polaritons across a rough surface, or the diffraction of light from a rough surface, can be simplified if the electromagnetic field in the medium is eliminated from the problem and only the field in the medium is considered. The mathematical tools that help us do this are the vectorial form of Kirchhoff's integral [2] and the extinction theorem [3].

To obtain the consequences of these tools for the situation of interest to us, let us denote by V the region $x_3 < \zeta(\vec{x}_\parallel)$ occupied by the dielectric medium. Maxwell's equations for the time-independent electric field in this region take the form

$$\left(\nabla^2 + \varepsilon(\omega)\,\frac{\omega^2}{c^2}\right) E_\alpha^<(\vec{x}|\omega) = 0 \qquad \alpha = 1, 2, 3. \tag{A.1}$$

In writing Eq. (A.1) we have used the condition $\nabla\cdot\vec{E}(\vec{x}|\omega) = 0$ that follows from the absence of external charges and the spatial homogeneity of the dielectric medium.

We introduce a scalar Green's function $G_\varepsilon(\vec{x};\vec{x}'|\omega)$ that is the solution of

$$\left(\nabla^2 + \varepsilon(\omega)\,\frac{\omega^2}{c^2}\right) G_\varepsilon(\vec{x};\vec{x}'|\omega) = -\,4\pi\delta(\vec{x}-\vec{x}') \tag{A.2}$$

and satisfies outgoing wave conditions at infinity. We next use Green's theorem [47] in the form

$$\int_V d^3x\{E_\alpha^<(\vec{x}|\omega)(\nabla^2+\varepsilon(\omega)\frac{\omega^2}{c^2})G_\varepsilon(\vec{x};\vec{x}'|\omega) -$$

$$- G_\varepsilon(\vec{x};\vec{x}'|\omega)(\nabla^2+ \varepsilon(\omega)\frac{\omega^2}{c^2})E_\alpha^<(\vec{x}|\omega)\}$$

$$= \int_V d^3x\{E_\alpha^<(\vec{x}|\omega)\nabla^2 G_\varepsilon(\vec{x};\vec{x}'|\omega) - G_\varepsilon(\vec{x};\vec{x}'|\omega)\nabla^2 E_\alpha^<(\vec{x}|\omega)\}$$

$$= \int_\Sigma dS\{E_\alpha^<(\vec{x}|\omega)\frac{\partial}{\partial n}G_\varepsilon(\vec{x};\vec{x}'|\omega) - G_\varepsilon(\vec{x};\vec{x}'|\omega)\frac{\partial}{\partial n}E_\alpha^<(\vec{x}|\omega)\}, \tag{A.3}$$

where Σ is the surface bounding the volume V. It consists of the surface $x_3 = \zeta(\vec{x}_\parallel)$ that we denote by S, and a hemispherical cap of infinite radius in the lower half-space. The latter does not contribute to the integral over Σ either because we seek electric fields that vanish as $x_3 \to -\infty$ or because of the radiation conditions satisfied by both $E_\alpha^<(\vec{x}|\omega)$ and $G_\varepsilon(\vec{x};\vec{x}'|\omega)$ as $x_3 \to -\infty$. The derivative $\partial/\partial n$ appearing in Eq. (A.3) is the derivative along the normal to the surface Σ directed away from the volume V. Thus, by the use of Eqs. (A.1) and (A.2) we can transform Eq. (A.3) into

$$\vec{E}^<(\vec{x}|\omega)\theta_V(\vec{x}) = \frac{1}{4\pi}\int_S dS'\{G_\varepsilon(\vec{x};\vec{x}'|\omega) \times$$

$$\times (\hat{n}'\cdot\nabla')\vec{E}^<(\vec{x}'|\omega) - (\hat{n}'\cdot\nabla' G_\varepsilon(\vec{x};\vec{x}'|\omega))\vec{E}^<(\vec{x}'|\omega)\}, \tag{A.4}$$

where

$$\theta_V(\vec{x}) = \int_V d^3x\ \delta(\vec{x}-\vec{x}') \tag{A.5}$$

equals unity if $\vec{x}\in V$ and vanishes if $\vec{x}\notin V$. In writing Eq. (A.4) we have used the fact that $G_\varepsilon(\vec{x};\vec{x}'|\omega)$ is symmetric in $\vec{x}$ and $\vec{x}'$.

We now rewrite the expression in braces in the integrand in Eq. (A.4) in such a way that only quantities that enter the boundary conditions on the surface S appear in it. We begin by writing this expression as

$$\{\ \} = G_\varepsilon(\vec{x};\vec{x}'|\omega)(\hat{n}'\cdot\nabla')\vec{E}^<(\vec{x}'|\omega) -$$

$$- E^<(\vec{x}'|\omega)(\hat{n}'\cdot\nabla')G_\varepsilon(\vec{x};\vec{x}'|\omega)$$

$$= (\hat{n}'\cdot\nabla')(G_\varepsilon(\vec{x};\vec{x}'|\omega)\vec{E}^<(\vec{x}'|\omega)) - \tag{A.6}$$

$$- 2\vec{E}^<(\vec{x}'|\omega)(\hat{n}'\cdot\nabla' G_\varepsilon(\vec{x};\vec{x}'|\omega)) .$$

We next introduce the two identities

$$\hat{n}' \times (\vec{E}^<(\vec{x}'|\omega) \times \nabla' G_\varepsilon(\vec{x};\vec{x}'|\omega)) \tag{A.7}$$

$$= \vec{E}^<(\vec{x}'|\omega)(\hat{n}'\cdot\nabla' G_\varepsilon(\vec{x};\vec{x}'|\omega) - (\hat{n}'\cdot\vec{E}^<(\vec{x}'|\omega))\nabla' G_\varepsilon(\vec{x};\vec{x}'|\omega)$$

and

$$\nabla' G_\varepsilon(\vec{x};\vec{x}'|\omega) \times (\hat{n}' \times \vec{E}^<(\vec{x}'|\omega))$$

$$= \hat{n}'(\vec{E}^<(\vec{x}'|\omega)\cdot\nabla' G_\varepsilon(\vec{x};\vec{x}'|\omega)) - \vec{E}^<(\vec{x}'|\omega)(\hat{n}'\cdot\nabla' G\ (\vec{x};\vec{x}'|\omega)). \tag{A.8}$$

We add the negative of Eq. (A.7) to Eq. (A.8) and obtain after a simple rearrangement that

$$-2\vec{E}^<(\vec{x}'|\omega)(\hat{n}'\cdot\nabla' G_\varepsilon(\vec{x};\vec{x}'|\omega)) = -\hat{n}'\times(\vec{E}^<(\vec{x}'|\omega)\times\nabla' G_\varepsilon(\vec{x};\vec{x}'|\omega)) +$$

$$+\nabla' G_\varepsilon(\vec{x};\vec{x}'|\omega)\times(\hat{n}' \times \vec{E}^<(\vec{x}'|\omega)) - (\hat{n}'\cdot\vec{E}^<(\vec{x}'|\omega))\nabla' G_\varepsilon(\vec{x};\vec{x}'|\omega) -$$

$$- \hat{n}'(\vec{E}^<(\vec{x}'|\omega)\cdot\nabla' G_\varepsilon(\vec{x};\vec{x}'|\omega)) \ . \tag{A.9}$$

When this result is used in Eq. (A.6) the latter becomes

$$\{\ \} = (\hat{n}'\cdot\nabla')(G_\varepsilon(\vec{x};\vec{x}'|\omega)\vec{E}^<(\vec{x}'|\omega)) -$$

$$- \hat{n}' \times (\vec{E}^<(\vec{x}'|\omega) \times \nabla' G_\varepsilon(\vec{x};\vec{x}'|\omega)) -$$

$$- (\hat{n}' \times \vec{E}^<(\vec{x}'|\omega)) \times \nabla' G_\varepsilon(\vec{x};\vec{x}'|\omega) -$$

$$- (\hat{n}'\cdot\vec{E}^<(\vec{x}'|\omega))\nabla' G_\varepsilon(\vec{x};\vec{x}'|\omega) -$$

$$- \hat{n}'(\vec{E}^<(\vec{x}'|\omega)\cdot\nabla' G_\varepsilon(\vec{x};\vec{x}'|\omega)) \ . \tag{A.10}$$

From the result that

$$\nabla' \times (G_\varepsilon(\vec{x};\vec{x}'|\omega)\vec{E}^<(\vec{x}'|\omega))$$

$$= G_\varepsilon(\vec{x};\vec{x}'|\omega)(\nabla' \times \vec{E}^<(\vec{x}'|\omega)) - \vec{E}^<(\vec{x}'|\omega) \times (\nabla' G_\varepsilon(\vec{x};\vec{x}'|\omega)) \tag{A.11}$$

we find that

$$\hat{n}' \times \nabla' \times (G_\varepsilon(\vec{x};\vec{x}'|\omega)\vec{E}^<(\vec{x}'|\omega)$$

$$= G_\varepsilon(\vec{x};\vec{x}'|\omega)\hat{n}' \times (\nabla' \times \vec{E}^<(\vec{x}'|\omega)) -$$

$$- \hat{n}' \times (\vec{E}^<(\vec{x}'|\omega) \times (\nabla' G_\varepsilon(\vec{x};\vec{x}'|\omega))) \ . \tag{A.12}$$

At the same time, we have that

$$\nabla' \cdot (G_\varepsilon(\vec{x};\vec{x}'|\omega)\vec{E}^<(\vec{x}'|\omega)) = \vec{E}^<(\vec{x}'|\omega) \cdot (\nabla' G_\varepsilon(\vec{x};\vec{x}'|\omega)) \tag{A.13}$$

because $\nabla' \cdot \vec{E}^<(\vec{x}'|\omega) = 0$, as has been noted above. If we use Eqs. (A.12) and (A.13) in Eq. (A.10) we obtain finally that

$$\begin{aligned}\{\ \} = &(\hat{n}' \cdot \nabla')(G_\varepsilon(\vec{x};\vec{x}'|\omega)\vec{E}^<(\vec{x}'|\omega) + \\ &+ \hat{n}' \times \nabla' \times (G_\varepsilon(\vec{x};\vec{x}'|\omega)\vec{E}^<(\vec{x}'|\omega)) - \\ &- \hat{n}'\nabla' \cdot (G_\varepsilon(\vec{x};\vec{x}'|\omega)\vec{E}^<(\vec{x}'|\omega)) - \\ &- (\hat{n}' \cdot \vec{E}^<(\vec{x}'|\omega))\nabla' G_\varepsilon(\vec{x};\vec{x}'|\omega) - \\ &- \hat{n}' \times \vec{E}^<(\vec{x}'|\omega) \times \nabla' G_\varepsilon(\vec{x};\vec{x}'|\omega) - \\ &- G_\varepsilon(\vec{x};\vec{x}'|\omega)\hat{n}' \times (\nabla' \times \vec{E}^<(\vec{x}'|\omega)) \quad .\end{aligned} \tag{A.14}$$

When Eq. (A.14) is substituted into Eq. (A.4) the surface integral of the first three terms vanishes [2]. Therefore, we obtain the result that

$$\begin{aligned}\vec{E}^<(\vec{x}|\omega)\theta_V(\vec{x}) = -\frac{1}{4\pi}\int_S dS'\{&\frac{i\omega}{c}(\hat{n}' \times \vec{H}^<(\vec{x}'|\omega))G_\varepsilon(\vec{x};\vec{x}'|\omega) + \\ &+ (\hat{n}' \times \vec{E}^<(\vec{x}'|\omega)) \times \nabla' G_\varepsilon(\vec{x};\vec{x}'|\omega) + \\ &+ (\hat{n}' \cdot \vec{E}^<(\vec{x}'|\omega))\nabla' G_\varepsilon(\vec{x};\vec{x}'|\omega)\} \quad ,\end{aligned} \tag{A.15}$$

where we have used the result that $\nabla' \times \vec{E}^<(\vec{x}'|\omega) = (i\omega/c)\vec{H}^<(\vec{x}'|\omega)$. We now use the boundary conditions at the surface S in the forms

$$\hat{n}' \times \vec{H}^<(\vec{x}'|\omega)|_{S'} = \hat{n}' \times \vec{H}^>(\vec{x}'|\omega)|_{S'} \tag{A.16a}$$

$$\hat{n}' \times \vec{E}^<(\vec{x}'|\omega)|_{S'} = \hat{n}' \times \vec{E}^>(\vec{x}'|\omega)|_{S'} \tag{A.16b}$$

$$\hat{n}' \cdot \vec{E}^<(\vec{x}'|\omega)|_{S'} = \frac{1}{\varepsilon(\omega)}\hat{n}' \cdot \vec{E}^>(\vec{x}'|\omega)|_{S'} , \tag{A.16c}$$

where the superscript ">" denotes the fields in the vacuum region $x_3 > \zeta(\vec{x}_\parallel)$, together with the relation $(i\omega/c)\vec{H}^>(\vec{x}'|\omega) = \nabla' \times \vec{E}^>(\vec{x}'|\omega)$. In this way we obtain the vectorial equivalent of the Kirchhoff integral:

$$\vec{E}^{<}(\vec{x}|\omega)\theta_V(\vec{x}) = -\frac{1}{4\pi}\int_S dS'\{(\hat{n}'\times(\nabla'\times\vec{E}^{>}(\vec{x}'|\omega)))G_\varepsilon(\vec{x};\vec{x}'|\omega) +$$
$$+ (\hat{n}'\times\vec{E}^{>}(\vec{x}'|\omega))\times\nabla' G_\varepsilon(\vec{x};\vec{x}'|\omega) +$$
$$+ \frac{1}{\varepsilon(\omega)}(\hat{n}'\cdot\vec{E}^{>}(\vec{x}'|\omega))\nabla' G_\varepsilon(\vec{x};\vec{x}'|\omega)\} . \qquad (A.17)$$

We now assume that the point of observation $\vec{x}$ is in the vacuum region. The left-hand side of Eq. (A.17) vanishes. This is a form of the extinction theorem [3]. In this way we obtain finally the boundary conditions at the surface $x_3 = \zeta(\vec{x}_\parallel)$ satisfied by the electric field in the vacuum:

$$\int_S dS'\{(\hat{n}'\times(\nabla'\times\vec{E}^{>}(\vec{x}'|\omega)))G_\varepsilon(\vec{x};\vec{x}'|\omega) +$$
$$+ (\hat{n}'\times\vec{E}^{>}(\vec{x}'|\omega))\times\nabla' G_\varepsilon(\vec{x};\vec{x}'|\omega) +$$
$$+ \frac{1}{\varepsilon(\omega)}(\hat{n}'\cdot\vec{E}^{>}(\vec{x}'|\omega))\nabla' G_\varepsilon(\vec{x};\vec{x}'|\omega)\} = 0 \qquad x_3 > \zeta(\vec{x}_\parallel). \qquad (A.18)$$

We conclude this Appendix by pointing out that a convenient representation for the Green's function $G_\varepsilon(\vec{x};\vec{x}'|\omega)$ is

$$G_\varepsilon(\vec{x};\vec{x}'|\omega) = \int\frac{d^2k_\parallel}{(2\pi)^2}\frac{2\pi i}{\alpha(k_\parallel\omega)}e^{i\vec{k}_\parallel\cdot(\vec{x}_\parallel-\vec{x}'_\parallel)}e^{i\alpha(k_\parallel\omega)|x_3-x'_3|}, \qquad (A.19)$$

where

$$\alpha(k_\parallel\omega) = [\varepsilon(\omega)\frac{\omega^2}{c^2} - k_\parallel^2]^{1/2}, \quad \mathrm{Re}\alpha(k_\parallel\omega) > 0,\ \mathrm{Im}\alpha(k_\parallel\omega) > 0. \qquad (A.20)$$

Appendix B. The Smoothing Method

Many times in the text we encounter equations that can be written in the abstract form

$$(H-V)R = f , \qquad (B.1)$$

where H is a non-stochastic operator, while V is a stochastic operator and f is a stochastic vector. What is required is $\langle R\rangle$, the value of R averaged over the ensemble characteristizing the stochastic nature of V and f. A systematic way of obtaining the equation satisfied by $\langle R\rangle$ is provided by the so-called smoothing method [85], which we outline and elaborate on here.

We begin by introducing the smoothing operator P that averages everything that stands to the right of it, e.g. $PR \equiv \langle R\rangle$. We also introduce the operator $Q = 1-P$ that projects out the fluctuating part of any object on which it acts. We then apply the operators P and Q in turn to Eq. (B.1), and use the identity $R = PR + QR$ to obtain the pair of equations

$$HPR - PV(PR+QR) = Pf \tag{B.2a}$$

$$HQR - QV(PR+QR) = Qf \quad . \tag{B.2b}$$

We solve Eq. (B.2b) for QR,

$$QR = (H-QV)^{-1}QVPR + (H-QV)^{-1}Qf \quad , \tag{B.3}$$

and substitute this solution into Eq. (B.2a) to obtain an equation for $PR \equiv \langle R\rangle$ that can be written in the form

$$(H-\langle M\rangle)\langle R\rangle = \langle F\rangle \quad , \tag{B.4}$$

where the (unaveraged) self-energy M is the solution of

$$M = V + VG_oQM \tag{B.5a}$$

$$= V + MG_oQV \quad , \tag{B.5b}$$

while F is given by the solution of

$$F = f + VG_oQF \quad . \tag{B.6}$$

In Eqs. (B.5) and (B.6) G_o is a Green's function defined by

$$G_o = H^{-1} \quad . \tag{B.7}$$

The results given by Eqs. (B.4) - (B.7) constitute the results of the smoothing method. We now proceed to rewrite them in a different fashion [86], which represents a resummation of the terms in iterative solutions of Eqs. (B.5) and (B.6).

We begin by introducing the Green's function G by

$$G = (H-V)^{-1} \quad , \tag{B.8}$$

or

$$G = G_o + G_oVG \tag{B.9a}$$

$$= G_o + GVG_o \quad . \tag{B.9b}$$

It is straightforward to show, by the methods that led from Eq. (B.1) to Eqs. (B.4) - (B.7), that the averaged Green's function $\langle G\rangle$ is the solution of

$$\langle G\rangle = G_o + G_o\langle M\rangle\langle G\rangle \tag{B.10a}$$

$$= G_o + \langle G\rangle\langle M\rangle G_o \ , \tag{B.10b}$$

where M is defined by Eqs. (B.5).

We now rewrite Eq. (B.9a) as

$$G = G_o + G_oV(PG+QG) \ , \tag{B.11a}$$

or

$$(I - G_oVQ)G = G_o + G_oV\langle G\rangle \ , \tag{B.11b}$$

whence it follows that

$$G = (I-G_oVQ)^{-1}G_o + (I-G_oVQ)^{-1}G_oV\langle G\rangle \ . \tag{B.12}$$

With the aid of the identities

$$(I-G_oVQ)^{-1}G_o = G_o \tag{B.13a}$$

$$(I-G_oVQ)^{-1}G_oV\langle G\rangle = G_o(I-VG_oQ)^{-1}V\langle G\rangle \ , \tag{B.13b}$$

that follow from the definition of the operator Q and the fact that G_o is non-stochastic, we transform Eq. (B.12) into

$$G = G_o + G_o(I-VG_oQ)^{-1}V\langle G\rangle \ . \tag{B.14}$$

In view of Eq. (B.5a) this becomes

$$G = G_o + G_oM\langle G\rangle \ . \tag{B.15}$$

If we compare this equation with Eq. (B.9a) we obtain the useful result that

$$VG = M\langle G\rangle \ . \tag{B.16}$$

We now return to Eq. (B.9a) and rewrite it as

$$G = G_o + G_o(V_1+V_2)G \tag{B.17}$$

and define a Green's function G_1 by

$$G_1 = G_o + G_oV_1G_1 \tag{B.18a}$$

$$= (I-G_oV_1)^{-1}G_o \ . \tag{B.18b}$$

Then the equation for G, Eq. (B.17), becomes

$$G = G_1 + G_1V_2G \ . \tag{B.19}$$

Now let

$$V_1 = \langle M\rangle \tag{B.20a}$$

$$V_2 = V - \langle M\rangle \equiv v \ . \tag{B.20b}$$

Then Eq. (B.18a) becomes

$$G_1 = G_o + G_o\langle M\rangle G_1 , \tag{B.21}$$

and a comparison of Eqs. (B.10a) and (B.21) reveals that $G_1 = \langle G\rangle$. Equation (B.19) now becomes

$$G = \langle G\rangle + \langle G\rangle vG \tag{B.22a}$$

$$= \langle G\rangle + Gv\langle G\rangle \quad . \tag{B.22b}$$

We next multiply Eq. (B.22b) from the left by V,

$$VG = V\langle G\rangle + VGv\langle G\rangle \tag{B.23}$$

and use Eq. (B.16) to obtain

$$M\langle G\rangle = V\langle G\rangle + M\langle G\rangle v\langle G\rangle \quad . \tag{B.24}$$

It follows from Eq. (B.24) that

$$M = V + M\langle G\rangle v \tag{B.25a}$$

$$= V + M\langle G\rangle(V-\langle M\rangle) \quad . \tag{B.25b}$$

This is the equation for M that we seek. In contrast with Eqs. (B.5) it contains the averaged Green's function $\langle G\rangle$ instead of the unperturbed Green's function G_o. In the language of diagrammatic perturbation theory it corresponds to the calculation of $\langle M\rangle$ by the use of "skeleton" diagrams.

We solve Eq. (B.24) for M formally in powers of V,

$$M = M_1 + M_2 + M_3 + \dots \tag{B.26}$$

and find that

$$M_1 = V \tag{B.27a}$$

$$M_2 = V\langle G\rangle V - V\langle G\rangle\langle V\rangle \tag{B.27b}$$

$$M_3 = V\langle G\rangle V\langle G\rangle V - V\langle G\rangle\langle V\langle G\rangle V\rangle - V\langle G\rangle\langle V\rangle\langle G\rangle V - V\langle G\rangle V\langle G\rangle\langle V\rangle + $$

$$+ 2V\langle G\rangle\langle V\rangle\langle G\rangle\langle V\rangle , \tag{B.27c}$$

so that finally

$$\langle M_1\rangle = \langle V\rangle \tag{B.28a}$$

$$\langle M_2\rangle = \langle V\langle G\rangle V\rangle - \langle V\rangle\langle G\rangle\langle V\rangle \tag{B.28b}$$

$$\langle M_3\rangle = \langle V\langle G\rangle V\langle G\rangle V\rangle - \langle V\rangle\langle G\rangle\langle V\langle G\rangle V\rangle -$$

$$- \langle V\langle G\rangle\langle V\rangle\langle G\rangle V\rangle - \langle V\langle G\rangle V\rangle\langle G\rangle\langle V\rangle +$$

$$+ 2\langle V\rangle\langle G\rangle\langle V\rangle\langle G\rangle\langle V\rangle \quad . \tag{B.28c}$$

Since, from Eq. (B.10) we see that

$$\langle G\rangle = \left(G_o^{-1} - \langle M\rangle\right)^{-1} , \tag{B.29}$$

the substitution of Eqs. (B.28) and (B.29) into the average of Eq. (B.26) yields a highly nonlinear integral equation for $\langle M \rangle$.

We return now to the function F defined by Eq. (B.6). We note from Eqs. (B.1) and (B.8) that the solution of the former can be written formally as

$$R = Gf \quad . \tag{B.30}$$

Thus, we multiply Eq. (B.22a) from the right by f to obtain

$$Gf = \langle G \rangle f + \langle G \rangle vGf \quad , \tag{B.31}$$

whence it follows that

$$Gf = (I - \langle G \rangle v)^{-1} \langle G \rangle f \tag{B.32a}$$

$$= \langle G \rangle (I - v \langle G \rangle)^{-1} f \quad . \tag{B.32b}$$

From Eq. (B.4) we find that

$$\langle R \rangle = \langle G \rangle \langle F \rangle \quad . \tag{B.33}$$

From Eqs. (B.30), (B.32b), and (B.33) we find that

$$\langle F \rangle = \langle (I - v \langle G \rangle)^{-1} f \rangle \quad . \tag{B.34}$$

This is the expression for $\langle F \rangle$ in "skeletonized" form.

References

1. Lord Rayleigh, Phil. Mag. 14, 70 (1907); Theory of Sound, 2nd. ed. (Dover, New York, 1945), vol. II, p. 89.
2. J. D. Jackson, Classical Electrodynamics (Wiley, New York, 1962), pp. 283-285.
3. E. Wolf, in Coherence and Quantum Optics, eds. L. Mandl and E. Wolf (Plenum, New York, 1973), p. 339.
4. R. F. Millar, Proc. Camb. Phil. Soc. 65, 773 (1969).
5. R. Petit and M. Cadilhac, C. R. Acad. Sci. B262, 468 (1966).
6. R. F. Millar, Proc. Camb. Phil. Soc. 69, 175, 217 (1971); Radio Sci. 8, 785 (1973).
7. N. R. Hill and V. Celli, Phys. Rev. B17, 2478 (1978).
8. P.M. van den Berg and J. T. Fokkema, J. Opt. Soc. Am. 69, 27 (1979); Radio Sci. 15, 723 (1980).
9. B. Laks, D. L. Mills, and A. A. Maradudin, Phys. Rev. B23, 4965 (1981).
10. N. E. Glass and A. A. Maradudin, Phys. Rev. B24, 595 (1981).
11. N. E. Glass, R. Loudon, and A. A. Maradudin, Phys. Rev. B24, 6843 (1981).
12. N. E. Glass and A. A. Maradudin, Electronics Lett. 17, 773 (1981).
13. R. E. Camley, N. E. Glass, and A. A. Maradudin, J. Appl. Phys. 53, 3170 (1970).
14. F. O. Goodman, J. Chem. Phys. 66, 976 (1977).
15. R. H. Ritchie, Phys. Rev. 106, 874 (1957).

16. H. Raether, in Physics of Thin Films, vol. 9 (Academic Press, New York, 1977), p. 145.
17. H. Raether, in Surface Polaritons, eds. V. M. Agranovich and D. L. Mills (North-Holland, Amsterdam, 1982), p. 331.
18. A. A. Maradudin, in Surface Polaritons, eds. V. M. Agranovich and D. L. Mills (North-Holland, Amsterdam, 1982), p. 405.
19. P. Mazur and D. L. Mills, J. Appl. Phys. 54, 3735 (1983).
20. R. G. Lee, Jr., G. A. Farias, A. A. Maradudin, and B. Mühlschlegel (unpublished work).
21. A. A. Maradudin (unpublished work).
22. H. Davies, Proc. Inst. Elec. Engrs. (London) 101, Pt. IV, 209 (1954).
23. H. E. Bennett, J. Opt. Soc. Am. 53, 1389 (1963).
24. J. O. Porteus, J. Opt. Soc. Am. 53, 1394 (1963).
25. J. Bramberg and H. Raether, Phys. Rev. Lett. 15, 882 (1965); Z. Physik 199, 118 (1967).
26. J. Bosenberg, Phys. Lett. 26A, 74 (1967).
27. E. A. Stern, Phys. Rev. Lett. 19, 1321 (1967).
28. R. H. Ritchie and R. E. Wilems, Phys. Rev. 178, 372 (1969); 184, 254 (1969).
29. J. Crowell and R. H. Ritchie, J. Opt. Soc. Am. 60, 794 (1970).
30. E. Kretschmann, T. L. Ferrel, and J. C. Ashley, Phys. Rev. Lett. 42, 1312 (1979).
31. J. G. Endriz and W. E. Spicer, Phys. Rev. B4, 4144, 4159 (1971).
32. A. Daudé, A. Savary, and S. Robin, J. Opt. Soc. Am. 62, 1 (1972).
33. T. F. Gesell, E. T. Arakawa, M. Williams, and R. N. Hamm, Phys. Rev. B7, 5141 (1972).
34. F. U. Hillebrecht, J. Phys. D13, 1625 (1980).
35. M. Rasigni, G. Rasigni, and J. P. Palmari, J. Opt. Soc Am. 71, 1230 (1981).
36. G. Rasigni, F. Varnier, M. Rasigni, J. P. Palmari, and A. Llebaria, Phys. Rev. B25, 2315 (1982).
37. P. Dobberstein, Phys. Lett. 31A, 307 (1970).
38. J. Eastman and P. W. Baumeister, Opt. Commun. 12, 418 (1974).
39. M. Rasigni, G. Rasigni, C. T. Hua, and J. Pons, Opt. Lett. 1, 126 (1977).
40. F. Jansen and R. W. Hoffman, Surf. Sci. 83, 313 (1979).
41. M. Rasigni and G. Rasigni, Phys. Rev. B19, 1915 (1979).
42. G. Rasigni, F. Varnier, M. Rasigni, and J. P. Palmari, Opt. Commun. 43, 409 (1982).
43. G. Rasigni, F.Varnier, M. Rasigni, J. P. Palmari, and A. Llebaria, Phys. Rev. B27, 819 (1983).
44. W. Kaspar and U. Kreibig, Surf. Sci. 69 619 (1977).
45. G. Rasigni, F. Varnier, M. Rasigni, and A. Llebaria, Phys. Rev. B29, 5931 (1984).
46. M. G. Kendall and A. Stuart, The Advanced Theory of Statistics (Griffin, London, 1963), Vol. I, p. 234.
47. Ref. 2, pp. 14-15.
48. G. A. Farias and A. A. Maradudin, Phys. Rev. B28, 5675 (1983).
49. G. C. Brown, V. Celli, M. Haller, and A. Marvin, Surf. Sci. 136, 381 (1984).

50. F. Toigo, A. Marvin, V. Celli, and N. R. Hill, Phys. Rev. B15, 5618 (1977).
51. R. W. Rendell, D. J. Scalapino, and B. Mühlschlegel, Phys. Rev. Lett. 41, 1746 (1978).
52. R. W. Rendell and D. J. Scalapino, Phys. Rev. B24, 3276 (1981).
53. D. W. Berreman, Phys. Rev. 163, 855 (1967).
54. R. Ruppin, Solid State Commun. 39, 908 (1981).
55. J. I. Gersten and A. Nitzan, J. Chem. Phys. 75, 1139 (1981).
56. P. C. Das and J. I. Gersten, Phys. Rev. B25, 6281 (1982).
57. A. G. Mal'shukov and Sh. A. Shekhmamet'ev, Fiz. Tver. Tela 25, 2623 (1983) [Soviet Physics - Solid State 25, 1509 (1983)].
58. A. A. Maradudin and W. M. Visscher, Zeit. Physik B (to appear).
59. See, for example, M. Abramowitz and I. A. Stegun, Handbook of Mathematical Functions (Dover, New York, 1965), p. 890. Note that in the expression for the weights w_i given in Section 25.4.45 of this reference the factor $(n+1)^2$ in the denominator must be replaced by $((n+1)!)^2$.
60. I am grateful to T. P. Shen for carrying out these calculations.
61. P. B. Johnson and R. W. Christy, Phys. Rev. B6, 4370 (1972).
62. I. Pockrand, Opt. Commun. 13, 311 (1975).
63. I. Pockrand and H. Raether, Opt. Commun. 17, 353 (1976); 18, 395 (1976).
64. P. F. Liao, J. G. Bergman, D. S. Chemla, A. Wokaun, J. Melngailis, A. M. Hawryluk, and N. P. Economou, Chem. Phys. Lett. 82, 355 (1981).
65. N. E. Glass, A. A. Maradudin, and V. Celli, Phys. Rev. B26 5357 (1982).
66. P. M. van den Berg and J. C. M. Borburgh, Appl. Phys. 3, 55 (1974).
67. T. S. Rahman and A. A. Maradudin, Phys. Rev. B21, 2137 (1980).
68. A. A. Maradudin, in Proc. Autumn School on Surface Physics, Varna, Bulgaria, October, 1980, ed. I. J. Lalov (Bulgarian Academy of Sciences, Sofia, 1981).
69. R. E. Palmer and S. E. Schnatterley, Phys. Rev. B4, 2329 (1971).
70. R. Kötz, M. J. Lewerenz, and E. Kretschmann, Phys. Lett. 70A, 452 (1979).
71. M. W. Williams, J. C. Ashley, E. Kretschmann, T. A. Callcott, M. S. Chung, and E. T. Arakawa, Phys. Lett. 73A, 231 (1979).
72. T. Lopez-Rios and Y. Borensztein, Phys. Rev. B31, 5507 (1985).
73. A. A. Maradudin and W. Zierau, Phys. Rev. B14, 484 (1976).
74. D. L. Mills, Phys. Rev. B12, 4036 (1975).
75. A. A. Maradudin, in Nonequilibrium Phonon Dynamics, ed. W. E. Bron (Plenum, New York, 1985), p. 395.
76. A. Wirgin and A. A. Maradudin, Phys. Rev. B31, 5573 (1985).

77. See, for example, B. Fischer and J. Lagois, in Topics in Current Physics, vol. 14: Excitons, ed. K. Cho (Springer-Verlag, Berlin, 1979), ch. 4.
78. A. A. Maradudin, Surf. Sci. 81, 15 (1979).
79. N. J. M. Horing, in Proc. 7th Intern. Vacuum Congr. and 3rd Intern. Conf. on Solid Surfaces, eds. R. Dobrozemsky, F. Rudenauer, F. P. Viebock and A. Breth (Berger, Vienna, 1977), p. 419.
80. A. A. Maradudin and D. L. Mills, Phys. Rev. B7, 2787 (1973).
81. D. N. Pattanayak and J. L. Birman, Solid State Commun. 28, 927 (1978).
82. R. Englman and R. Ruppin, J. Phys. C1, 614, 1515 (1968).
83. R. Ruppin and R. Englman, J. Phys. C1, 630 (1968).
84. L. Genzel, T. P. Martin, and U. Kreibig, Z. Physik B21, 339 (1975).
85. See. for example, U. Frisch, in Probabilistic Methods in Applied Mathematics, ed. A. T. Barucha-Reid, vol. I (Academic Press, New York, 1968), p. 76.
86. G. Brown, V. Celli, M. Haller, A. A. Maradudin, and A. M. Marvin, Phys. Rev. B31, 4993 (1985).

Poles of the Reflectance Function and Their Relation to Leaky Waves*

F.Y. Kou and T. Tamir

Department of Electrical Engineering and Computer Science, Polytechnic Institute of New York, Brooklyn, N.Y. 11201, USA

Recent work on optical waveguides has revealed the presence of novel types of leaky waves along thin-film layers bounded by two open (substrate and cover) regions. The new field varieties include waves that radiate selectively into only one of the two open regions[1] and waves having backward propagation characteristics[2]. However, although a few such novel leaky waves have been identified analytically, the actual excitation of these waves by realistic sources has not yet been reported.

We therefore examine the general problem of field reflection by a dielectric layer bounded by two open regions having different permittivities. For this purpose, we assume the presence of a localized source and describe its field in terms of an integral in a complex wavenumber plane consisting of four Riemann sheets. By using suitable contour deformations, it is possible to express the total field in terms of two continuous wave spectra (given by branch-cut integrals) and a discrete set of leaky waves (given by pole residues). These leaky waves are therefore given directly by the poles of the reflectance function of the layered structure. The presence of an experimentally identifiable leaky wave depends on its field being measured as a distinct, and preferably dominant, constituent of the total field. We therefore choose a field representation that clearly separates leaky waves from the continuous wave spectra. With this choice, the excitable leaky waves are those represented by residues due to poles that appear in certain selected portions of the complex Riemann plane. All the other leaky-wave poles yield no residues, and therefore they do not represent fields that can be physically identified. We then find that, out of all possible leaky-wave varieties, only a few of them can be excited. However, as some of these excitable fields are of novel types, the results suggest the possibility of field configurations that had not been previously recognized.

*This work was supported by the National Science Foundation and by the Joint Services Electronics Program

[1] C.W. Hsue and T. Tamir, J. Opt. Soc. Amer. A, 1, 923, 1984
[2] G.I. Stegeman et al., Opt. Lett. 8, 383, 1983

Surface Polaritons in Semiconductor Superlattices

N. Raj and D.R. Tilley

Physics Department, University of Essex, Colchester CO4 3SQ, U.K.

We use previously derived results [1] for transverse electromagnetic waves propagating in an infinite superlattice where the z-axis is taken as normal to the layers which have thicknesses d_1, d_2 and isotropic dielectric functions ε_1, ε_2, both frequency-dependent. Taking the long wavelength limit $\lambda >> \Lambda$, where λ is the optical wavelength and $\Lambda = d_1 + d_2$ the superlattice period, we show that the superlattice has the optical properties of a conventional uniaxial medium with principal values of the dielectric tensor ε_{zz} along the uniaxis, $\varepsilon_{zz}^{-1} = (\varepsilon_1^{-1} d_1 + \varepsilon_2^{-1} d_2)/(d_1+d_2)$ and ε_{xx} in the transverse plane, $\varepsilon_{xx} = (\varepsilon_1 d_1 + \varepsilon_2 d_2)/(d_1+d_2)$. This result is used to derive the dispersion equations for polaritons at single and double interfaces between superlattices and ordinary media. The dispersion relation for a single interface surface polariton is

$$q_x^2 = \frac{\omega^2}{c^2} \frac{\varepsilon_m \varepsilon_{zz} (\varepsilon_{xx} - \varepsilon_m)}{\varepsilon_{xx} \varepsilon_{zz} - \varepsilon_m^2}$$

where ε_m is the dielectric constant of the bounding medium and q_x is the wavenumber in the surface plane. Dispersion curves are shown in Fig. 1 [2].

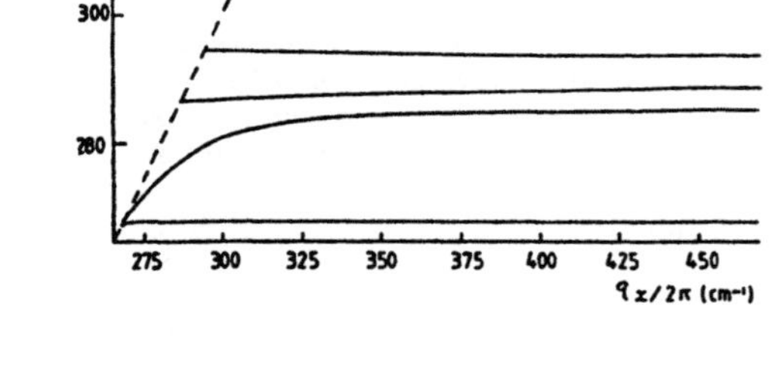

Fig. 1. Dispersion curves, ω vs. q_x, for the interface between a GaAs/$Ga_{.86}Al_{.14}As$ superlattice and vacuum.

References

1. P. Yeh, A. Yariv and C.S. Hong, J. Opt. Soc. Am. 67, 423 (1977)
2. N. Raj and D.R. Tilley, Sol. St. Comms., Vol. 55, No.4, 373-376 (1985)

Plasmon Modes in Semiconductor Superlattices

N.C. Constantinou and M.G. Cottam

Physics Department, University of Essex, Colchester CO4 3SQ, U.K.

We have performed calculations for the bulk and surface plasmon modes of a superlattice consisting of two-dimensional electron-gas layers. Applications are discussed for multiple quantum wells in semiconductor superlattices.

Previous calculations for the bulk plasmon modes (e.g. see [2]) have considered electron-gas layers separated by media of the same thickness and dielectric constant. We have extended the theory to the case of alternating ...ABABAB... structures where media A and B may have different thicknesses and/or dielectric constants. Details of the calculations are presented in [1], the main features being the prediction of 'acoustic' and 'optic' branches in the dispersion relation. These modes should correspond to two peaks in the light scattering spectrum. Numerical calculations are given for the case of GaAs/$Al_xGa_{1-x}As$ superlattices. Retardation effects have been taken into account, but these are shown to be generally unimportant for the modes and frequencies of interest.

We have also evaluated the dispersion relations of surface plasmons in semi-infinite superlattices [1]. In doing this we have extended previous calculations (e.g. see [3]) by employing a more general boundary condition at the surface of the semi-infinite superlattice, allowing for the areal electron concentration at the superlattice surface to be different from the bulk value. Results for the surface plasmon dispersion relation are given in an explicit analytic form, including the special case studied in [3]. For non-alternating superlattices it is found, as in [3], that the surface branch may occur above or below the bulk continuum,depending on the ratio of the dielectric constants outside and inside the superlattice. No surface plasmon is predicted if the in-plane wavenumber is less than a critical value, which depends on the above ratio (as in [3]) and also on the surface-charge density. The surface modes in the alternating structure have similarly been evaluated [1]. In this case, it is found that there may be two surface plasmon branches, and for certain in-plane wavenumbers,one of the surface plasmons may occur between the two bulk mode regions. The spectral intensities associated with the various plasmon excitations can be conveniently evaluated using linear response theory, and we shall report on this shortly.

References

1. N.C. Constantinou and M.G. Cottam: submitted to J. Phys. C (1985)
2. S. Das Sarma and J.J. Quinn: Phys. Rev. B25, 7603 (1982)
3. G.F. Giuliani and J.J. Quinn: Phys. Rev. Lett. 51, 919 (1983)

Surface Phonons and Reconstruction in Diamond Structure Crystals

W. Goldammer, W. Zierau, and W. Ludwig

Institut für Theoretische Physik II, Universität Münster, Domagkstrasse 75, D-4400 Münster, F.R.G.

The surface phonon spectrum has been investigated for a series of crystals with diamond structure for the occurrence of soft modes. Those modes are an indication of a structural phase transition leading to the formation of a new superstructure. Calculations base on a force constant model being fitted to experimental bulk phonon frequencies. The (111) surface of C, Si, Ge and α-Sn is treated by a Green's function method (for details see [1]). We then studied the influence of the change of surface force constants on the surface phonon dispersion.

A soft mode was found in case of Si, Ge [1,2] and α-Sn [3]. However, no softening was observed for C [3]. Analyzing the positions of the soft modes, the modes can be attributed to superstructures being (7x7) for Si, (8x8) for Ge and (3x3) for α-Sn. In addition, we propose a model of the reconstructed surface being calculated from the eigenvectors associated with the soft mode [1,2]. Fig. 1 shows our results for Si(111) and its comparison to an experimental TED pattern [4].

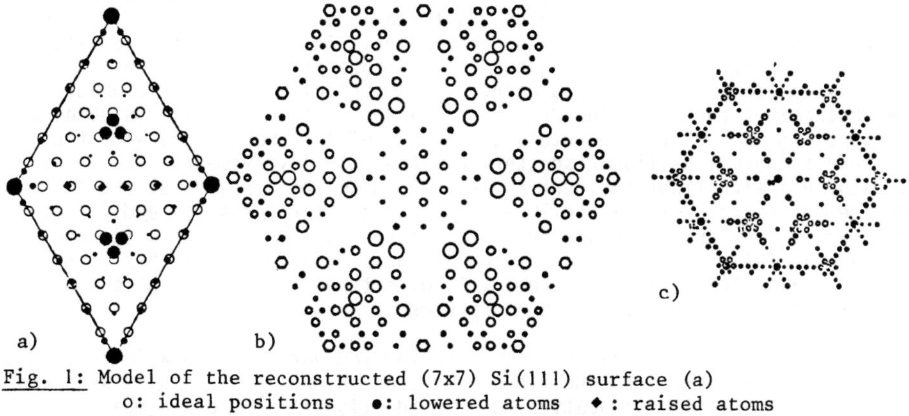

Fig. 1: Model of the reconstructed (7x7) Si(111) surface (a)
o: ideal positions ●: lowered atoms ◆: raised atoms
and its comparison with TED pattern (b) calculated from (a) and experimentally determined (c) [4].

The idea of our calculation is to show the intimate relation between the dynamical behavior of a surface and its structure. It seems that the electronic structure being indirectly contained in our model (via the bulk force constants) plays an important role in the formation of the different reconstruction patterns.

1. W. Goldammer, W. Ludwig, W. Zierau and C. Falter: Surface Sci 141, 139, (1984).
2. W. Zierau, W. Goldammer, C. FAlter and W. Ludwig: Superlattices and Microstructures 1, 55 (1985).
3. W. Goldammer, W. Zierau and W. Ludwig, to be published.
4. E.G. McRae and P.M. Petroff: Surface Sci 147, 385 (1984).

Surface Plasmon – Surface Acoustic Wave Interactions on Thin Silver Films*

W.M. Robertson, A.L. Moretti, and R. Bray

Physics Department, Purdue University, West Lafayette, IN 47907, USA

The high surface fields associated with surface plasmons (SP's) are responsible for contributing to the strong enhancement observed for a number of non-linear optical phenomena such as Raman scattering by chemical adsorbates and second harmonic generation. We have demonstrated experimentally [1] that such an enhancement can also be observed when SP's are used as an intermediate agent for Brillouin scattering from thermal equilibrium surface acoustic waves (SAW's) of thin (450 Å) silver films.

In our experiments the SP's were generated by means of the Kretschmann geometry, with the films deposited in vacuum on the flat face of a pyrex hemisphere. Light incident at an appropriate angle through the hemisphere, can resonantly generate SP's with well-defined k-vectors at the metal/air interface. Two modes of interaction of the SP's with the SAW's were studied: (1) the SP could be scattered by a SAW directly into light on the air side of the film; (2) the initial SP can be scattered into another SP which can then reverse couple to light through the hemisphere. In each case the resultant scattered light, shifted in frequency by the absorption or emission of a phonon, was collected and analyzed using a 5-pass Fabry-Perot interferometer. The enhancement factor in the Brillouin scattering cross-section is obtained by comparing the scattering intensity in each of the above cases with that obtained in normal Brillouin scattering without SP involvement. The latter was achieved with light incident and scattered on the air side of the film. Enhancement factors of 25x and 500x were obtained for cases (1) and (2) respectively. The spatial distribution of scattered light in cases (1) and (2) must be carefully analyzed in arriving at these enhancement factors. The experimental values obtained agreed well with theoretical calculations using the parameters appropriate for our experiment.

This technique is capable of demonstrating the various SP-SAW interactions, and of sizeably increasing the signal obtained in the intrinsically weak process of surface Brillouin scattering. In addition, it provides a means of measuring the enhancement due to the SP's alone. This is in contrast to Raman scattering from chemical adsorbates, where the enhancement is due to a combination of electromagnetic (SP) and chemical factors.

References

1. A.L. Moretti, W.M. Robertson and Ralph Bray, Phys. Rev. B31, 3361 (1985).

*Work supported by NSF-MRL Program DMR80-20249 at Purdue University.

Part II

Electromagnetic Probes of Surface Excitations

Surface Brillouin and Raman Scattering

F. Nizzoli

Department of Physics and G.N.S.M., University of Modena,
I-41100 Modena, Italy

1. Introduction

First order light scattering from collective excitations in transparent materials (fluids, solids) has been studied since the beginning of the century [1,2]. More recently (since 1965) light scattering has been used to investigate acoustic and optic surface vibrations. Two different points of view can be adopted. In fact, surface phonons can be detected which are either externally excited with a variety of different techniques or naturally present in the sample at a finite temperature T due to thermal excitations (thermal noise). The two different methods have been developed rather independently.

The **external generation of surface modes** is a technique which has grown explosively in the framework of surface wave technology. Applications to electronics and nondestructive testing are well established. The most important applications have been in the area of signal processing, where long wavelength surface waves in the frequency range 1 MHz- 1 GHz are employed. Detection of generated surface waves occurs both through interdigital transducers [3] and the interaction of light with the travelling surface wave. Many techniques are used, all based on the analysis of the diffracted electromagnetic field and the subsequent electronic processing of the electrical signals. This area has been reviewed in detail by STEGEMAN [4] and is excluded from the present discussion, which is confined to scattering of light from long wavelength phonons in thermal equilibrium. As an exception, we will consider light scattering from generated surface waves involving surface plasmon polaritons.

With respect to the case of generated phonons, the area of **thermal excited phonons** has been mainly developed in the framework of basic solid state research. As it is well known, the scattering from acoustic and optic phonons is named after Brillouin and Raman respectively. The Brillouin scattering probes waves in the frequency range 1-100 GHz while in Raman scattering optic phonons of frequency greater than 10 THz are detected. In light scattering from thermal noise the modes are excited by statistical fluctuations, and there is no net transport of energy via these modes from one part of the sample to another. Hence these modes occur as standing waves. A second major difference, with respect to externally generated phonons, is that it is not the propagation distance that one can detect experimentally, but rather the phonon lifetime. Since these modes are excited by the fluctuations, they correspond to the true normal modes of the sample. In this way the detail of the phonon density of states contributes to the information obtained from a Brillouin or Raman scattering experiment.

Before facing the problem of light scattering from surface phonons, we briefly review classical results on light scattering from bulk collective modes in transparent media (fluids and solids). In fact, many concepts in light scattering and the nomenclature itself are derived from this well-developed field.

2. Light Scattering in Fluids

The approach used in this case is via thermodynamic fluctuation theory. In fact in this case the atoms are free to move over large distances, there is no long range order and the fluid can be considered as a continuous dielectric medium. The pressure fluctuations δp can be decomposed in density fluctuations $\delta\rho$ and in entropy fluctuations δs according to the following equation

$$\delta p=\left[\frac{\partial p}{\partial \rho}\right]_s \delta\rho + \left[\frac{\partial p}{\partial s}\right]_\rho \delta s. \qquad (1)$$

The first term on the right-hand side is related to the existence in the fluid of longitudinal sound waves (periodic density fluctuations) while the second is responsible for entropy fluctuations. In their very elegant theory LANDAU and PLACZEK [5] showed that a monochromatic light of frequency ω_0 is inelastically scattered by the longitudinal sound waves,giving rise to the **Brillouin doublet** of frequencies ω_0-vq, ω_0+vq, where v and q are the propagation velocity and the wavevector of the waves. The two Brillouin lines are called Stokes and anti Stokes components respectively. The entropy fluctuations turn out to produce in the spectrum of the scattered light a central unshifted peak (**Rayleigh line**) of frequency ω_0, which is therefore called 'quasi-elastic' component. MOUNTAIN [6] has expliciteIy derived the structure factor $S(\omega)$ for scattering from a simple fluid in terms of macroscopic constants, as the sound absorption coefficient Γ, the thermal diffusivity χ and the specific heats at constant pressure and volume c_p and c_v. A typical structure factor is shown in Fig.1. The lineshape of each line is a lorentzian. In the theory of Landau and Placzek the ratio between the scattering intensities of the quasi elastic component and of the two Brillouin components is given by $(c_p-c_v)/c_v$.

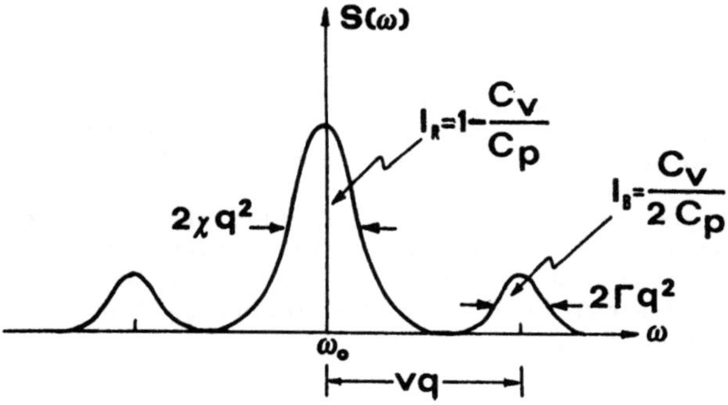

Fig.1 The structure factor $S(\omega)$ for scattering from a fluid showing the full width at half maximum of each spectral line.

3. Brillouin Scattering in Transparent Solids

Three-dimensional crystalline solids are conveniently described in terms of translational invariance through discrete lattice vectors and of Bloch wavefunctions. In quantum mechanics,first order light scattering is described as creation and annihilation of phonons of wavevector **q** and frequency Ω, with energy and momentum conservation. In the long wavelength limit,where umklapp process are not important,we have

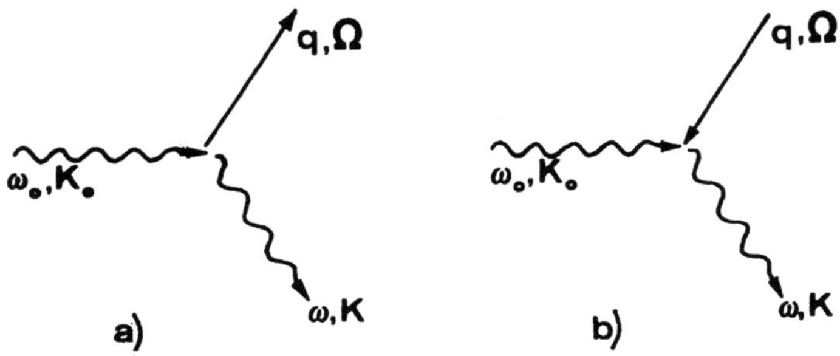

Fig.2 Kinematics of first order Stokes (a) and anti-Stokes (b) scattering events

phonon creation (Stokes events) $\begin{cases} \omega=\omega_0-\Omega \\ \mathbf{K}=\mathbf{K}_0-\mathbf{q} \end{cases}$

phonon annihilation (anti-Stokes events) $\begin{cases} \omega=\omega_0+\Omega \\ \mathbf{K}=\mathbf{K}_0+\mathbf{q} \end{cases}$

as indicated in Fig.2. ω_0 and $\mathbf{K}_0$ are the frequency and wavevector of the incident light, whereas ω and $\mathbf{K}$ are the corresponding quantities of the scattered light.

It is clear from energy and wavevector conservation that the scattering geometry completely specifies the acoustic wavevector $\mathbf{q}$, whose modulus is given by

$$q \simeq 2\,K_0\,\sin(\theta/2)$$

where θ is the scattering angle. Therefore in general,the Brillouin spectrum from bulk sound waves in a solid is composed of three Brillouin doublets, one for each acoustic branch. Obviously,symmetry reasons may cause doublets to disappear or to be degenerate.

The physical reason for light scattering is to be found in the inhomogeneities of the medium due to fluctuations in the dielectric constant, which in turn are induced by the thermally excited phonons. In fact the fluctuations of the dielectric function $\delta\varepsilon_{\alpha\beta}$ are linear combinations of the strains $\eta_{\gamma\delta}$ through the elasto optic (Pockels) coefficients $k_{\alpha\beta\gamma\delta}$

$$\delta\varepsilon_{\alpha\beta} = \sum_{\gamma\delta} k_{\alpha\beta\gamma\delta}\,\eta_{\gamma\delta}. \tag{2}$$

The strains $\eta_{\gamma\delta}$, at least in the simple case of optically isotropic materials [7], are symmetric combinations of the displacement gradients

$$\eta_{\gamma\delta} = \frac{1}{2}\left[\frac{\partial u_\gamma}{\partial x_\delta}+\frac{\partial u_\delta}{\partial x_\gamma}\right]. \tag{3}$$

The detailed derivation of the scattering cross-section may be found in LANDAU and LIFSHITZ [8] and here is only briefly outlined. As derived in [8] the electric field E of the scattered light, at large distance R from the scattering volume V and to first order Born approximation, is given by

$$E=-\frac{e^{iKR}}{4\pi\varepsilon R}\mathbf{K}\times\left[\mathbf{K}\times\int_V \delta\varepsilon E_0 e^{-i(\mathbf{K}-\mathbf{K}_0).\mathbf{r}}d^3r\right] \tag{4}$$

where the second rank tensor $\delta\varepsilon$ is defined by (2) and (3). Integrals of the type (4) immediately lead to momentum conservation. In fact, for a given excitation of wavevector $\mathbf{q}$, $\delta\varepsilon_{\alpha\beta}$ turns out to be proportional to $\exp(i\mathbf{q}.\mathbf{r})$. If V is large compared to q^{-1}, the integral (4) is proportional to $\delta(\mathbf{K}-\mathbf{K}_0\pm\mathbf{q})$. The conservation of momentum is therefore recovered.

The mean square displacement per mode q is related to the polarisation vector $\mathbf{e}$ by the relationship

$$<|\mathbf{u}|^2> = \frac{k_B T}{V\rho v^2 q^2}|\mathbf{e}|^2 \tag{5}$$

where ρ is the mass density of the medium and v is the velocity of the mode considered. The power scattered into the unit solid angle per unit incident power and unit interaction volume is given by

$$\left[\frac{dP}{d\Omega}\right] = \frac{R^2}{VE_0^2}E^2 = \frac{k_B T\pi^2}{4\lambda^4\rho v^2}|g|^2 \tag{6}$$

where λ is the wavelength of the light. $\mathbf{g}$ is a vector expressed in terms of the elasto optic coefficients $k_{\alpha\beta}$, of the electric field $\mathbf{E}_0$ of the incident light, of the polarisation vector $\mathbf{e}$ and of the wavevector $\mathbf{q}$ of the phonon

$$g_\alpha=\frac{1}{qE_0}[(k_{11}-k_{12})(e_\alpha\mathbf{q}.\mathbf{E}_0+q_\alpha\mathbf{e}.\mathbf{E}_0)+2k_{12}\delta_{\alpha\beta}E_\beta\mathbf{e}.\mathbf{q}]. \tag{7}$$

This expression is valid for an isotropic medium. Equation (7) contains the selection rules for Brilloiun scattering from bulk waves. In particular, the cross-section depends on the polarisations of the incoming and scattered light beams, according to the following cases:

a) s→s (TE→TE) scattering (from longitudinal waves)
$|\mathbf{g}|^2=4k_{12}^2$
This term comes from the diagonal contribution $\mathbf{e}.\mathbf{q}$ to (7)

b) s→p (TE→TM) scattering (from transverse waves, polarised normally to the scattering plane)
$|\mathbf{g}|^2=(k_{11}-k_{12})^2\cos\theta/2$
This term comes from $\mathbf{e}.\mathbf{E}_0$ in (7)

c) p→s (TM→TE) scattering (from transverse waves, polarised normally to the scattering plane)
$|\mathbf{g}|^2=(k_{11}-k_{12})^2\cos\theta/2$
This term comes from $\mathbf{q}.\mathbf{E}_0$ in (7)

d) p→p (TM→TM) scattering (from longitudinal waves)
$|\mathbf{g}|^2=4|(k_{11}-k_{12})\cos^2\theta/2 + k_{12}\cos\theta|^2$
Here there are nonvanishing contributions from $\mathbf{e}.\mathbf{q}$, $\mathbf{q}.\mathbf{E}_0$ and $\mathbf{e}.\mathbf{E}_0$.

Notice that backscattering ($\theta=\pi$) occurs via longitudinal waves only. Transverse phonons depolarise the light, while scattering by longitudinal phonons is fully polarised. For anisotropic crystals the selection rules become more complicated. Useful results for anisotropic crystals can be found in [9] and [10].

4. Brillouin Scattering from Bulk Modes in Opaque Solids

The recent progress on Brilloiun scattering from opaque materials and thin films has been made possible by advances in instrumentation. In fact, in scattering from opaque materials, the light reflected from the surface in any space direction due to scratches, surface imperfections and roughness gives rise to very strong spurious signals at the incident laser frequency. In addition, the scattering volume is very small, i.e. at most an acoustic wavelength deep, and the signals are very weøk compared with Brilloiun scattering in transparent media. We postpone the discussion of the instrumentation to the next section. Here we want to describe the main **effects of opacity.**

First of all **opacity limits the scattering** to a volume close to the sample surface. Therefore the argument presented in the previous section to show that (4) is proportional to $\delta(\mathbf{K}-\mathbf{K}_0\pm\mathbf{q})$ is no longer valid. Instead $(\mathbf{K}-\mathbf{K}_0)=\mathbf{K}'-i\mathbf{K}''$ is now complex, so that we are left with integrals of the type

$$F = \int \eta_{\alpha\beta}\, e^{-i(\mathbf{K}'-i\mathbf{K}'')\cdot\mathbf{r}}\, d^3r. \tag{8}$$

By considering a phonon of wavevector $\mathbf{q}$, $\eta_{\alpha\beta}$ is proportional to $\exp(i\mathbf{q}\cdot\mathbf{r})$. Carrying out the integration over the surface coordinates x,y we have

$$F = \int_{-\infty}^{0} e^{-[i(q_z-K_z')z-K_z''z]}\, dz \propto \frac{1}{[i(q_z-K_z')-K_z'']} \tag{9}$$

where it has been considered a semiinfinite medium extending from $z=-\infty$ to $z=0$. Let us assume, for simplicity's sake, a backscattering geometry with $\mathbf{K}'-i\mathbf{K}''$ perpendicular to the surface. Under this simplifying hypothesis $\mathbf{K}'-i\mathbf{K}''= 2\mathbf{K}_0(n+ik)$, where $n+ik$ is the complex refractive index of the medium. We obtain

$$F^2 \propto \frac{2kK_0}{[(q-2nK_0)^2+(2kK_0)^2]}. \tag{10}$$

This equation shows that the presence of optical absorption allows phonon exchange not only at the frequency $\Omega=2nK_0v$ but rather within a finite frequency range. From (10) **the broadening** has a full width at half maximum of $\Delta\omega=4kK_0v$. An effect of this nature was observed experimentally by PINE [11] and SANDERCOCK [12]. In Fig.3 the results for Si and Ge taken from the latter reference are reproduced. In particular the increase of the bulk phonon linewidth in Ge is explicitely shown, by lowering the wavelength of the laser light and therefore increasing the extinction coefficient k. These data were used to deduce the complex refractive indices which were responsible for the opaque character of the medium.

An additional important message is present in Fig.3c. The lineshape of the longitudinal bulk phonons is not simply symmetric, as given by (10), but rather asymmetric. This effect was first observed by DRESSELHAUS and PINE [13] and described in quantitative terms by DERVISCH and LOUDON [14]. The idea is simple. In opaque materials the scattering occurs in a region close to the surface, so that the effect of **phonon reflection** at the boundary must be considered. For example, an anti-Stokes event involves now not only a phonon with $\mathbf{q}$ but also the reflected phonon with $-\mathbf{q}$, the phonon reflection at a boundary being a coherent process. Therefore the scattering form factor F contains, in addition to the term (8), another term obtained with $\eta_{\alpha\beta}(-\mathbf{q})\propto-\exp(-i\mathbf{q}\cdot\mathbf{r})$. The cross-section turns out to be proportional to

$$F^2 \propto \left| \frac{1}{[i(q-K')-K'']} - \frac{1}{[i(-q-K')-K'']} \right|^2 \propto \frac{q^2}{[(q^2-4n^2K_0^2+4k^2K_0^2)^2+(8nkK_0^2)^2]} \tag{11}$$

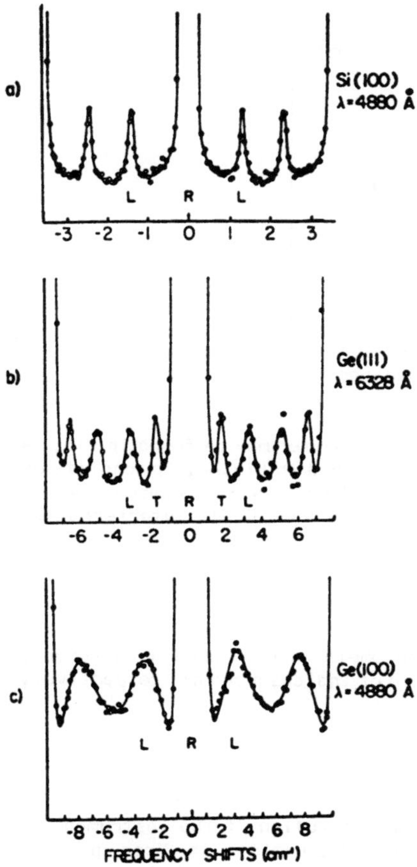

Fig.3 Brillouin spectra of silicon and germanium measured on different crystal faces and with different incident wavelengths. After SANDERCOCK [12]

Notice the **asymmetric lineshape** due to the term q^2 in the numerator. In addition also the maximum of the phonon line is shifted from $\Omega=2nK_0v$ to $\Omega=2K_0v(n^2+k^2)^{1/2}$.

In this section we have considered the effects of the opacity on the Brillouin spectra of bulk acoustic modes. However, there are even more dramatic consequences of opacity on the Brillouin spectroscopy: the appearance of new excitations associated with the surface and the scattering of light from the dynamical corrugation induced by surface excitations in highly opaque materials. These arguments will be illustrated in detail in section 6. The next section is devoted to the description of the pieces of apparatus which have made possible the detection of complex surface Brillouin spectra in opaque materials.

5. Instrumentation for Surface Brillouin Scattering

As already remarked in the previous section, Brillouin scattering from opaque materials is a difficult experiment, since the detection of very week signals requires long data acquisition times and a very high rejection of the stray light signal. Enough resolution is usually achieved with a plane parallel Fabry-Perot interferometer, used as a scanning spectrometer. In special applications, where the highest resolution is required, a spherical Fabry-Perot must be used [15]: in the latter case a resolution as high as 1 MHz is possible.

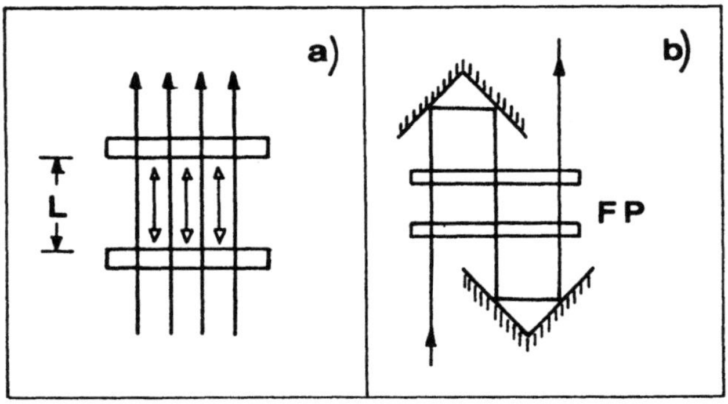

Fig.4 (a) Fabry-Perot interferometer; (b) three-pass interferometer

A general description of a Fabry-Perot interferometer may be found in [16]. A plane parallel Fabry-Perot interferometer is sketched in Fig.4a. It consists of two flat mirrors mounted accurately parallel to each other with a spacing L which may be varied, typically through piezoelectric transducers. The interferometer transmits light of wavelength λ_n if the spacing L is such that

$$L = \frac{n\lambda_n}{2} \qquad (12)$$

where n is an integer. Therefore the interferometer can be used as a tunable filter. Two neighboring orders of interference of indices n and n+1 respectively are separated in wavenumber by the quantity

$$\frac{1}{\lambda_{n+1}} - \frac{1}{\lambda_n} = \frac{1}{L}\left[\frac{n+1}{2} - \frac{n}{2}\right] = \frac{1}{2L} \qquad (13)$$

which is called **free spectral range** (FSR). The **resolution** of the instrument is determined by the width of the transmission peak. The ratio of the free spectral range to the width is called **finesse** F. In usual application $F \approx 100$. Another important parameter of the interferometer is the **contrast** C, defined as the ratio of maximum to minimum transmission, given by

$$C = 1 + \frac{4}{\pi^2} F^2 \qquad (14)$$

As a matter of fact, for measurements in opaque materials the elastically scattered component of the scattered light exceeds the Brillouin component by a factor 10^4-10^5. Therefore a simple Fabry-Perot interferometer is inadequate for such a situation. SANDERCOCK [17] was the first to show that passing light more than once through the same Fabry-Perot interferometer substantially increased the contrast, reaching values greater than 10^9. Fig.4b shows the design of a three-pass Fabry-Perot interferometer. A **multipass interferometer** requires an active stabilization system to maintain the alignment. By adopting piezoelectric scanning, a dynamic stabilization can be obtained, whereby the instrument is rapidly scanned and continuously realigned to mantain correct mirror spacing and parallelness.

A typical apparatus is shown in Fig.5. The argon ion laser is operated in a stabilised single longitudinal mode configuration with power levels not exceeding 1 W. The heart of the apparatus, the Fabry-Perot multipass interferometer, is normally configured to pass the light three times between the same interferometer plates. Light passed by the Fabry-Perot is detected by a

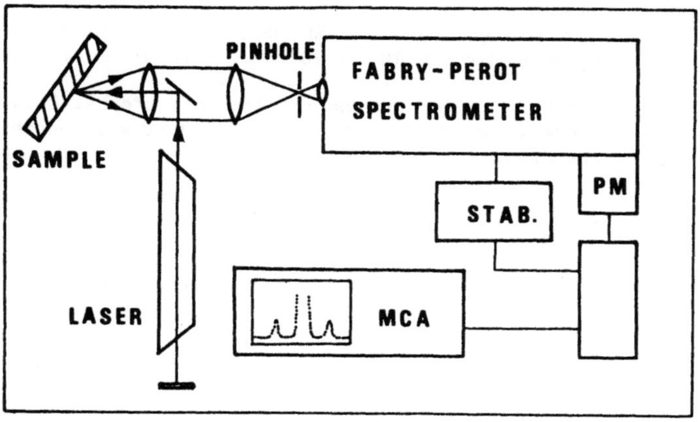

Fig.5 Typical Brillouin scattering apparatus for measurement of opaque materials in backscattering geometry.

photomultiplier tube (PM). The resulting sequence of electrical pulses is stored in a multi-channel analizer (MCA) or computer. As already remarked, the stabilization system (STAB) is a necessary requirement. Since 1969 various other schemes for Fabry-Perot scanning, stabilization and data acquisition have been realised. The reader may refer to recent reviews [18,19].

The next important advance in Brillouin spectroscopy was the development of **tandem interferometers**, in order to increase the FSR. In fact, spectra for opaque materials may contain several features (surface modes, thin film Lamb/Love modes, bulk modes) extending up to 100 GHz. In a single Fabry-Perot interferometer the interference transmission pattern is repeated, so that spectra of contiguous orders extending more than half the FSR may overlap.

The principle upon which tandem interferometers are based is the following. Consider a pair of Fabry-Perot interferometers, with different plate separation L_1 and L_2, which could be scanned simultaneously. The two interferometers transmit wavelengths respectively

$$\lambda_1 = \frac{2L_1}{n_1} \quad ; \quad \lambda_2 = \frac{2L_2}{n_2} \tag{15}$$

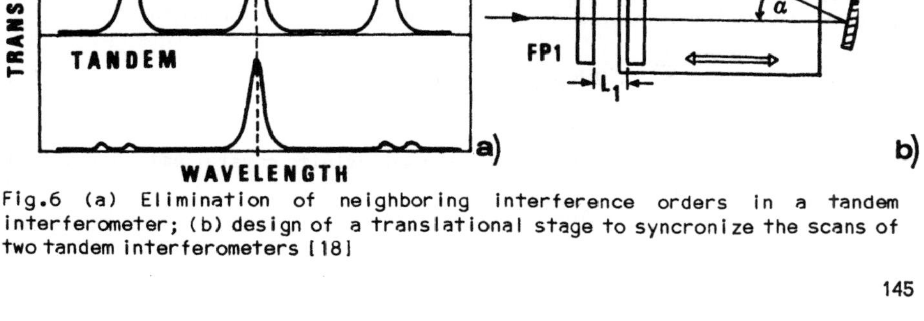

Fig.6 (a) Elimination of neighboring interference orders in a tandem interferometer; (b) design of a translational stage to syncronize the scans of two tandem interferometers [18]

where n_1 and n_2 are integers. Light will be transmitted through the combination only if $\lambda_1=\lambda_2$. Because $L_1 \neq L_2$, contiguous transmission peaks are suppressed and the FSR may be increased by a factor of 10 to 20 over that of a single interferometer.

Fig.6a illustrates the elimination of neighboring interference orders. Only small ghosts of neighboring orders remain. To scan the transmitted wavelength the mirror spacings L_1 and L_2 are incremented by δL_1 and δL_2

$$\delta\lambda_1 = \frac{2\,\delta L_1}{n_1} \quad ; \quad \delta\lambda_2 = \frac{2\,\delta L_2}{n_2} \tag{16}$$

with the condition $\delta\lambda_1=\delta\lambda_2$. This condition, together with (15) and $\lambda_1=\lambda_2$, requires

$$\frac{\delta L_1}{\delta L_2} = \frac{L_1}{L_2} \tag{17}$$

The scan syncronization is difficult to be achieved in practice. For the kind of experiments discussed below two methods have turned out to be successful. DIL et al. [20] used electronic feedback to lock the scan together of two piezoelectrically scanned interferometers. SANDERCOCK [18] realised a mechanical scan synchronization of two interferometers. In this set up, illustrated in Fig.6b, the scanning mirrors of two interferometers are mounted on the same scanning stage, one with the mirror axis parallel to the scan direction, the other offset by an angle α. For a spacing change δL_1 of the former interferometer, the spacing change of the latter is $\delta L_2=\delta L_1 \cos\alpha$. From (17) the syncronization condition is simply given by $L_2=L_1 \cos\alpha$.

6. Normal Modes of Vibration in a Semiinfinite Medium and in a Supported Film

Brillouin scattering probes acoustic phonons in the long wavelength limit, where the theory of elasticity is fully adequate to describe the dispersion relations and the vibration amplitudes of the modes. Surface waves in solids are reviewed in detail in many books and articles, e.g. in AULD [21], FARNELL [22] and FARNELL and ADLER [23]. In the classical literature emphasis is placed on genuine surface modes, localised at a surface or in a thin supported film. More recently the measurement of Brillouin spectra, directly related to the phonon density of states, has rised new interest upon the so-called "mixed modes" (combination of evanescent waves and bulk modes) which form a continuous spectrum. The pseudo Rayleigh waves well known in acoustics [22] actually belong to this part of the spectrum. In this section a derivation of the surface acoustic modes is presented referring to both the discrete and the continuous spectra.

Let us consider a semiinfinite medium in the half space z<0 with the surface in the x-y plane. The surface coordinate is labelled by $\mathbf{R}$. ρ is the mass density of the medium and L is the side of the surface. Because of the translational symmetry in the surface plane, the component of the phonon wavevector Q parallel to the surface is real, whereas the normal component q_z can assume either real or complex values. The displacement field can be expressed in normal modes

$$\mathbf{u}(\mathbf{R},z,t)=\frac{1}{L\rho^{1/2}}\sum_{\mathbf{Q},\Omega_n} e^{i\mathbf{Q}\cdot\mathbf{R}}\,\mathbf{w}(z,\mathbf{Q},\Omega_n)\,A(\mathbf{Q},\Omega_n,t) \tag{18}$$

where $\mathbf{w}$ is the polarisation vector in the semiinfinite crystal and A the normal coordinate defined as

$$A(\mathbf{Q},\Omega_n,t)=\left[\frac{h}{4\pi\Omega_n}\right]^{1/2}[a^+(-\mathbf{Q},n)\exp(-i\Omega_n t)+a(\mathbf{Q},n)\exp(i\Omega_n t)]. \qquad (19)$$

a^+ and a are the creation and the annihilation operators respectively. The sum in (18) is over all the phonon states of parallel wavevector Q and energy Ω_n. The polarisation vector **w** can be found with the **partial wave method** [21] by assuming that **w** is written as a linear combination of the polarisation vectors **e** of the infinite medium. This assumption is perfectly valid in the long wavelength limit, where the bulk phonons "feel" the surface only as a boundary condition. , being not affected by microscopic changes of the interatomic forces close to the surface. The polarisation vector **e** of the infinite medium is solution of the equation of elasticity [21] written for the Fourier component $(\mathbf{Q},q_z)$

$$[\rho\,\Omega^2\,\delta_{\alpha\beta}-c_{\gamma\alpha\beta\delta}\,q_\gamma\,q_\delta]\,e_\beta(\mathbf{Q},q_z,\Omega)=0. \qquad (20)$$

The $c_{\gamma\alpha\beta\delta}$ are the components of the elastic tensor referred to the surface frame. Usually the above equation is solved at a fixed wavevector $(\mathbf{Q},q_z)$ and gives three eigenvalues Ω_n corresponding to as many acoustic branches. Instead, in our case, we fix **Q** and Ω in order to find the (6 in general) values of q_z which satisfy the secular problem derived from [20]. In the following the solutions q_z are labelled by λ. For bulk modes the $q_z{}^\lambda$ are real and represent transverse and longitudinal travelling waves. In surface problems,however, we have to consider transverse and longitudinal cut-off frequencies (Ω_T and Ω_L respectively) below which [20] allows solutions with complex $q_z{}^\lambda$. This corresponds to evanescent waves localised at the surface and decaying into the bulk, i.e. with $\mathrm{Im}(q_z{}^\lambda)<0$. Exponential waves with $\mathrm{Im}(q_z{}^\lambda)>0$ must be discarded because they grow exponentially inside the medium and do not satisfy the required boundary conditions. By considering for simplicity a single transverse wave (the other one is decoupled in many practical applications) we have the following cases of superposition of partial waves:

a) $\Omega<\Omega_T<\Omega_L$
Allowed $q_z{}^\lambda$: 1 transverse with $\mathrm{Im}(q_z{}^T)<0$
1 longitudinal with $\mathrm{Im}(q_z{}^L)<0$.
A solution in this frequency range represents obviously a surface mode (**Rayleigh wave**) because its amplitude goes to zero for $z\to-\infty$.

b) $\Omega_T<\Omega<\Omega_L$
Allowed $q_z{}^\lambda$: 2 transverse with $\mathrm{Im}(q_z{}^T)=0$ (bulk waves with $q_z{}^T,-q_z{}^T$)
1 longitudinal with $\mathrm{Im}(q_z{}^L)<0$.
A solution in this frequency range is a superposition of bulk and localised modes. For this reason these modes are called "mixed modes".

c) $\Omega_T<\Omega_L<\Omega$
Allowed $q_z{}^\lambda$: 2 transverse with $\mathrm{Im}(q_z{}^T)=0$ (bulk waves with $q_z{}^T,-q_z{}^T$)
2 longitudinal with $\mathrm{Im}(q_z{}^L)=0$ (bulk waves with $q_z{}^L,-q_z{}^L$).
This solution is bulk-like,and the only effect of the surface is to modify the density of states through the boundary conditions.

The superposition of partial waves may be considered as the equivalent of a reflection problem at the boundary. Fig.7 shows the physical meaning of the scattering in the three cases.

The effect of the surface is to couple the modes so that the resulting displacement field satisfies the surface boundary condition. The superposition of partial waves can be written as

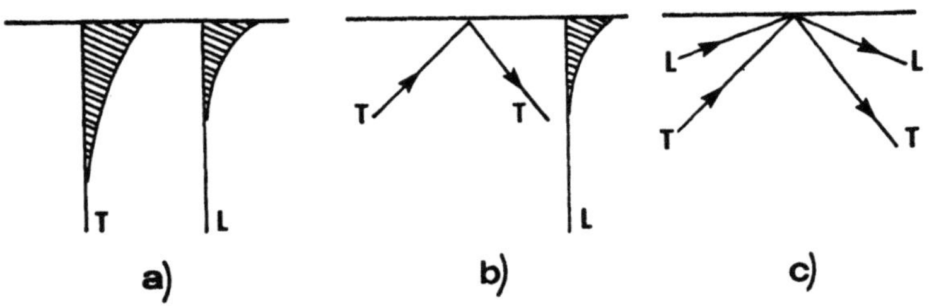

Fig.7 (a) superposition of two evanescent waves at the surface; (b) reflection of a T mode at the surface with coupling to an evanescent L mode; (c) reflection of coupled T and L waves

$$\mathbf{w}(z,\mathbf{Q},\Omega_n)=\sum_{\lambda} a(\mathbf{Q},q_z{}^{\lambda},\Omega_n)\exp(iq_z{}^{\lambda}z)\,\mathbf{e}(\mathbf{Q},q_z{}^{\lambda},\Omega_n) \qquad (21)$$

where the $a(\mathbf{Q},q_z{}^{\lambda},\Omega_n)$ are coefficients to be determined by imposing the boundary conditions. Usually [21] stress-free surface boundary conditions are imposed:

$$c_{z\beta\mu\nu}\,\eta_{\mu\nu}\Big|_{z=0}=0 \qquad (\beta=x,y,z) \qquad (22)$$

where the strain components $\eta_{\mu\nu}$ are defined by (3). (22) may be also written as

$$c_{z\beta\mu\nu}\,q_\mu w_\nu\Big|_{z=0}=0. \qquad (\beta=x,y,z) \qquad (23)$$

In the simple case considered above (coupling between one transverse and one longitudinal phonon branch) the condition (23) forms a set of two linear equations in the unknowns $a(\mathbf{Q},q_z{}^{\lambda},\Omega_n)$. With reference to the frequency ranges a), b), c) introduced above, we have three regimes of solution.

a) $\Omega<\Omega_T<\Omega_L$
(23) is an homogeneous set of two equations with two unknowns, because there are only two partial waves in the expansion (21). The solution corresponds to the frequency Ω_R of the Rayleigh wave. The spectrum is discrete.

b) $\Omega_T<\Omega<\Omega_L$
(23) becomes an inhomogeneous set of equations, the partial waves and the corresponding unknowns being three. We have a solution for every frequency in this range and the spectrum (of mixed modes) is continuous.

c) $\Omega_T<\Omega_L<\Omega$
Similar to case b), but the unknowns are four and hence there are two linearly independent solutions.

The above conclusions can be generalised to the case of two transverse branches which couple with the longitudinal one. Numerical solutions are required but for the isotropic case, where the implicit Stoneley-Viktorov equation for the Rayleigh wave is known [21] and the continuous spectrum of mixed modes has been calculated analytically [24-27]. A variety of solutions for the Rayleigh wave in different crystals is presented in the general references [20,21] where piezoelectric crystals are also considered. Numerical results obtained for the continuous spectrum of anisotropic cubic crystals have been reviewed by STEGEMAN and NIZZOLI [19]. As an example we present here a calculation of $|w_\alpha|^2/\Omega^2$ for transverse and longitudinal displacements, in the case of an isotropic solid.
Fig.8 shows both the Rayleigh wave and the continuous spetrum. Notice the typical behaviour of the transverse component that forms a sort of plateau between Ω_T and Ω_L while the longitudinal component is peaked close to Ω_L. Although the detail of

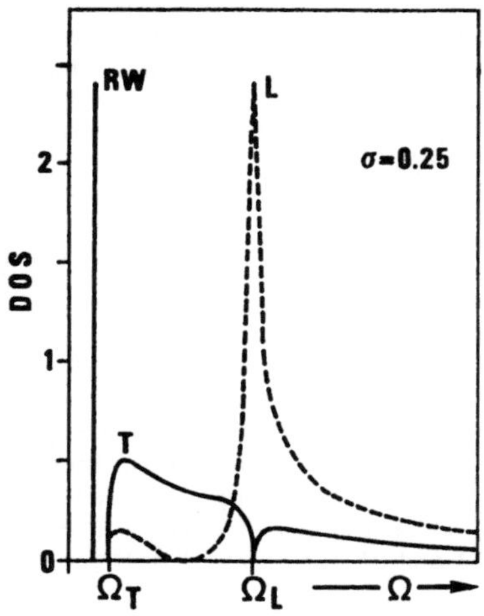

Fig.8 Density of states for shear vertical (T) and londitudinal (L) displacements in an isotropic solid with Poisson ratio $\sigma=1/4$. The vertical lines mark the Rayleigh wave.

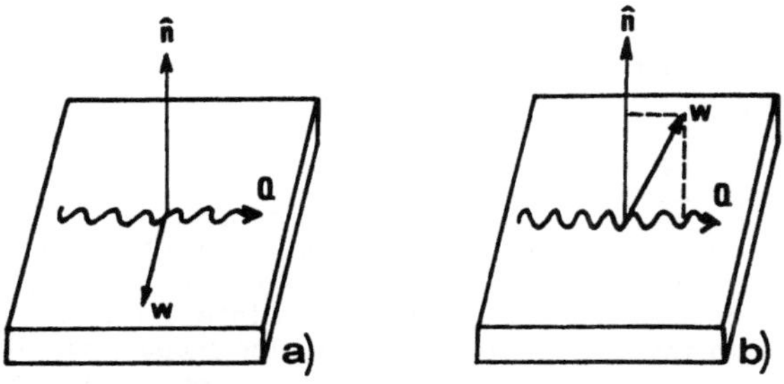

Fig.9 Polarisation **w** of Love (a) and Lamb (b) waves in a free plate

the density of states depends on the particular case, we will see later that the lineshapes shown in Fig.8 determine the general behaviour of the Brillouin cross section in opaque materials.

Many features are known of the vibrational modes of free and supported films [20,22], mainly derived from the early work in geophysics. Fig.9 shows Love modes (polarised normally to the sagittal plane) and Lamb modes (polarised in the sagittal plane) of a free plate. The sagittal plane is defined by **Q** and the surface normal $\hat{\mathbf{n}}$.

Generalised Love and Lamb modes exist in **supported films** when the transverse threshold frequency of the film is lower than the transverse threshold frequency of the substrate. In this case the appropriate boundary conditions are given by the continuity of displacements and stresses at the interface, in addition to zero stresses at the film-vacuum surface.

7. Theory of Surface Brillouin Scattering

Brillouin scattering from surface and bulk phonons in opaque media on reflection has been treated with different mathematical methods by many authors who have reached essentially the same conclusions [24-31]. Theoretical work on scattering from thin films on reflection and trasmission has also been reported [32-35]. Brillouin scattering using guided waves in dielectric films has been treated in [36-37].

We outline here the theory of Brillouin scattering in an opaque material. The generalisation to supported films, although rather complicated to be explicitly carried out, does not involve new concepts. The presence of the phonons in thermal equilibrium modifies the dielectric tensor in the following way

$$\epsilon_{\alpha\beta} = \begin{cases} \epsilon_0\delta_{\alpha\beta} + \delta\epsilon_{\alpha\beta} & z<0 \quad \text{(medium)} \\ \delta_{\alpha\beta} & z>0 \quad \text{(vacuum)} \end{cases} \qquad (24)$$

where ϵ_0 is the dielectric function of the medium in absence of phonons and $\delta\epsilon_{\alpha\beta}$ the dielectric function modulated by the phonons and given by (2). $\delta\epsilon_{\alpha\beta}$ can be expanded in normal coordinates by using the expression (18) for the displacement field. The scattering problem consists essentially in calculating the scattered field in vacuum by taking into account the **elasto optic coupling** in the medium and the **dynamical corrugation of the surface** induced by the phonons (**ripple effect**).

To fix the ideas, let us consider p(TM) incident light. In this case the derivation of the scattering electromagnetic field is more easly carried out in terms of the magnetic field **B**. The Maxwell equations written in terms of **B** can be obtained by expressing $(\varepsilon_{\alpha\beta})^{-1}$ in terms of ε_0 and $\delta\varepsilon_{\alpha\beta}$ [29] as follows

$$\varepsilon_0\left[\nabla^2 - \frac{\varepsilon_0}{c^2}\frac{\partial^2}{\partial t^2}\right]\mathbf{B} = -\nabla\times(\delta\varepsilon\ \nabla\times\mathbf{B}) \qquad z<0 \tag{25}$$

$$\left[\nabla^2 - \frac{1}{c^2}\frac{\partial^2}{\partial t^2}\right]\mathbf{B} = 0. \qquad z>0 \tag{26}$$

The solution of (25) and (26) to zeroth order in the displacement field simply gives the well known Fresnel coefficients for refraction of light at the surface of a dielectric medium. We are interested here in a solution for **B** to first order in the displacement field **u**. The proper solution will be the sum of a particular solution of the inhomogeneous Maxwell equation (25) and of the general solution of the homogeneous equations (obtained with $\delta\varepsilon_{\alpha\beta}=0$) to satisfy the boundary conditions:

a) continuity of **B** at the surface
b) continuity at the surface of the tangential component of the electric field **E**, e.g. of $\varepsilon^{-1}\nabla\times\mathbf{B}$.

The electromagnetic boundary conditions are to be written on the corrugated surface, whose profile is just $u_Z(\mathbf{R},0,t)$ as given by (18). By assuming a small corrugation, an expansion in u_Z can be made and only first order terms can be retained. The general form of the ratio of the scattered field in vacuum $\mathbf{B_K}$ to the incoming field $\mathbf{B_{K_0}}$ is given by

$$\left|\frac{\mathbf{B_K}}{\mathbf{B_{K_0}}}\right| = \left|\frac{1}{L\rho^{1/2}}\sum_n \int \frac{dt}{2\pi}\exp[i(\omega-\omega_0)t]\, A(\mathbf{K}-\mathbf{K}_0,\Omega_n,t)\, D^{t\to r}(\mathbf{K}-\mathbf{K}_0,\Omega_n)\right| \tag{27}$$

where **K** and $\mathbf{K}_0$ are the parallel wavevectors of the scattered and incoming light respectively and $D^{t\to r}$ is a scattering amplitude which depends on the polarisations of the scattered (r) and incoming light (t) and on the scattering geometry. Long calculations are needed to explicitly find the expression of $D^{t\to r}$, expecially for a supported film. Analytical expressions may be found in the general references given at the beginning of the section.
Once the scattered field in vacuum (27) is known, the thermal average of the scattered Poynting vector may be evaluated

$$S = \left\langle \frac{c}{8\pi}\mathrm{Re}[E(r,t)\times B(r,t)^*]\right\rangle. \tag{28}$$

From (27) and (28) we get the for $t\to r$ scattering

$$S = \frac{c}{8\pi L^2\rho}|\mathbf{B_{K_0}}|^2\,\mathrm{Re}\sum_{\mathbf{K},\omega,n} \mathbf{k}\left|D^{t\to r}(\mathbf{K}-\mathbf{K}_0,\Omega_n)\right|^2$$

$$\times\left[\int\frac{dt}{2\pi}\exp[i(\omega-\omega_0)t]\,\langle A(\mathbf{K}-\mathbf{K}_0,\Omega_n,t)A^+(\mathbf{K}-\mathbf{K}_0,\Omega_n,0)\rangle\right] \tag{29}$$

where **k** is the total wavevector of the scattered light. Since the modes of the perfect crystal are harmonic, the integration in (29) gives

$$\frac{h}{4\pi\Omega_n}\left[\delta[\Omega_n-(\omega-\omega_0)]N(\Omega_n)+\delta[\Omega_n+(\omega-\omega_0)][N(\Omega_n)+1]\right] \tag{30}$$

which includes anti-Stokes and Stokes processes. $N(\Omega)$ is the Bose occupation number. Let $N^*=N$, $N+1$ for anti-Stokes and Stokes events respectively. We use now the transformation

$$\frac{1}{L^2}\sum_{\mathbf{K}} \rightarrow \frac{1}{4\pi^2}\left[\frac{\omega^2}{c^2}\right]\int\cos\theta\, d\Omega_a \tag{31}$$

where $d\Omega_a$ is the differential solid angle. θ_0 is the angle between the surface normal and the impinging laser beam, while θ is the corresponding angle for the scattered light. Projecting (29) on the z-axis and dividing by the projected Poynting vector of the incident field [38] we get the differential cross-section

$$\left[\frac{d^2\sigma}{d\Omega_a d\omega}\right]_{t\rightarrow r}=\frac{\omega^2 h}{(2\pi)^3c^2}\,\frac{\cos^2\theta}{\cos\theta_0}\,\frac{N^*(|\omega-\omega_0|)}{2\rho|\omega-\omega_0|}\sum_n\left|D^{t\rightarrow r}(\mathbf{K}-\mathbf{K}_0,\Omega_n)\right|^2\delta[\Omega_n-|\omega-\omega_0|]. \tag{32}$$

The summation index n in (32) is a discrete one in the case of localised surface modes, as Rayleigh and Lamb modes. For the continuous spectrum the expression (32) is still valid with the substitution

$$\sum_n \delta[\Omega_n-|\omega-\omega_0|] \rightarrow \left[\frac{dn}{d\Omega_n}\right]_{\Omega_n=|\omega-\omega_0|}\sum_j \tag{33}$$

where on the right-hand side j runs over the linearly independent solutions of the continuous spectrum ($j=1,2,3$ in the most general case for $\Omega>\Omega_L$).

Before ending the section it is useful to write the cross-section for TM→TM in-plane scattering in the limit of large ε_0. As we will see in the next section many surface Brillouin scattering experiments in opaque materials can be interpreted on the basis of this formula:

$$\left[\frac{d^2\sigma}{d\Omega_a d\omega}\right]_{p\rightarrow p}\simeq\frac{\omega^4 k_B T}{2\pi^2c^4\rho(\omega-\omega_0)^2}\,\frac{\cos^2\theta\cos\theta_0}{[(\cos\theta+\varepsilon_0^{-1/2})(\cos\theta_0+\varepsilon_0^{-1/2})]^2}$$

$$\times\sum_n |\varepsilon_0^{-3/2}(k_{1111}-k_{1133}c_{1133}/c_{3333})\,(\sin\theta-\sin\theta_0)\,w_L(0,\mathbf{Q},\Omega_n)/2$$

$$+(1-\sin\theta\sin\theta_0)\,w_T(0,\mathbf{Q},\Omega_n)|^2\,\delta[\Omega_n-|\omega-\omega_0|]. \tag{34}$$

Notice that the ripple contribution is proportional to the transverse displacement component w_T while, at least in this approximation, the elasto optic contribution is simply proportional to the longitudinal displacement component w_L. It must be also remarked that (34) is only suitable for surface Brillouin scattering, the bulk resonances being washed out by the approximation $\varepsilon\rightarrow\infty$.

8. Experimental Results in Brillouin Scattering from Surface Acoustic Waves

Three **experimental geometries** have been used to study surface and thin film acoustic phonons, as well as bulk phonons near a surface. They are schematically shown in Fig.10.

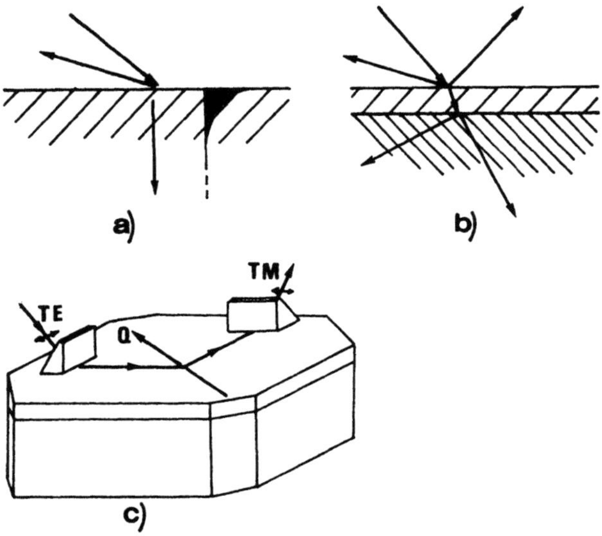

Fig.10 The surface Brillouin scattering geometries studied to date. The explanation is given in the text

We have the following possibilities.

a) Scattering on reflection from a surface [39]. This technique works well for opaque materials as metals and semiconductors. Brillouin scattering occurs from the surface and from within the optical skin depth. The same approach can be used also for supported films, either the film or the substrate (or both) being opaque. In this geometry the scattering from bulk waves is usually weak or even absent, as in metals.

b) Light is incident onto a trasparent film on a transparent substrate and the Brillouin scattered light is detected either in reflection or in trasmission. The incident light does penetrate into the substrate and hence intense Brillouin scattering from bulk waves can occur. This kind of experiment is possible if the acoustic properties of the film and substrate media are very different and the Brillouin spectra are well separated. Polimer films on glass substrate have been investigated with this technique [20].

c) This geometry relies on electromagnetic waves which are guided parallel to the surface [36]. For example, for an integrated optics waveguide, a dielectric film of higher refractive index than the two surrounding media is required. The same incident field continuosly resample the phonon modes and therefore the scattering phenomenon is enhanced. Since the optical field decays exponentially into both media, Brillouin scattering can be studied in the surface region of the bounding media as well as in the film.

In the last decade many experimental results have been accumulated. Our purpose is not to review the subject [18,19] but rather to present a few representative results which have been obtained. In Fig.11 the Brillouin spectra of aluminium is presented. It is shown in the inset that the data have been satisfactory explained in terms of the ripple effect [24]. In fact in a metal $|\varepsilon_0|$ is large so that in (34) the elasto-optic term is completely washed out and the ripple term dominates. The latter term is proportional to the shear horizontal transverse displacement component w_T. Therefore the cross-section is quite similar to the density of states for vertical displacements shown in Fig.8.
In particular the dip of the continuous spectrum for $\Omega=\Omega_L$ allows us to measure directly the longitudinal velocity v_L. The transverse velocity v_T may be

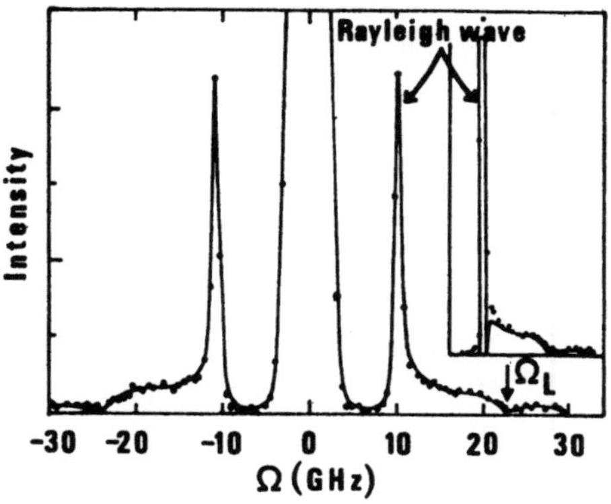

Fig.11 Measured backscattering Brillouin spectrum for polycrystalline aluminium [39]. The inset shows the comparison between theory [24] and experiment. After SANDERCOCK [18]

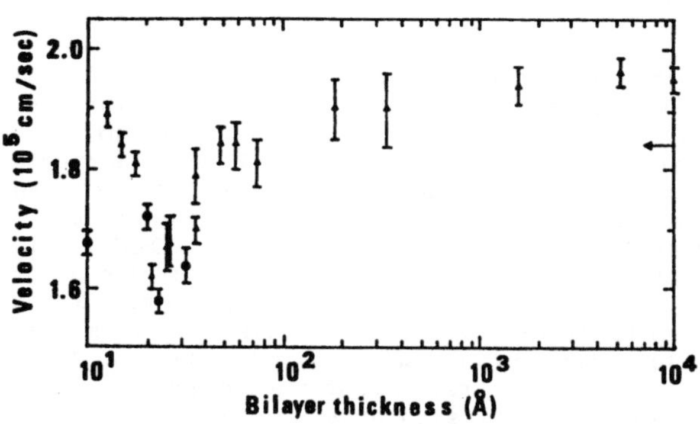

Fig.12 Surface wave velocity measured by Brillouin scattering from Nb-Cu superlattices as a function of bilayer thickness. The arrow indicates the calculated velocity for films with bilayer thickness $\leqslant$100 Å. After KUENY et al. [40]

evaluated from the velocity of the surface mode, that contributes to the spectrum with a well-defined peak.

An interesting application of Brillouin scattering to highly absorbing media has been developed in the field of metallic superlattices [40]. As shown in Fig.12 the Rayleigh velocity measured from the Brillouin spectra in a Nb-Cu superlattice exhibits a marked softening of the shear modulus for a bilayer thickness of about 20 Å. It is interesting to notice that, although Brillouin scattering probes macroscopic properties, in this particular case it reveals a microscopic effect,due to the presence of periodically repeated interfaces.

The crystalline surfaces of silicon and gallium arsenide have been investigated [39]. Fig.13 shows the backscattering spectrum taken for (110) GaAs by John Sandercock. It may be diveded in three parts: the Rayleigh wave gives rise to the narrow low-frequency spectral line; the mixed modes discussed in section 6 contribute to the spectral features in the frequency range 15-30 GHz; two bulk resonances (T), (L) are present at higher frequencies exhibiting broad lineshapes. The calculated cross-section in Fig.13 has been fitted to the experimental data having as the only free parameters the elasto-optic

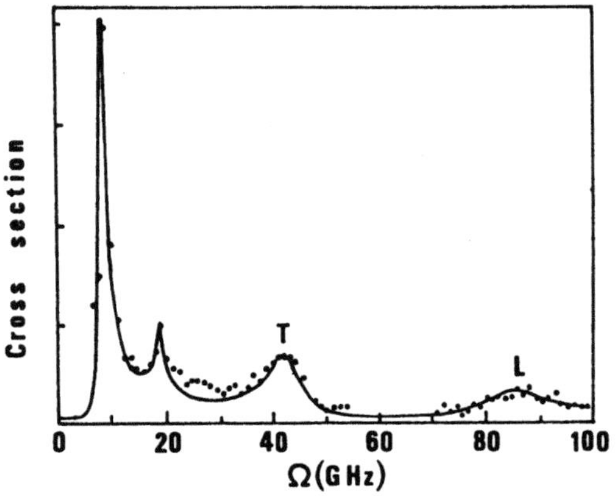

Fig.13 Experimental and calculated Brillouin spectra for the (110) surface of GaAs with Q along [1$\bar{1}$0]. The elasto-optic constants are k_{11}=42.5, k_{12}=51 and k_{44}=25.5. After MARVIN et al. [30]

coefficients $k_{\alpha\beta}$ of GaAs, which have been determined in this way. Notice that an absolute determination of the elasto-optic coeffients may be possible via surface Brillouin scattering in opaque material, because the spectra contain at the same time comparable contributions from the elasto-optic and the ripple couplings. In particular, in Fig.13 the Rayleigh wave scatters mainly via the ripple effect, while the bulk resonances are completely due to the elasto-optic effect. It is of special interest the narrow peak located at $\Omega=\Omega_L\simeq 19$ GHz. It is mainly due to the elasto-optic coupling of the light with the continuum of longitudinal modes already shown in Fig.8 and which contributes to the scattering intensity according to the formula (34). From (34) it is clear that the longitudinal modes close to Ω_L show up in the spectra when ε_0 is not too large and the tensor quantity $k_{1111}-k_{1133}c_{1133}/c_{3333}$, referred to the surface scattering frame, is large. The (110) surface of GaAs turns out to satisfy these conditions. A typical backscattering spectrum for an opaque supported film is shown in Fig.14. Many spectral lines are present, representing scattering from Rayleigh, Sezawa, Lamb modes plus smaller contributions from the mixed modes. Due to the presence of a metallic coating, the computed cross-section has been calculated by

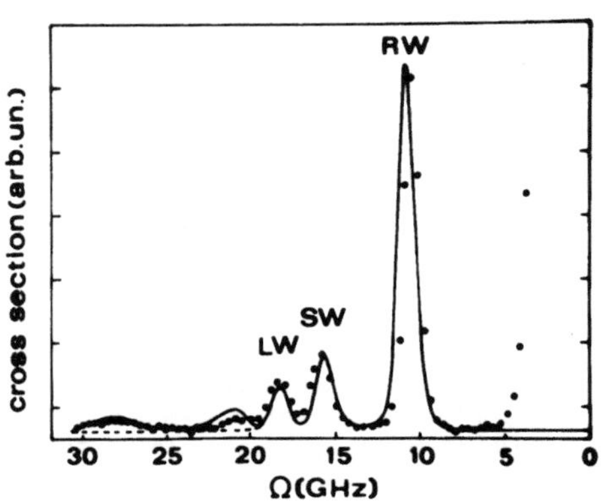

Fig.14 Experimental and computed Brillouin cross-section for a 2000 Å film of aluminium deposited on silicon. The broken line shows the spectrum without the contribution of the continuum of mixed modes. After SANDERCOCK et al. [42]

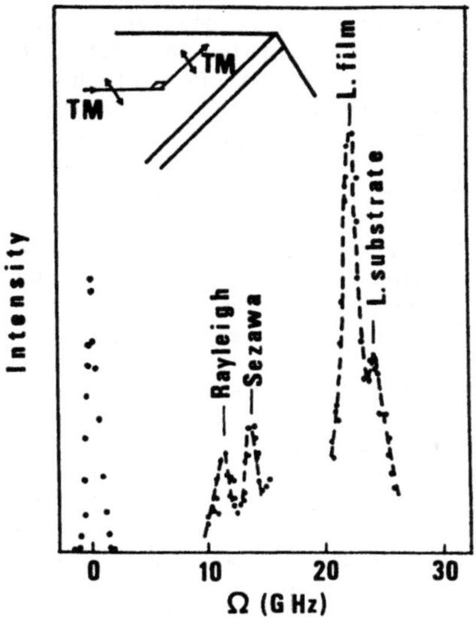

Fig. 15 The Brillouin spectrum obtained from a thin glass film on a transparent substrate. Both the incident and scattered light were propagated as TM polarised guided waves. The various peaks observed are labelled. After ROWELL et al. [43]

retaining only the ripple term at the film-air surface,and it turns out to reproduce quite satisfactory the scattering strength of the various waves. Measurements and cross-section calculations for transparent films on opaque substrates in backscattering geometry have been also reported. Results for SiO_2 on silicon [41] demonstrated that the terms in the cross-section due to the ripple scattering at both interfaces and to the elasto-optic coupling are equally important, so that strong interference effects may occur.

An example of Brillouin scattering in guided wave geometry (c) is shown in Fig.15. These measurements [43] gave the first experimental estimate of the surface phonon lifetime via Brillouin scattering. Peaks due to surface phonons (Rayleigh, Sezawa) are visible, together with bulk resonances. For the film thickness used, the Rayleigh wave amplitude is negligible at the film-substrate interface and the mode is essentially that associated with the film material. Instead,the Sezawa wave is guided by both interfaces. It was found that, when compared to the same acoustic wavevector, the lifetime of the Rayleigh wave was comparable with those of the bulk phonons in the film. The lifetime of the Sezawa mode was found to be substantially shorter, presumably due to scattering roughness and material imperfections at the film-substrate interface.

9. Light Scattering from Surface Waves Involving Surface Plasmons

Many contributions to these Proceedings (for example by F. Abeles, R.F. Wallis) extensively deal with surface plasmon polaritons. In particular, it is well known that surface plasmon polaritons (or surface electromagnetic waves: SEW) can be excited at the air-metal interface of an air-metal-prism system by using the attenuated total reflection (ATR) method in the so-called Kretschmann configuration [44]. The electric field associated with these polaritons is strongly enhanced with respect to the incident field [44]. This property has been widely applied to several kinds of light spectroscopies, as surface enhanced Raman scattering and non-linear phenomena [45].

FUKUI et al. [46] have derived a theory of enhanced Brillouin scattering from metal films involving surface plasmons. These authors calculated the scattered field in backscattering geometry (on the prism side in the ATR Kretschmann configuration) varying the incidence and scattering angles around the condition of surface plasmon coupling. Bulk and surface acoustic waves in thermal equilibrium in the metal-prism system were assumed. Two main conclusions were drawn. First the **enhancement** obtained for scattering from both bulk and surface phonons when surface plasmon are involved is typically 10^3 over the usual backscattering case. Then the enhancement is more effective for the **ripple scattering**, the elasto-optic coupling in the metal film being weak. From the experimental point of view there has been no attempt to verify the theory of FUKUI et al. [46], neither to take advantage from the enhancement to obtain a stronger Brillouin signal.

In contrast, both experimental [47] and theoretical [48] work has been carried out in the special case of light scattering in Kretschmann configuration when a surface acoustic wave (SAW) acts as a **grating to produce diffracion.** TALAAT et al. [47] excited a SAW in a 706 Å thick silver film deposited on a $LiNbO_3$ plate, placed in optical contact with the base of a glass prism (Fig.16a). The intensities of the reflected and transmitted beams were measured versus the incidence angle, together with the intensities of the backward and forward (first-order) diffracted beams. Coupling of volume electromagnetic waves (VEW) and SEW, and viceversa, occurs when the usual ATR condition $(\omega/c)n_p \sin\theta_{ATR}=K_S$ is verified, where n_p is the refractive index of the prism, K_S is the SEW wavevector, ω is the frequency of the light and θ_{ATR} the ATR incidence angle. The periodic corrugations at the two metal film interfaces (of amplitude ≃ 5 Å) produce an appreciable first-order diffraction of VEW into VEW and VEW into SEW, and viceversa, whenever the momentum and energy conservation is satisfied. As a consequence, prism coupling (Kretschmann) processes and diffraction events by a SAW coherently interfere and give rise to maxima and minima in the diffracted light. A crucial role is played by the magnitude and phase differences of the SAW at the two metal film interfaces. In the experiment considered, the SAW is almost exactly the same at any depth in the film. Therefore, due to the different form of the scattering intensity for backward and forward diffraction, minima and maxima

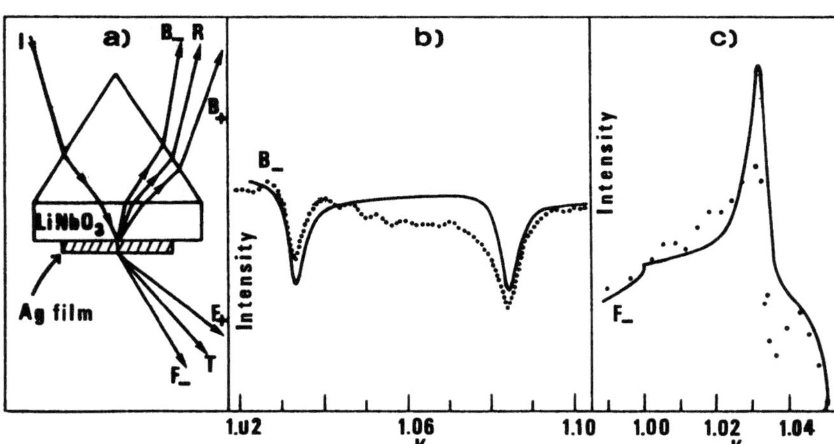

Fig.16 (a) Combined Kretschmann prism and diffraction from a SAW dynamic grating. Incoming (I), reflected (R), transmitted (T), forward diffracted (F_-,F_+) and backward diffracted (B_-,B_+) light; (b) experimental and calculated B_- spectrum; (c) experimental and calculated F_- spectrum. The parallel component of the wavevector in the prism $K_{\parallel}$ is expressed in unit of ω/c. After MARVIN et al. [48]

are obtained respectively when three-wave mixing of VEW, SEW and SAW occurs. This is shown in the examples given in Fig.16b and Fig.16c. Minima and maxima take place when the parallel wavevector $K_{\parallel}$ satisfies the ATR condition, plus or minus a SAW wavevector.

Finally, it is worth mentioning that the same kind of experiment discussed above has been very recently performed also with the thermally excited SAW in the silver film [49]. The forward diffracted light has been analysed and an enhancement has been found in agreement with the theory [48]. The ratio of the Brillouin signal in ATR condition with respect to the usual backscattering signal from the Rayleigh wave turns out to be about 25.

10. Surface Raman Scattering

This section is devoted to a short summary of the most important results obtained in the Raman scattering from surface optical phonons. Enhanced Raman scattering and non-linear interactions involving surface polaritons [45] are completely excluded from this brief review.

In surface Brillouin scattering, studied in sections 6 to 8, we considered the coupling between light and acoustic waves. In that case the interaction does not modify the dispersion relations themselves, because there is no mode crossing between the surface elecromagnetic wave and the acoustic phonons, due to the great difference in the propagation velocity (light versus sound). On the contrary, the surface optic lattice vibrations are coupled in the long wavelength regime to the surface electromagnetic waves, giving rise to surface polariton branches [50]. The form of these branches depends on the geometry of the sample. In any case, the dielectric permittivities of the media must obey certain conditions to allow surface polaritons to propagate [50]. We may distinguish three cases:

a) single interface, typically a semi-infinite medium. Only a single branch of surface polaritons is expected, extending in the frequency range between the tansverse (Ω_T) and the longitudinal (Ω_L) phonon frequencies.
b) double interface geometry, typically a film of thickness h deposited on a semiinfinite substrate. In the limit of a thick film two branches are found, with a dispersion depending on the properties of the two non-interacting interfaces. In a thin supported film, the two polaritons interfere and the mode frequencies move further apart, giving rise to an upper mode (UM) and to a lower mode (LM) whose separation depends on the film thickness and the dielectric properties of the media. The UM and the LM exist in the gap between Ω_T and Ω_L of the supported film.
c) The modes previously considered are localised at the boundaries. We may also consider guided-wave polaritons decaying exponentially outside a supported film and exhibiting an oscillatory behaviour across the film [51]. Many branches of guided wave exist outside the gap defined by Ω_T and Ω_L. The number of modes increases with $\mathbf{Q}.\mathbf{h}$ and in the limit of $\mathbf{Q}.\mathbf{h}\to\infty$ two continua are generated. Notice the analogy with the Lamb modes of a slab discussed in section 6. Guided-wave polaritons are either TM or TE polarised.

The theory of surface Raman scattering has been developed during the 70's and is well established. Essentially the light scattering cross section of surface polaritons is derived in the linear response theory by evaluating the electromagnetic field generated by the polarisation due to the coupling between the light and the crystal excitations. The resulting cross-section formulae are long and complicated and are not given here. The reader may find theoretical references in the recent review paper of USHIODA and LOUDON [50]. A typical experimental geometry for (Stokes) Raman scattering is depicted in Fig.17. The same conservation equations as in Brillouin scattering (section 3) hold for

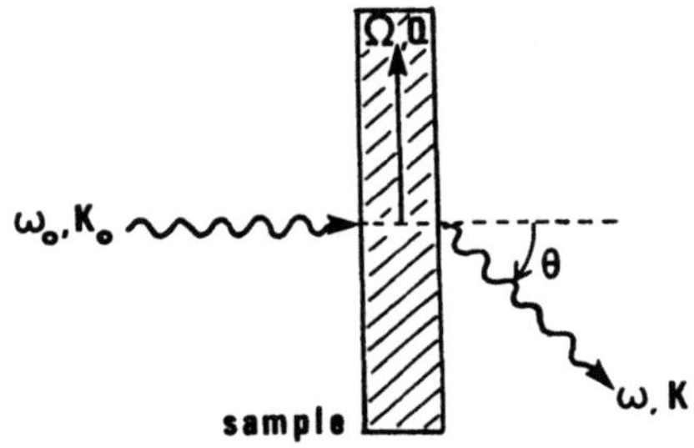

Fig.17 Experimental geometry for surface polariton scattering

Raman scattering. The kinematics of the scattering is that shown in Fig.2. In Raman scattering the phonon frequency Ω is not negligible compared with ω_0, so that an experiment detects surface polaritons not at fixed parallel wavevector **Q** (as in Brillouin scattering) but rather according to the kinematical relation

$$\frac{c^2Q^2}{\varepsilon\omega_0^{\,2}} = \frac{\Omega^2}{\omega_0^{\,2}} + 4\left[1-\frac{\Omega}{\omega_0}\right]\sin^2\frac{\theta}{2} \tag{35}$$

where ε is the dielectric function of the medium. By assuming ε to be constant, [35] defines for any scattering angle θ a curve in the Ω,Q plane. Therefore by changing θ, different regions of the polariton dispersion relation can be probed.

Usually the most interesting region for surface Raman scattering corresponds to small angle scattering ($\theta<4°$). As it can be seen in Fig.17 small angle scattering means forward scattering. Actually it has been shown theoretically [52] that the intensity for forward scattering overcomes by a factor of 10^2 to 10^3 the intensity for backward scattering. Therefore only the forward geometry is used for surface Raman scattering.

Measurements representative of the three cases a), b), c) mentioned above are now presented. For what concerns the case a) Fig.18 shows the room-temperature

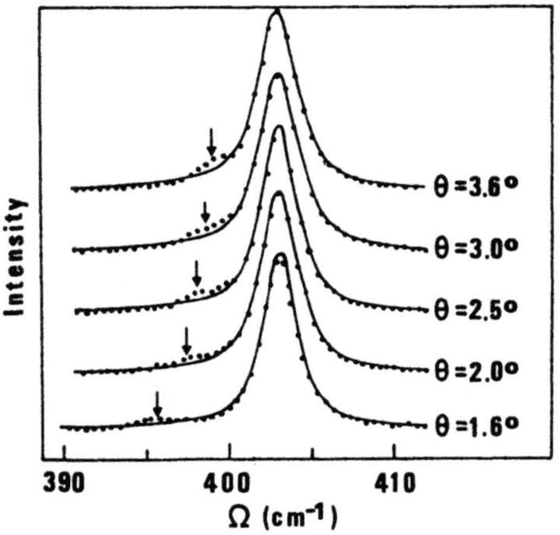

Fig.18 Raman spectra of surface polaritons in a (111) slab of GaP (20 µm). Data collected using the 5145 Å line of an argon ion laser. After VALDEZ and USHIODA [54].

Raman spectra of a 20 μm slab of GaP for several scattering angles in the range 1.6°-3.6°. The large peak at 403 cm^{-1} is the LO phonon peak of bulk GaP and the small peak indicated by an arrow is the surface polariton peak. It is apparent that quite a sophisticated apparatus is needed to detect the feeble surface signal, approximately 2÷3 photon counts per second. Thus the main effort is directed towards reduction of the elastically scattered light, minimizing the photomultiplier dark noise and maximizing the Raman signal. A long collection time is needed, which requires a spectrometer interfaced to a minicomputer as the experiment controller and data accumulator. Such a system has been developed by USHIODA et al. [53]. The surface polariton peak of Fig.18 shifts to lower frequencies for smaller scattering angles. The measured surface polariton dispersion derived from Fig.18 is in good agreement with the theoretical calculation based on the known dielectric data of GaP. A single smooth polariton branch is obtained.

Typical Raman spectra obtained for a supported film (case b) are shown in Fig.19a. As it is expected, the backward spectrum shows only bulk features due to creation of bulk transverse and longitudinal optical phonons. The forward spectrum contains the surface contribution. The LM peak is well separated from the bulk peak [55]. The UM is not directly visible because is very close to the LO bulk frequency. By subtracting the backward spectrum from the forward spectrum also the UM peak clearly appears. The subtracting procedure requires spectra measured simultaneously and in the same experimental conditions, so that a suitable scattering geometry based on the use of moving mirrors has been designed for this purpose [56]. The experimental data for the system air/GaAs/sapphire obtained with the subtraction procedure are shown in Fig.19b. When the air is substituted with benzene the UM lowers and separates from the LO bulk frequency. Therefore its detection is made easier and more measurements are possible, as shown in Fig.19b.

The guided-wave polaritons (case c) were also investigated experimentally. Figure 20 shows the mode dispersion for the guided-wave branches below the TO phonon frequency in a 30 μm GaP slab. Thicker samples (e.g. 125 μm) exhibit a continuum of modes instead of a discrete spectrum. USHIODA and LOUDON [50] also report measurements of guided-wave polaritons above the LO phonon frequency.

Also, the effects of surface roughness on the propagation of surface polaritons have been studied. The surface roughness changes both the linewidth

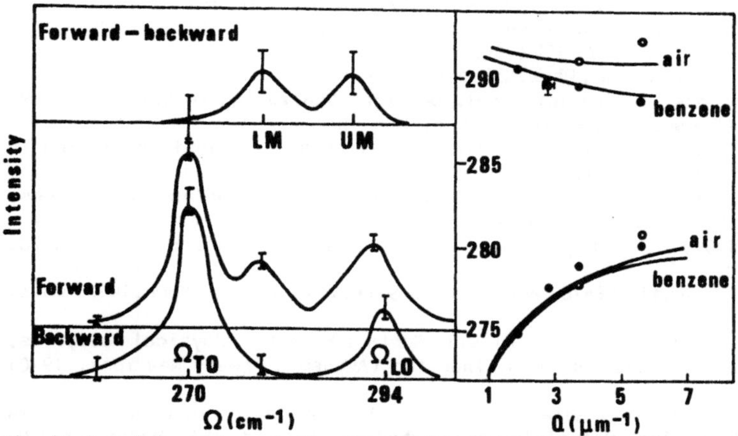

Fig.19 (a) Extraction of the UM and LM peaks by subtracting the forward and backward spectra. Film of 2500 Å of GaAs on sapphire; (b) Dispersion curves for the UM and LM in air/GaAs/sapphire (o) and benzene/GaAs/sapphire (•). After PRIEUR and USHIODA [56]

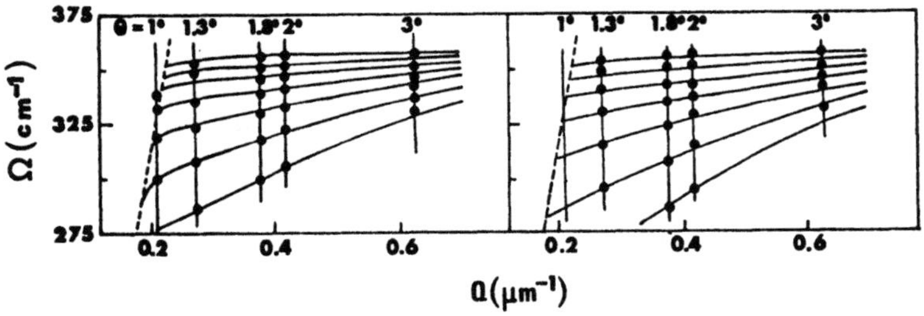

Fig.20 Experimental dispersion curves of guided-wave polaritons in a 30 µm GaP slab. After USHIODA and LOUDON [50]

and the frequency of the peaks in the Raman spectra of surface phonon polaritons [50].

References

1. A. Einstein: Ann. Physik (Germany) **33**, 1275 (1910)
2. L. Brillouin: Ann. Phys. (Paris) **17**, 88 (1922)
3. R.M. White and F.M. Voltmer: Appl. Phys. Lett. **7**, 314 (1965)
4. G.I. Stegeman: J. Appl. Phys. **47**, 1712 (1976); IEEE Trans. Sonics and Ultrason. **SU-23** (1976)
5. L.D. Landau and G. Placzek: Phys. Z. Sovietunion **5**, 172 (1934)
6. R.D. Mountain: Rev. Mod. Phys. **38**, 205, (1966)
7. D.F. Nelson and M. Lax: Phys. Rev. Lett. **24**, 379 (1970)
8. L.D. Landau and E.M. Lifshitz : **Electrodinamics of Continuous Media** (Pergamon Press, New York, 1958) chap. 14
9. H.Z. Cummins and P.E. Schoen: "Linear Scattering from Thermal Fluctuations" in **Laser Handbook**, Ed. F.T. Arecchi and O.E. Schulz-DuBois (North-Holland, Amsterdam, 1972) p. 1031
10. R. Vacher and L. Boyer: Phys. Rev. **B6**, 639 (1972)
11. A.S. Pine: Phys. Rev. **B5**, 3003 (1972)
12. J.R. Sandercock: Phys. Rev. Lett. **28**, 237 (1972)
13. G. Dresselhaus and A.S. Pine: Solis St. Commun. **16**, 1001 (19775)
14. A. Dervisch and R. Loudon: J. Phys. C 9, L669 (1976)
15. M. Hercher: Appl. Opt. **7**, 951 (1968)
16. P. Jacquinot: Rep. Prog. Phys. **23**, 268 (1960)
17. J.R. Sandercock: in **2nd Int. Conf. on Light Scattering in Solids**, Ed. M. Balkanski (Flammarion, Paris, 1969) p. 9
18. J.R. Sandercock: "Trends in Brillouin Scattering" in **Light Scattering in Solids III**, Ed. M. Cardona and G. Guntherodt (Springer, Berlin, 1982) p. 173
19. G.I. Stegeman and F. Nizzoli: "Surface Vibrations" in **Surface Excitations**, Ed. V.M. Agranovich and R. Loudon (North-Holland, Amsterdam, 1984) chap. 4
20. J.G. Dil, N.C.J.A. van Hijningen, F. van Dorst and R.M. Aarst: Appl. Opt. **20**, 1374 (1981)
21. B.A. Auld: **Acoustic Fields and Waves in Solids** (Wiley, New York, 1972) vols. 1&2
22. G.W. Farnell: "Properties of Elastic Surface Waves" in **Physical Acoustics**, vol.6, Ed. W.P. Mason and R.N. Thurston (Academic Press, New York, 1970) pag.109
23. G.W. Farnell and E.L. Adler: "Elastic Wave Propagation in Thin Layers" in **Physical Acoustics, Principles and Methods**, vol.9, Ed. W.P. Mason and R.N. Thurston (Academic Press, New York, 1972) pag.35
24. R. Loudon: Phys. Rev. Lett. **40**, 581 (1978)
25. R. Loudon: J. Phys. C **11**, 403 (1978)

26. K.R. Subbaswamy and A.A.Maradudin: Phys. Rev. **B18**, 4181 (1978)
27. N.L. Rowell and G.I. Stegeman: Phys. Rev. **B18**, 2598 (1978)
28. R. Loudon and J.R. Sandercock: J. Phys. C **13**, 2609 (1980)
29. A.M. Marvin, V. Bortolani and F. Nizzoli: J. Phys. C **13**, 299 (1980)
30. A.M. Marvin, V. Bortolani, F. Nizzoli and G. Santoro: J. Phys. C **13**, 1607 (1980)
31. V.R. Velasco and F. Garcia-Moliner: Solid St. Commun. **33**, 1 and 2237 (1980)
32. V. Bortolani, F. Nizzoli, G. Santoro and J.R. Sandercock: Phys. Rev. **B25**, 3442 (1982)
33. V. Bortolani, A.M. Marvin, F. Nizzoli and G. Santoro: J. Phys. C **16**, 219 (1983)
34. E.L. Albuquerque: J. Phys. C **13**, 2623 (1980)
35. E.L. Albuquerque, R. Loudon and D.R. Tilley: J. Phys. C **13**, 1775 (1980)
36. N.L. Rowell and G.I. Stegeman: Phys. Rev. Lett. **41**, 970 (1978)
37. N.L. Rowell and G.I. Stegeman: Can. J. Phys. **60**, 1788 and 1804 (1982)
38. A.M. Marvin, F. Toigo and V. Celli: Phys. Rev. **B11**, 2777 (1975)
39. J.R. Sandercock: Solid St. Commun. **26**, 547 (1978)
40. A. Kueny, M. Grimsditch, K. Miyano, I. Banerjee, C. Falco and I.K. Schuller: Phys. Rev. Lett. **48**, 166 (1982)
41. V. Bortolani, F. Nizzoli, G. Santoro and J.R. Sandercock: Phys. Rev. **B25**, 3442 (1982)
42. J.R. Sandercock, F. Nizzoli, V. Bortolani, G. Santoro and A.M. Marvin: in **Proc. 3rd Europ. Conf. on Surface Science** (Societe du Vide, Paris 1980) p.754
43. N.L. Rowell, V.C-Y. So, R. Normandin and G.I. Stegeman: Can. J. Phys. **60**, 1821 (1982)
44. E. Kretschmann: Z. Phys. **227**, 412 (1969); **241**, 313 (1971)
45. Y.J. Chen and E. Burstein: "Three-Wave Nonlinear Interactions Involving Surface Polaritons; Raman Scattering, Light Diffraction and Parametric Mixing" in **Surface Polaritons** , Ed. V.M. Agranovich and D.L. Mills (North-Holland, Amsterdam, 1982) pag. 587
46. M. Fukui, O. Toda, V.C-Y. So and G.I. Stegeman: Solid St. Commun. **36**, 995 (1980); J. Phys. C **14**, 5591 (1981)
47. H. Talaat, W.P. Chen and E. Burstein: J. Phys. C **15**, 3261 (1982)
48. A. Marvin, V. Bortolani, F. Nizzoli, G. Santoro and V. Celli: J. Phys. C **15**, 3273 (1982)
49. A.L. Moretti, W.M. Robertson, B. Fisher and R. Bray: Phys. Rev. **B31**, 3361 (1985)
50. S. Ushioda and R. Loudon: "Raman Scattering by Surface Polaritons" in **Surface Polaritons** , Ed. V.M. Agranovich and D.L. Mills (North-Holland, Amsterdam, 1982) pag. 535
51. K.L. Kliewer and R. Fuchs: Phys. Rev. **144**, 495 (1966)
52. Y.J. Chen, E. Burstein and D.L. Mills: Phys. Rev. Lett. **34**, 1516 (1975)
53. S. Ushioda, J.B. Valdez, W.H. Ward and A.R. Evans: Rev. Sci. Instrum. **45**, 479 (1974)
54. J.B. Valdez and S. Ushioda: Phys. Rev. Lett. **38**, 1098 (1977)
55. D.J. Evans, S. Ushioda and J.D. McMullen: Phys. Rev. Lett. **31**, 372 (1973)
56. J.Y. Prieur and S. Ushioda: Phys. Rev. Lett. **34**, 1012 (1975)

Response Functions at Surfaces

R. Del Sole

Dipartimento di Fisica, II Università di Roma, v. O. Raimondo, I-00173 Roma, Italy

1 Introduction

The dielectric response of solid surfaces is a growing field of interest because it provides a unified description of many phenomena of experimental interest in surface physics, as optical properties, static screening, charge-surface interaction, electromagnetic excitations, and so on [1]. Much experimental work has been carried out in the last decade and earlier, but theory is still developing. The dielectric formulation of surface response is indeed much more difficult from the mathematical and computational point of view than that of bulk response, because of the reduced surface symmetry. Most of the theoretical work has been made for the simplest surface, i.e. the jellium surface[2], while very little has been made to take into account the microscopic structure of crystal surfaces.

In these lectures I want to outline the formal development of the theory of surface response and its use in interpreting experiments. I will take into account some features characterizing crystal surfaces, as the anisotropy, while other aspects of crystal structure, as local-field effects, will be neglected for sake of simplicity. The definition and the methods of calculation of the surface dielectric tensor, which completely determines the response of the surface, will be considered in Section 2. In Section 3 we treat the response to transverse fields, i.e. optical properties, for the general case of an anisotropic surface. The simpler case of the jellium surface is then recovered, and contact is made with the response functions relevant in that case. In Section 4 the response to longitudinal fields is considered, in particular static screening and electron-energy loss (EEL) experiments.

In Section 5 I shall describe some calculations of optical and EEL spectra carried out for a semiconductor surface, Si(111)-2x1, chosen as a test case to illustrate the application of the dielectric theory. The atomic structure of the 2x1 reconstruction had not been known until a few years ago. Calculations of optical properties [3], carried out for one of the proposed models, i.e. the π-bonded chain model[4], predict a large anisotropy of reflectance within the surface plane, in very good agreement with experiment [5]. This and other experiments [6,7] have recently given great evidence in favour of the chain model, which is now generally accepted. The surface dielectric tensor, extracted from optical experiments, allows interpretation of EEL spectra [8,9]. Energy losses related to surface states are explained in terms of excitations of collective surface modes and of electron-hole pairs.

2 Dielectric Response of Surfaces

In this Section we consider the dielectric response of a semi-infinite solid to a perturbation. For sake of simplicity, we shall treat the subject within the self-consistent field approximation, often referred to also as Random Phase Approximation (RPA) [10]. This is an independent-particle approximation, since each electron is supposed to feel, in addition to the field of ions and of external particles (the perturbation), the self-consistent Coulomb potential generated by the electrons. Important effects, as the electron-hole interaction, are left aside this approximation[11]. We shall also neglect local-field effects, i.e. the microscopic fluctuations of the potential, induced, even in response to a long wave length disturbance, by the microscopic structure of the crystal, which in principle might be considered within RPA[11]. In spite of these approximations, realistic calculations of surface response are very difficult; on the other hand, even results based on simple models are relevant, in view of the lack of more detailed calculations.

2.1 Dielectric Tensor

We consider a system of non-interacting electrons subject to a perturbation, characterized by a vector potential $\underline{A}(\underline{r},t)$ and by a scalar potential $V(\underline{r},t)$. The perturbed electrons yield an induced current:

$$J_i^{in}(\underline{r},t) = \Sigma_j \int_{-\infty}^{t} dt' \int d^3r' \sigma_{ij}(\underline{r},\underline{r}';t-t') E_j(\underline{r}',t') \quad , \tag{1}$$

where $E_j(\underline{r}',t')$ is the perturbing electric field and $\sigma_{ij}(\underline{r},\underline{r}';t-t')$ is the complex conductivity tensor. This is the most general linear relation between the two vectors $\underline{J}^{in}$ and $\underline{E}$. The induced current at the time t is determined by the values of the field at preceding times t', because of causality. Before proceeding to quote the expression for σ_{ij} which obtains from linear response theory [10], it is convenient to consider time-Fourier transforms of the type:

$$\sigma_{ij}(\underline{r},\underline{r}';t-t') = (2\pi)^{-1} \int d\omega \exp[-i\omega(t-t')] \sigma_{ij}(\underline{r},\underline{r}';\omega) \quad . \tag{2}$$

Then (1) becomes[10]:

$$J_i^{in}(\underline{r},\omega) = \Sigma_j \int d^3r' \sigma_{ij}(\underline{r},\underline{r}';\omega) E_j(\underline{r}',\omega) \quad , \tag{3}$$

where

$$\sigma_{ij}(\underline{r},\underline{r}';\omega) = -(i/\omega) [\chi_{ij}^{jj}(\underline{r},\underline{r}';\omega) - (e/m) \delta_{ij} \rho(\underline{r}) \delta(\underline{r}-\underline{r}')] \quad . \tag{4}$$

Here $\rho(\underline{r})$ is the ground-state electron density and $\chi_{ij}^{jj}(\underline{r},\underline{r}';\omega)$ is the Fourier transform of the retarded current-current correlation function

$$\chi_{ij}^{jj}(\underline{r},\underline{r}';t-t') = i\theta(t-t') \langle [J_i(\underline{r},t), J_j(\underline{r}',t')] \rangle . \tag{5}$$

$J_i(\underline{r},t)$ is the i-component of the current operator in the Heisemberg representation of the unperturbed crystal, $\theta(t)$ is the step function accounting for causality, and $<[\ ,\ ]>$ means the expectation value on the ground state (we restrict ourselves to zero temperature) of the commutator of the quantum mechanical operators appearing within square brackets. The current-current correlation function $\chi^{jj}_{ij}(\underline{r},\underline{r}';t-t')$, which is a property of the crystal, independent of the perturbation, describes how a current perturbation, occurring at point $\underline{r}'$ and time t', propagates to the position $\underline{r}$ at time t. One may readily verify that χ^{jj}_{ij} is proportional to the retarded Green's function [12]. It depends on the time difference t-t', rather than on t and t' separately, because of translational invariance in time.

The response function (4) completely describes the electron response to the perturbation. In fact, the induced charge can be computed from the current via the continuity relation

$$\mathrm{div}\ \underline{J}^{in}(\underline{r},\omega)-i\omega\rho^{in}(\underline{r},\omega) = 0 \quad . \tag{6}$$

A different way of looking at electron response is to consider the polarization induced by the perturbation, defined as:

$$\underline{P}(\underline{r},t) = \int_{-\infty}^{t}dt'\underline{J}^{in}(\underline{r},t') \tag{7}$$

or, Fourier-transformed:

$$\underline{P}(\underline{r},\omega) = (i/\omega)\underline{J}^{in}(\underline{r},\omega) = \int d^3r'\ \underline{\underline{\alpha}}(\underline{r},\underline{r}';\omega)\cdot\underline{E}(\underline{r}',\omega) \quad , \tag{8}$$

where the polarizability is:

$$\alpha_{ij}(\underline{r},\underline{r}';\omega) = [\chi^{jj}_{ij}(\underline{r},\underline{r}';\omega)-(e/m)\delta_{ij}\rho(\underline{r})\delta(\underline{r}-\underline{r}')]/\omega^2 \quad . \tag{9}$$

From this we get the electric displacement

$$\underline{D} = \underline{E} + 4\pi\underline{P} \tag{10}$$

which is related to the total field $\underline{E}$ via the dielectric tensor $\epsilon_{ij}(\underline{r},\underline{r}';\omega)$:

$$D_i(\underline{r},\omega) = \Sigma_j \int d^3r'\ \epsilon_{ij}(\underline{r},\underline{r}';\omega)E_j(\underline{r}',\omega) \quad . \tag{11}$$

This can be computed from one-electron eigenfunctions within RPA [10]:

$$\epsilon_{ij}(\underline{r},\underline{r}';\omega) = \delta_{ij}\delta(\underline{r}-\underline{r}')\ [1-(4\pi e/m\omega^2)\rho(\underline{r})] \quad + \\ +(4\pi/\omega^2)\Sigma_{if}\left[\frac{J_i^{if}(\underline{r})J_j^{fi}(\underline{r}')}{\omega+\omega_{if}+i\eta} - \frac{J_j^{if}(\underline{r}')J_i^{fi}(\underline{r})}{\omega-\omega_{if}+i\eta}\right] \quad , \tag{12}$$

where i and f label filled and empty electron states respectively, $\hbar\omega = E_i - E_f$,

$$J_i^{if}(\underline{r}) = (e\hbar/2mi)[\psi_i^*(\underline{r})\frac{\partial}{\partial r_i}\psi_f(\underline{r})-\psi_f(\underline{r})\frac{\partial}{\partial r_i}\psi_i^*(\underline{r})] \tag{13}$$

is the matrix element of the current operator between i and f states, and $\eta \to 0^+$.

2.2 Dielectric Function

In the static case ($\omega=0$) or whenever it is possible to neglect retardation ($\omega/c \to 0$), the electric field is longitudinal, i.e. can be described by a scalar potential. In this case the dielectric response simplifies, since it is sufficient to consider the induced charge (a scalar quantity) instead of the induced current (a vector):

$$\rho^{in}(\underline{r},\omega) = -\int d^3r'\,\chi_{\rho\rho}(\underline{r},\underline{r}';\omega)V(\underline{r}',\omega) \quad , \tag{14}$$

where $\chi_{\rho\rho}(\underline{r},\underline{r}';\omega)$ is the Fourier-transform of the retarded density-density correlation function:

$$\chi_{\rho\rho}(\underline{r},\underline{r}';t-t') = i\theta(t-t')\langle[\rho(\underline{r},t),\rho(\underline{r}',t')]\rangle, \tag{15}$$

and $\rho(\underline{r},t)$ is the charge density operator in the Heisemberg representation. The induced potential is easily obtained from the induced charge:

$$V^{in}(\underline{r},\omega) = V(\underline{r},\omega) - V^{ex}(\underline{r},\omega) = -\int d^3r'\,v(\underline{r}-\underline{r}')\int d^3r''\chi_{\rho\rho}(\underline{r}',\underline{r}'';\omega)V(\underline{r}'',\omega), \tag{16}$$

where $v(\underline{r})$ is the Coulomb potential. The external potential is therefore related to the total potential by the dielectric function $\epsilon_p(\underline{r},\underline{r}',\omega)$:

$$V^{ex}(\underline{r},\omega) = \int d^3r'\,\epsilon_p(\underline{r},\underline{r}';\omega)V(\underline{r}',\omega) \tag{17}$$

with

$$\epsilon_p(\underline{r},\underline{r}';\omega) = \delta(\underline{r}-\underline{r}') + \int d^3r''v(\underline{r}-\underline{r}'')\,\chi_{\rho\rho}(\underline{r}'',\underline{r}';\omega) \quad . \tag{18}$$

The dielectric function can be obtained from the dielectric tensor, as shown below. To this end, it is better to perform spatial Fourier-transforms according to:

$$\underline{E}(\underline{k},\omega) = (2\pi)^{-3}\int d^3k\,\exp(-i\underline{k}\cdot\underline{r})\underline{E}(\underline{r},\omega) \quad , \tag{19}$$

$$\epsilon_p(\underline{k},\underline{k}';\omega) = (2\pi)^{-3}\int d^3k\int d^3k'\exp(-i\underline{k}\cdot\underline{r}+i\underline{k}'\cdot\underline{r}')\epsilon_p(\underline{r},\underline{r}';\omega) \quad . \tag{20}$$

The constitutive relation (11) in this representation becomes:

$$\underline{D}(\underline{k},\omega) = \int d^3k'\,\underline{\underline{\epsilon}}(\underline{k},\underline{k}';\omega)\cdot\underline{E}(\underline{k}',\omega) \quad . \tag{21}$$

Let us define the longitudinal component of $\underline{E}$ as:

$$E^L(\underline{k},\omega) = (\underline{k}/k)\cdot\underline{E}(\underline{k},\omega) \quad , \tag{22}$$

i.e. the component of $\underline{E}$ along $\underline{k}$. The Maxwell equation

$$\mathrm{div}\underline{D} = 4\pi\rho^{ex} \qquad (23)$$

which is Fourier-transformed to:

$$ikD^{L}(\underline{k},\omega) = 4\pi\rho^{ex}(\underline{k},\omega) \quad , \qquad (24)$$

implies that $D^{L}(\underline{k},\omega) = E^{ex,L}(\underline{k},\omega)$, (25)
since the latter is related to ρ^{ex} by the same equation. Taking the longitudinal component of (21) in the non-retarded limit $\omega/c \to 0$, when

$$\underline{E}(\underline{k},\omega) = -i\underline{k}\, V(\underline{k},\omega) \quad , \qquad (26)$$

we get

$$E^{ex,L}(\underline{k},\omega) = \int d^3k'\, \varepsilon_{LL}(\underline{k},\underline{k}';\omega) E^{L}(\underline{k}',\omega) \quad , \qquad (27)$$

where

$$\varepsilon_{LL}(\underline{k},\underline{k}';\omega) = (\underline{k}/k)\cdot \underline{\underline{\varepsilon}}(\underline{k},\underline{k}';\omega)\cdot(\underline{k}'/k') \quad . \qquad (28)$$

ε_{LL} relates the longitudinal external field to the longitudinal total field, which is all we need in the quasi-static case. The relation between potentials is

$$V^{ex}(\underline{k},\omega) = \int d^3k'(k'/k)\varepsilon_{LL}(\underline{k},\underline{k}';\omega)V(\underline{k}',\omega) \qquad (29)$$

from which we identify

$$\varepsilon_{p}(\underline{k},\underline{k}';\omega) = (k'/k)\,\varepsilon_{LL}(\underline{k},\underline{k}';\omega) = [\,\underline{k}\cdot\underline{\underline{\varepsilon}}\,(\underline{k},\underline{k}';\omega)\cdot\underline{k}'\,]/k^{2} \quad . \qquad (30)$$

In a homogeneous system, as for instance in free space or in a homogeneous electron gas (jellium), the presence of translation invariance in space makes the response functions depend only on the difference $\underline{r}'-\underline{r}$, rather than on $\underline{r}$ and $\underline{r}'$ separately. As a consequence, their space Fourier transforms performed according to (20) are non-vanishing only for $\underline{k}'=\underline{k}$. This is approximately true also for systems with weakly inhomogeneous charge density, as infinite simple metals or small gap semiconductors. In this case, the dielectric function $\varepsilon_p(\underline{k},\omega)$ is the same as $\varepsilon_{LL}(\underline{k},\omega)$, the part of the dielectric tensor relating longitudinal fields. Moreover, in cubic crystals for $k \to 0$ the dielectric tensor becomes a scalar:

$$\varepsilon_{ij}(\underline{k},\omega) = \delta_{ij}\,\varepsilon(\underline{k},\omega) \qquad (k \to 0) \qquad (31)$$

called dielectric constant. It follows that the longitudinal part of the dielectric tensor, ε_{LL}, the dielectric function relating potentials, ε_p and the dielectric constant ε are all equal. In systems with less symmetry, as for instance for a semi-infinite crystal, one must pay attention to distinguish between them.

The dielectric function can be computed directly from eigenfunctions according to (18)[10]:

$$\varepsilon_p(\underline{k},\underline{k}';\omega) = \delta(\underline{k}-\underline{k}') +$$

$$(4\pi e^2/k^2)\,\Sigma_{if}\left[\frac{A^{if}(\underline{k})A^{fi}(-\underline{k}')}{\omega_{fi}-\omega-i\eta} + \frac{A^{if}(-\underline{k}')A^{fi}(\underline{k})}{\omega_{fi}+\omega+i\eta}\right] , \qquad (32)$$

where $$A^{if}(\underline{k}) = \int d^3r\, \psi_i^*(\underline{r})\exp(-i\underline{k}\cdot\underline{r})\psi_f(\underline{r}) , \qquad (33)$$

and $4\pi e^2/k^2$ is the Fourier-transform of the Coulomb potential.

2.3 Semi-Infinite Crystals

Dealing with surfaces, it is convenient to use the partial Fourier-transform, performed only on the surface coordinate $\underline{r}_{\parallel}$:

$$\underline{E}(\underline{k}_{\parallel},z;\omega) = \int d^2r_{\parallel}\exp(-i\underline{k}_{\parallel}\cdot\underline{r}_{\parallel})\underline{E}(\underline{r},\omega) , \qquad (34)$$

$$\varepsilon_{ij}(\underline{k}_{\parallel},\underline{k}_{\parallel}';z,z';\omega) =$$

$$= (2\pi)^{-2}\int d^2r_{\parallel}\int d^2r_{\parallel}'\exp(-i\underline{k}_{\parallel}\cdot\underline{r}_{\parallel}+i\underline{k}_{\parallel}'\cdot\underline{r}_{\parallel}')\,\varepsilon_{ij}(\underline{r},\underline{r}';\omega) , \qquad (35)$$

to take advantage of the two-dimensional periodicity parallel to the surface plane. The constitutive relation becomes:

$$\underline{D}(\underline{k}_{\parallel},z;\omega) = (4\pi^2/A)\Sigma_{\underline{G}_{\parallel}}\int dz'\ \underline{\underline{\varepsilon}}(\underline{k}_{\parallel},\underline{k}_{\parallel}+\underline{G}_{\parallel};z,z';\omega)\cdot\underline{E}(\underline{k}_{\parallel}+\underline{G}_{\parallel},z';\omega) , \qquad (36)$$

(A being the surface area), where the integral over $\underline{k}'_{\parallel}$ has become a sum over the reciprocal lattice vectors, because of the two-dimensional periodicity. We do not get simply $\underline{k}_{\parallel}=\underline{k}'_{\parallel}$ as in jellium, because of the inhomogeneous microscopic structure of the crystal. This is the so-called local-field effect. However, if we can neglect it, i.e. if the off-diagonal $\underline{G}_{\parallel}\neq 0$ components of the dielectric tensor are small, we finally get:

$$\underline{D}(\underline{k}_{\parallel},z;\omega) = \int dz'\ \underline{\underline{\varepsilon}}(\underline{k}_{\parallel};z,z';\omega)\cdot\underline{E}(\underline{k}_{\parallel},z';\omega) \qquad (37)$$

where

$$\varepsilon_{ij}(\underline{k}_{\parallel};z,z';\omega) = \delta_{ij}\ \delta(z-z')[1-(4\pi e/m\omega^2)\bar{\rho}(z)] +$$

$$-(4\pi/\omega^2 A)\Sigma_{if}\left[\frac{J_i^{if}(k_{\parallel},z)J_j^{fi}(-k_{\parallel},z')}{\omega+\omega_{if}+i\eta} - \frac{J_j^{if}(-k_{\parallel},z')J_i^{fi}(k_{\parallel},z)}{\omega-\omega_{if}+i\eta}\right] , \qquad (38)$$

$$J_i^{if}(\underline{k}_{\parallel},z) = \int d^2r_{\parallel}\exp(-i\underline{k}_{\parallel}\cdot\underline{r}_{\parallel})J^{if}(\underline{r}) \qquad (39)$$

and the ground state charge density $\bar{\rho}(z)$ has been averaged on the surface plane.

In the same way, we get for the dielectric function:

$$\varepsilon_p(\underline{k}_{\parallel};z,z';\omega) = \delta(z-z')+(2\pi/k_{\parallel}A)\int dz''\exp(-k_{\parallel}|z-z''|) \cdot$$
$$\cdot\sum_{if}\left[\frac{A^{if}(\underline{k}_{\parallel},z'')A^{fi}(-\underline{k}_{\parallel},z')}{\omega_{fi}-\omega-i\eta}+\frac{A^{if}(-\underline{k}_{\parallel},z')A^{fi}(\underline{k}_{\parallel},z'')}{\omega_{fi}+\omega+i\eta}\right] \qquad (40)$$

where $A^{if}(\underline{k}_{\parallel},z) = \int d^2r_{\parallel}\psi_i^*(\underline{r})\exp(-i\underline{k}_{\parallel}\cdot\underline{r}_{\parallel})\,\psi_f(\underline{r})$ (41)

and $(2\pi/k_{\parallel})\exp(-k_{\parallel}|z|)$ is the partial Fourier-transform of the Coulomb potential.

3 Response to Transverse Fields: Optical Properties

In this Section we find expressions for the reflection coefficient of a semi-infinite crystal. We consider the propagation equations for the electric field, by eliminating the magnetic field from Maxwell's equations:

$$-\frac{d^2E_x(q_x,z)}{dz^2}+iq_x\frac{dE_z(q_x,z)}{dz} = (\omega^2/c^2)\sum_j\int dz'\varepsilon_{xj}(q_x;z,z')E_j(q_x,z') \qquad (42a)$$

$$-\frac{d^2E_y(q_x,z)}{dz^2}+q_x^2E_y(q_x,z) = (\omega^2/c^2)\sum_j\int dz'\varepsilon_{yj}(q_x;z,z')E_j(q_x,z') \qquad (42b)$$

$$iq_x\frac{dE_x(q_x,z)}{dz}+q_x^2E_z(q_x,z) = (\omega^2/c^2)\sum_j\int dz'\varepsilon_{zj}(q_x;z,z')E_j(q_x,z')\ , \qquad (42c)$$

where the surface component of the light wave vector $\underline{q}_{\parallel}$ has been taken parallel to x.

The dielectric tensor of the vacuum-crystal interface may be decomposed in this way:

$$\varepsilon_{ij}(z,z') = \varepsilon_0(z)\delta(z-z')+\Delta\varepsilon_{ij}(z,z')\ , \qquad (43)$$

where $\varepsilon_0(z) = 1$ for $z<0$ (vacuum)

$\varepsilon_0(z) = \varepsilon_b$ for $z>0$ (semiconductor) (44)

(the q_x-dependence has been here suppressed for sake of simplicity). The first term alone corresponds to the sharp step-like interface, giving rise to Fresnel formulas for s and p reflectivities as zero-order approximation [13]. The second term, $\Delta\varepsilon_{ij}(z,z')$, accounts for the microscopic properties of the interface. It cannot be treated as a perturbation, since it is of the same order as ε_b. However it is of short range, extending for a depth d much smaller than the light wave length λ. The solution to first order in d/λ has been given by BAGCHI and RAJAGOPAL [14] for jellium (where $\varepsilon_{ij}(q_x\to0;z,z')$ is diagonal [15]), and later generalized to anisotropic

crystal surfaces, involving off-diagonal components of $\Delta\epsilon_{ij}$ [16]. The solution proceeds in three steps:
i) first (42) are transformed into the integral equations

$$E_i(z) = E_i^0(z) - (\omega^2/c^2)\Sigma_{jk} \int dz' \int dz'' G_{ij}(z,z')\Delta\epsilon_{jk}(z',z'')E_k(z'') \quad , \qquad (45)$$

(where $E_i^0(z)$ is the solution of the zero-order equations) using the Green's function matrix $G_{ij}(z,z')$ of the zero-order Fresnel problem[16]. We quote here the component relevant for the s-wave, $G_{yy}(z,z')$:

$$G_{yy}(z,z') = (q_z+k_z)(4iq_z k_z)^{-1} G^{in}(z_m)G^{out}(z_M) \quad , \qquad (46)$$

where z_m and z_M are respectively the minimum and maximum value among z and z', and

$$G^{in}(z) = 2k_z \exp(-iq_z z)/(q_z+k_z) \quad , \quad (z<0)$$
$$G^{in}(z) = \exp(-ik_z z) - \exp(ik_z z)(q_z-k_z)/(q_z+k_z) \quad ; \quad (z>0) \qquad (47)$$

$$G^{out}(z) = \exp(iq_z z) + \exp(-iq_z z)(q_z-k_z)/(q_z+k_z) \quad , \quad (z<0)$$
$$G^{out}(z) = 2q_z \exp(ik_z z)/(q_z+k_z) \quad . \quad (z>0) \qquad (48)$$

Here $q_z = (\omega^2/c^2 - q_x^2)^{1/2}$ and $k_z = (\omega^2\epsilon_b/c^2 - q_x^2)^{1/2}$, with Im $k_z > 0$. The other non-vanishing components G_{xx}, G_{zz}, G_{xz} and G_{zx} are quoted in Ref.14.
ii) The relevant range of integration in the z''-integral in the right-hand side of (45) is near the surface. There E_x and E_y are expected to be slowly varying (they are continous in the zero-order solution) and can be replaced by their values at the surface. However E_z is discontinuous in the zero - order solution. We eliminate it by inverting the constitutive equation relating D_z to $\underline{E}$:

$$E_z(z) = \int dz' \epsilon^{-1}_{zz}(z,z')[D_z(z') - \int dz'' \sum_{j=x,y} \epsilon_{zj}(z',z'')E_j(z'')] \qquad (49)$$

where $\int dz'' \epsilon^{-1}_{zz}(z,z'')\epsilon_{zz}(z'',z') = \delta(z-z') \quad . \qquad (50)$

D_z is also classically continuous, so that it can be replaced by its surface value.
iii) Since we are interested in a solution to first order in d/λ, and the integral in (45) is itself of this order, we there replace the surface values of E_x, E_y and D_z by their zero-order values. This is appropriate, except when strong surface resonances of the dielectric tensor occur, as in the case of a strongly bound surface exciton, or when the zero-order solution vanishes, as in the case of normal modes supported by the excitations of the surface layer. If this is the case, the solution can be obtained with some more effort[17-19].

Now it is straightforward to calculate the field in vacuum, since only known quantities are involved in the integral in (45). The results for the reflection coefficients of s and p radiation are:

$$\Delta R_s/R_s^0 = 4(\omega/c)\cos\theta \; \mathrm{Im}\,[\Delta\tilde{\varepsilon}_{yy}/(\varepsilon_b-1)] \tag{51}$$

for incident s-light (which is polarized perpendicular to the plane of incidence, i.e. along y),

$$\Delta R_p/R_p^0 = 4(\omega/c)\cos\theta\,\mathrm{Im}\left[\frac{(\varepsilon_b-\sin^2\theta)\,\Delta\tilde{\varepsilon}_{xx} + \varepsilon_b^{\,2}\sin^2\theta\,\overline{\Delta\varepsilon_{zz}^{-1}}}{(\varepsilon_b-1)(\varepsilon_b\cos^2\theta-\sin^2\theta)}\right] \tag{52}$$

for incident p-light (polarized in the plane of incidence, xz), where

$$\Delta\tilde{\varepsilon}_{ii} = \int dz \int dz'[\,\Delta\varepsilon_{ii}(z,z') +$$

$$-\int dz'' \int dz''' \Delta\varepsilon_{iz}(z,z')\varepsilon_{zz}^{-1}(z',z'')\Delta\varepsilon_{zi}(z'',z''')] \tag{53}$$

for $i = x,y$,

$$\overline{\Delta\varepsilon_{zz}^{-1}} = \int dz \int dz'[\,\varepsilon_{zz}^{-1}(z,z') - \delta(z-z')/\,\varepsilon_0(z)] \quad , \tag{54}$$

and θ is the angle of incidence. The off-diagonal components of the dielectric tensor mix s and p waves, so that, in the case of incident s-wave (p-wave), the reflected light has a p- (s-) component, whose intensity is of second order in d/λ. Therefore such mixing can be neglected in the first order solution which is of interest here. A local approximation of the bulk dielectric tensor (the first term in the right hand side of (43)) has been used to derive (51) and (52). As a consequence, these results are not valid when bulk spatial dispersion (i. e. non-locality) becomes important, namely at frequencies higher than the plasma frequency [2].

If light comes from the interior of the crystal (internal reflection) analogous formulas are derived [16]. Also, transmission [20] and ellipsometry [21] can be computed in this way in terms of the optical surface response functions $\Delta\tilde{\varepsilon}_{ii}$ and $\overline{\Delta\varepsilon_{zz}^{-1}}$. Power absorption is then easily derived from energy balance. The behavior of the fields near the surface , which is an essential ingredient to compute the photoemission yield[2], is also obtained from the solution of (42).

In the case of the jellium surface, the dielectric tensor is diagonal because of symmetry [15]:

$$\begin{aligned} &\varepsilon_{xx}(z,z') = \varepsilon_{yy}(z,z') \neq \varepsilon_{zz}(z,z') \\ &\varepsilon_{ij}(z,z') = 0 \quad \text{for } i \neq j \end{aligned} \tag{55}$$

for $q_{\|} \to 0$. The surface response function $\Delta\tilde{\varepsilon}_{ii}(\omega)$ becomes

$$\Delta\tilde{\varepsilon}_{\|} = \int dz \int dz' \Delta\varepsilon_{ii}(z,z') \tag{56}$$

for any direction in the surface plane. In the original derivation of the response of the jellium surface by Feibelman [2], different response functions $d_{\|}(\omega)$ and $d_{\perp}(\omega)$ were defined:

$$d_{\parallel}(\omega) = \left[\int dz\, z \frac{d}{dz} \int dz' \varepsilon_{xx}(z,z')\right] / \left[\int dz \frac{d}{dz} \int dz' \varepsilon_{xx}(z,z')\right] \qquad (57)$$

and

$$d_{\perp}(\omega) = \left[\int dz\, z \frac{d}{dz} \int dz' \varepsilon_{zz}^{-1}(z,z')\right] / \left[\int dz \frac{d}{dz} \int dz' \varepsilon_{zz}^{-1}(z,z')\right] \quad . \qquad (58)$$

Since $\int dz' \varepsilon_{xx}(z,z')$ and $\int dz' \varepsilon_{zz}^{-1}(z,z')$ are steeply varying functions only near the surface, their derivatives are peaked there and vanish elsewhere, so that $d_{\parallel}$ and $d_{\perp}$ measure the effective depth of the surface. We now show that $d_{\parallel}$ is related to $\Delta\tilde{\varepsilon}_{\parallel}$ and $d_{\perp}$ to $\overline{\Delta\varepsilon_{zz}^{-1}}$. To this end, we integrate by parts (57) and (58) from a position Z_v in vacuum (where $\int \varepsilon_{ii}(Z_v,z')dz' \simeq 1$) and Z_b in the bulk (where $\int \varepsilon_{ii}(Z_b,z')dz' \simeq \varepsilon_b$), getting:

$$d_{\parallel}(\omega) = \left[Z_b \varepsilon_b - Z_v - \int dz \int dz' \varepsilon_{xx}(z,z')\right] / (\varepsilon_b - 1) \qquad (59)$$

$$d_{\perp}(\omega) = \left[Z_b / \varepsilon_b - Z_v - \int dz \int dz' \varepsilon_{zz}^{-1}(z,z')\right] / (\varepsilon_b^{-1} - 1) \quad . \qquad (60)$$

Recalling the definitions (54) and (56) , we get:

$$d_{\parallel}(\omega) = - \Delta\tilde{\varepsilon}_{\parallel}(\omega) / [\varepsilon_b(\omega) - 1] \qquad (61)$$

$$d_{\perp}(\omega) = - \overline{\Delta\varepsilon_{zz}^{-1}}(\omega) / [\varepsilon_b^{-1} - 1] \quad . \qquad (62)$$

4 Response to Longitudinal Fields

4.1 Static Screening

Given an external potential, generated, for instance, by a surface impurity, the problem is to find the total screened potential. Because of the relation (17) it is sufficient to invert the dielectric function ε_p i.e. to find ε_p^{-1} such that:

$$\int dz'' \varepsilon_p^{-1}(\underline{q}_{\parallel}; z, z'') \varepsilon_p(\underline{q}_{\parallel}; z'', z') = \delta(z - z') \quad . \qquad (63)$$

Then the total potential is:

$$V(\underline{q}_{\parallel}, z) = \int dz' \varepsilon_p^{-1}(\underline{q}_{\parallel}; z, z') V^{ex}(\underline{q}_{\parallel}, z') \quad . \qquad (64)$$

Some work has been performed using a macroscopic three-layers model[22] , or based on bulk screening modified to account for crystal termination [23] . However, the consideration of the microscopic structure of the surface, as for instance the contribution to screening of electrons in surface states, is still missing. The problem is even more complex if the field in the very neighborhood of the impurity is of interest, since in this case local-field effects are important. Only little work has been made on local fields [24] . In general, surface screening is one of the most relevant open questions of surface physics at present.

4.2 Electron Energy Loss

The field generated by an incoming electron is well described by a scalar potential, so that the same formulas as for static screening can be used to obtain the induced field. It is however simpler, since most of the losses occur when the electrons are outside the crystal [25,26], so that the perturbing field is slowly varying inside the crystal, while the induced field is of interest only outside, where microscopic fluctuations are not important. This is somehow similar to the case of optical properties, where long-range disturbances and responses are involved.

From standard dipole scattering theory, we find that the probability $P(k,k')d\Omega_{k'}d(\hbar\omega)$ that an incident electron is scattered into the range of energy loss between $\hbar\omega$ and $\hbar(\omega+d\omega)$ and into the solid angle $d\Omega_{k'}$ around the direction of $\underline{k}'$ for small momentum transfer $q_{\|}=|\underline{k}'_{\|}-\underline{k}_{\|}|<<k$, is [26]:

$$P(\underline{k},\underline{k}') = 2(me/\pi\hbar^2)^2(1/\cos\theta)(k'/k)q_{\|}\ \mathrm{Im}\ g(\underline{q}_{\|},\omega)/|q_{\|}^2+q_{\perp}^2|^2 \quad , \tag{65}$$

where θ is the angle of incidence. Here we make a contact between $g(\underline{q}_{\|},\omega)$ defined in Ref. 27 and the inverse dielectric function. In the treatment of Ref. 27, the crystal occupies the half-space $z<0$. An external charge is present outside, for $z>d$, giving rise to a potential

$$V^{ex}(\underline{q}_{\|},z) = \tilde{V}^{ex}(\underline{q}_{\|})\exp(-q_{\|}|z-d|) \quad . \tag{66}$$

$g(\underline{q}_{\|},\omega)$ is defined from the induced potential in vacuum, which is $(z>0)$:

$$V^{in}(\underline{q}_{\|},z) = g(\underline{q}_{\|},\omega)\tilde{V}^{ex}(\underline{q}_{\|})\exp[-q_{\|}(z+d)] \quad . \tag{67}$$

The relation between these potentials is given by the inverse dielectric function, so that we can identify:

$$g(\underline{q}_{\|},\omega)\exp(-q_{\|}z) = \int_{-\infty}^{\infty} dz'[\varepsilon_p^{-1}(q_{\|};z,z')-\delta(z-z')]\exp(q_{\|}z') \quad , \tag{68}$$

where the relevant range of integration over z' is $z'<0^+$, including the crystal surface. Then, we integrate over z from 0 to infinity (excluding the surface) and get

$$g(\underline{q}_{\|},\omega) = q_{\|}\int_{0^+}^{\infty}dz\int_{-\infty}^{0^+}dz'\varepsilon_p^{-1}(q_{\|};z,z')\exp(q_{\|}z') \quad . \tag{69}$$

Inserting (69) in (65), we express the loss probability in terms of the inverse dielectric function.

However, EEL experiments are usually analysed in terms of a simpler three-layers model [26]. We here describe a recent generalization of this model to the case of anisotropic surfaces [28]. We consider a system characterized by the local dielectric tensor :

$$\begin{aligned} &z<0 \quad \text{(vacuum)} && \varepsilon_{ij}=\delta_{ij} \\ &z>d \quad \text{(bulk)} && \varepsilon_{ij}=\delta_{ij}\varepsilon_b \\ &0<z<d \quad \text{(surface layer)} && \varepsilon_{xx}\neq\varepsilon_{yy}\neq\varepsilon_{zz}\ ;\ \varepsilon_{ij}=0 \text{ for } i\neq j \quad . \end{aligned} \tag{70}$$

The dielectric function might be obtained from the dielectric tensor performing Fourier-transforms according to (30). It is however simpler to solve Poisson's equation for the total potential inside each medium, and match the fields according to Maxwell's boundary conditions. To simplify notations, let us assume the scattering to occur in the yz-plane. The electric potential in the vacuum region must satisfy Laplace equation and can thus be written as:

$$V(q_y,z) = \exp(-q_y z) + g \exp(q_y z) \quad , \qquad z<0 \tag{71}$$

where the first term is the external potential and the second term is the induced potential. Equation (71) defines the surface response $g(q_y,\omega)$. Similarly, the electric potential in the bulk ($z>d$) is given by

$$V(q_y,z) = C \exp[-q_y(z-d)] \quad . \qquad z>d \tag{72}$$

Finally in the surface layer ($0<z<d$) V can be expressed as

$$V(q_y,z) = A \exp(iq_z z) + B \exp(-iq_z z) \quad , \qquad 0<z<d \tag{73}$$

where q_z is obtained from Maxwell's equation div $\underline{D} = 0$, i.e.

$$q_y^2 \varepsilon_{yy} + q_z^2 \varepsilon_{zz} = 0 \quad . \tag{74}$$

The unknown amplitudes A, B, C and g are determined by matching V and the normal component of the electric displacement across the boundaries. For the loss function we obtain

$$\operatorname{Im} g(q_y,\omega) = \operatorname{Im}\{-2/[1+\varepsilon_{eff}(q_y,\omega)]\} \tag{75}$$

where the effective dielectric function is

$$\varepsilon_{eff}(q_y,\omega) = \varepsilon\,[\,\varepsilon_b \cos(q_z d) - \varepsilon \sin(q_z d)]/[\varepsilon \cos(q_z d) + \varepsilon_b \sin(q_z d)] \quad , \tag{76}$$

and $q_z = q_y \varepsilon/\varepsilon_{zz} = q_y(-\varepsilon_{yy}/\varepsilon_{zz})^{1/2}$, (77)

$$\varepsilon = (-\varepsilon_{yy}\varepsilon_{zz})^{1/2} \quad . \tag{78}$$

In the limit $q_z d \to 0$, (76) becomes

$$\varepsilon_{eff} = \varepsilon_b + q_y d(\varepsilon_{yy} - \varepsilon_b^2/\varepsilon_{zz}) \quad . \tag{79}$$

This reduces to $\varepsilon_{eff} = \varepsilon_b + q_y d(\varepsilon_s - \varepsilon_b^2/\varepsilon_s)$ (80)

for an isotropic surface layer ($\varepsilon_{yy} = \varepsilon_{zz} = \varepsilon_s$), in agreement with [26].

The poles of the loss function, i.e. the zeroes of the denominator of (75), determine the electromagnetic excitations of the surface in the non-retarded limit. As $q_y \to 0$, they occur for $\varepsilon_{yy} \to -\infty$ and for $\varepsilon_{zz} \to 0^+$, as can be easily seen from (79).

5 Application to Si(111)-2x1

The application of the dielectric theory to crystal surfaces is hindered not only by mathematical difficulties but also by the ignorance of surface atomic structure, which is generally strongly perturbed with respect to the ideal crystal termination. As an example, the 2x1 reconstruction of Si(111), one of the most studied surfaces, was not understood until a few years ago, when Pandey's π-bonded chain model succeeded in explaining a number of experimental data [4]. The geometry of this model is characterized by the surface atoms being as close as bulk nearest neighbors ($d \simeq 2.35$ Å) and forming zig-zag chains along the [1$\bar{1}$0] direction, while interchain distances are rather large. This involves a big rearrangement of the first two layers with respect to the ideal crystal termination, where surface atoms are second nearest neighbors ($d = 3.85$ Å), in order to minimize the total energy[4,30]. Electronic surface states arise from the broken bonds (mainly p_z-like, usually called dangling bonds), which interact with their nearest neighbors within the chains to form π-bonds. These bonds are responsible for the stability of the 2x1 reconstruction. In view of the good present understanding of its structure, this surface has become a good test-case for the theory of the dielectric response of crystal surfaces. In this Section we assume from the beginning the validity of the π-bonded chain model, and try to give a unified description of optical [5,29] and of EEL[8,9] experiments in terms of the surface dielectric tensor.

5.1 Reflectivity

The calculation of the reflectivity of a semi-infinite crystal, including the contribution of the microscopic structure of the surface, is a difficult task from the computational point of view. The first realistic calculation for crystal surfaces has been recently carried out for Si(111)2x1 [28]. The reflectance for normally incident light has been calculated simulating the semi-infinite Si crystal by a slab of 12 atomic layers, whose electron wave functions have been computed within the semi-empirical tight-binding scheme of VOGL et al.[31]. The quantities which can be evaluated from wave functiona are the imaginary parts of the double integrals, over z and z', of the dielectric tensor components:

$$\mathrm{Im} \int dz \int dz' \varepsilon_{ij}(\underline{q}_{\|} \to 0; z,z') = (4\pi e^2/m\omega^2 A)\sum_{if} p_i^{if} p_j^{fi}\, \delta(\omega_{fi} - \omega) \tag{81}$$

following from (12), with

$$p_i^{if} = \int d^3r\, J_i(\underline{r}) = (\hbar/i)\int d^3r\, \psi_i^*(\underline{r}) \frac{\partial}{\partial r_i}\psi_f(\underline{r}) \quad . \tag{82}$$

Then, the real parts are determined using Kramers-Kronig relations[10]. The evaluation of (82) involves a double sum over basis function , of about ten thousands terms. This must be repeated for each of the relevant (if) transitions, which amount in our case to about fifty thousand Although very heavy, the calculation can be carried out using the best

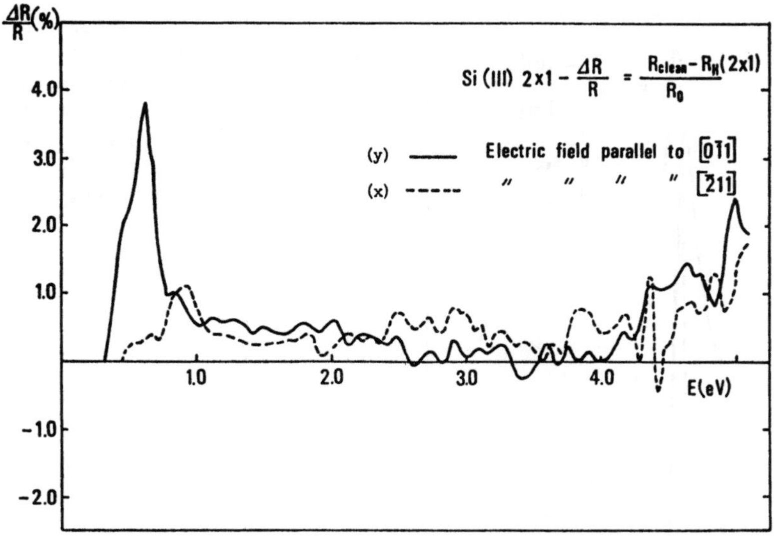

Fig. 1 - Differential reflectivity of Si(111)2x1 calculated as the difference between the clean surface reflectivity and the reflectivity after deposition of one monolayer of hydrogen, with the 2x1 structure kept unaltered

computers presently available. The fourfold integral in (53) has been estimated within a three-layers model, resulting anyway as small. The computed reflectivity is plotted in Fig.1, for light polarized along (y) and perpendicular to (x) the chains. To make evident the contribution of transitions from or to surface states , it is customary to subtract the reflectivity of the saturated surface ($1x1\text{-}O_2$ covered , in experiment [5,29]). For sake of simplicity, the calculation subtracts the reflectivity of the 2x1-H covered surface [28]. The differential reflectivity below 1 eV, arising from transitions between surface states, shows a very large anisotropy, resulting from the one-dimensional character of the chains [3]. Such anisotropy is in very good agreement with experiment [5]. This is a great evidence in favour of the chain model, since another most popular model, the buckling model, yields a polarization dependence opposite to that shown in Fig.1 [32]. The other structures can also be interpreted in terms of transitions between surface states at particular points of the Surface Brillouin Zone (SBZ).

5.2 EEL Spectra

EEL experiments [8,9] have also revealed a structure sensitive to surface contamination, which has been interpreted in terms of electronic transitions between surface states. The peak is, however, generally at higher energies (0.6 - 0.8 eV) with respect to the optical peak (0.45 eV). A weaker structure is also present at about 0.5 eV for low energies of the primary beam. These two structures have been attributed to transitions at

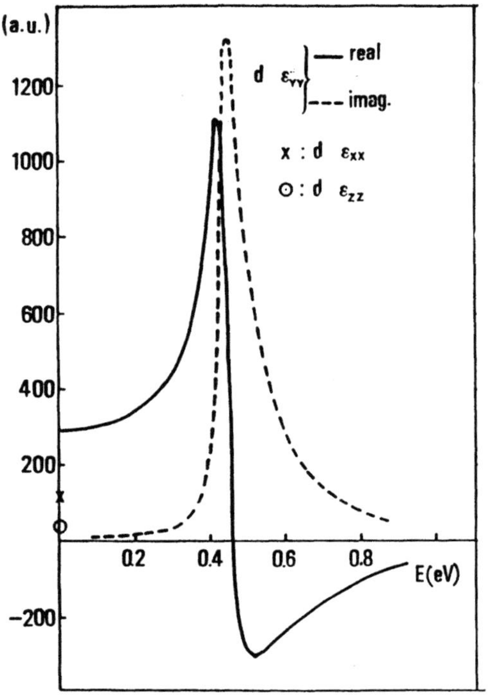

Fig. 2 - Dielectric tensor of the anisotropic surface layer of Si(111) 2x1 times its depth d. The nearly constant $\varepsilon_{xx}(\omega)$ and $\varepsilon_{zz}(\omega)$ are represented by their static values (the cross and the empty circle respectively)

different points of the SBZ [8] or to excitation of two collective modes of the surface [33]. A calculation of the loss function has been carried out within the framework of the anisotropic three-layers model described in Section 4 [28], showing that both previous interpretations were incorrect. The main peak of the EEL is a normal mode of the anisotropic surface, while the structure at lower energy is due to electron-hole pair excitations (as in the optical spectrum).

The dielectric tensor of the surface layer (for $\hbar\omega$<1eV) is obtained from a model two-dimensional calculation including only surface states [28], whose parameters were adjusted to fit optical experiments, or the computed optical properties (in particular the computed ε_{zz}, which is not known experimentally). $\varepsilon_{yy}(\omega)$ turns out to have a strong resonance at 0.45 eV for $q_y = 0$, and at higher energies for $q_y \neq 0$, while $\varepsilon_{zz}(\omega)$ is nearly constant and positive (see Fig.2). In contrast to the model where the surface dielectric tensor is described by an isotropic Lorentz oscillator [26], the number of solutions (surface normal modes) of the equation

$$1+\varepsilon_{eff}(q_y,\omega) = 0 \qquad (83)$$

depends on the value of q_y in our case. No solution exsists at very low q_y's (q_y<0.03 Å^{-1}), since the real part of ε_{yy} does not reach $-\infty$ as required by (79), while only one mode exists in the range of q_y relevant to most experiments (q_y<0.25 Å^{-1}). For larger values of q_y, the denominator of (75) has more than a single zero. q_z is real because ε_{yy} and ε_{zz} have

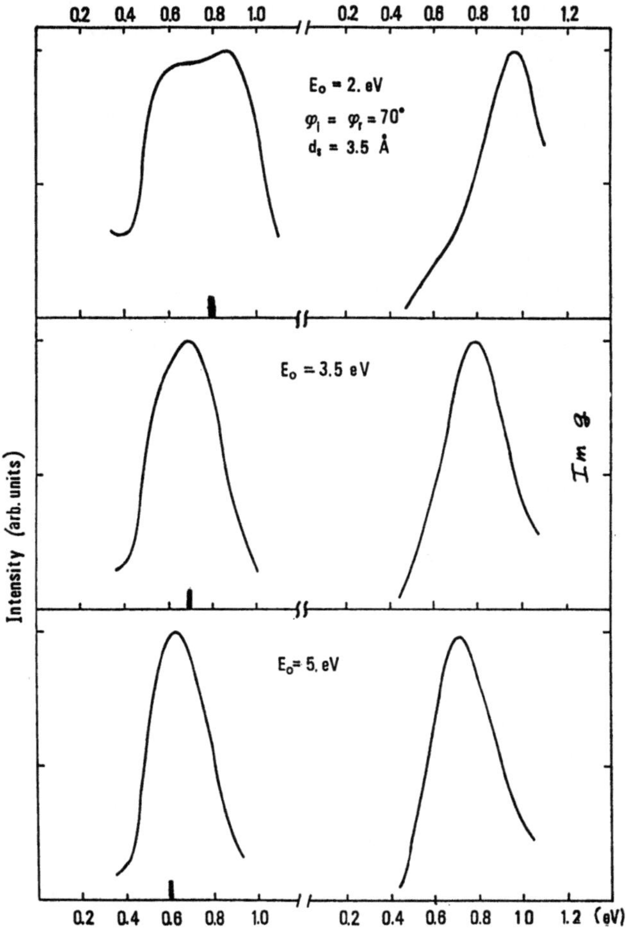

Fig. 3 - Loss spectra (curves on the left) and loss function (on the right) calculated for different primary energies E_o under specular reflection ($\phi_r = \phi_i = 70°$), with the incidence plane parallel to the chains of the Si (1$\bar{1}$1)2x1 surface and assuming a surface layer thickness d= 3.5 Å. Tickmarks denote the position of the main peak in the experimental spectra of Ref. 6

different signs (see (74)) and it is quantized in the surface layer. These modes are all longitudinal "bulk" modes of the surface layer. If d were increased at fixed q_y, an increasingly larger number of modes would appear, tending to the continuum of bulk modes of the semi-infinite anisotropic crystal.

The computed loss spectra are shown in Fig.3, where the set of curves on the left hand side refer to the loss probability P given by (65), and the set on the right to Im g. We shall consider P first. Both the main peak position and lineshape depend strongly on the primary energy E_0, in a way similar to that observed in experiment. In particular we observe the presence of a double structure in the $E_0 = 2$ eV spectrum, where the main peak is accompanied by a weaker structure at about 0.6 eV. The main peak occurs close to the zero of $Re(\epsilon_{eff}+1)$ and can thus be interpreted as a collective mode of the surface. The origin of the weaker low-energy structure can be best understood by looking at Im g on the right. This shows a well-pronounced low-energy tail, whose origin is the peak of $Im\,\epsilon_{eff}(\omega)$ in (75). It is this tail -enhanced by the kinematical term (i.e. the other factors in (65))- which gives rise to the shoulder in the calculated loss intensity.

In conclusion, the anisotropy of the surface dielectric tensor -originating from the chain-like structure of the reconstructed first layer- plays a major role in determining optical and EEL spectra of Si(111)2x1. The good agreement between theory and experiment is an important confirmation of the chain model.

It is a pleasure to thank A. Selloni for her collaboration in the research work discussed in the last Section.

References

1 M. Nakayama, J. Phys. Soc. Jpn. 39 , 265 (1975)
2 P. J. Feibelman, Progr. Surf. Sci. 12 , 287 (1982)
3 R. Del Sole and A. Selloni, Solid State Commun. 50 , 825 (1984)
4 K. C. Pandey, Phys. Rev. Lett. 47 , 1913 (1981); ibidem 49 , 223 (1982)
5 P. Chiaradia, A. Cricenti, S. Selci and G. Chiarotti, Phys. Rev. Lett. 52 , 1145 (1984)
6 F. J. Himpsel, P. Heimann and D. E. Eastman, Phys. Rev. B 24 , 2003 (1981); I. G. Uhrberg, G. U. Hansson, J. M. Nicholls and S. A. Flodstrom Phys. Rev. Lett. 48 , 1032 (1982)
7 R. M. Tromp, L. Smit and F. J. van der Veen, Phys. Rev. Lett. 51 , 1672 (1983)
8 R. Matz, H. Luth and A. Ritz, Solid State Commun. 46 , 343 (1983)
9 N. J. Di Nardo, J. E. Demuth, W. A. Thompson and Ph. Avouris, Phys. Rev. B 31 , 4077 (1985)
10 H. Ehrenreich, in The Optical Properties of Solids (Academic, New York, 1965)
11 W. Hanke, Adv. Phys. 27 , 287 (1978)
12 D. N. Zubarev, Usp. Fiz. Nauk. 71 , 71 (Sov. Phys. Uspekhi 3 , 320) (1960)
13 J. D. Jackson, Classical Electrodynamics (John Wiley, New York, 1962)
14 A. Bagchi and A. K. Rajagopal, Solid State Commun. 31 , 127 (1979)
15 A. Bagchi, Phys. Rev. B 15 , 3060 (1977)
16 R. Del Sole, Solid State Commun. 37 , 537 (1981)
17 J. E. Sipe, Phys. Rev. B 22 , 1589 (1980)
18 B. B. Dasgupta and R. Fuchs, Phys. Rev. B 23 , 3710 (1981)
19 R. G. Barrera and A. Bagchi, Phys. Rev. B 24 , 1612 (1981)
20 R. Del Sole, Surface Sci. 123 , 231 (1982)

21 R. Del Sole and E. Fiorino, J. Phys. Soc. Jpn. 49 , Suppl. A, 1133 (1980)
22 L. V. Keldysh, Pi'sma Zh. Eksp. Fiz. 29 , 716 (JETP Lett. 29 , 653) (1979)
23 F. Bechstedt, R. Enderlein and D. Reichardt, Phys. St. Sol. (b) 117 , 261 (1983)
24 C. H. Wu and W. Hańke, Solid State Commun. 23 , 829 (1977)
25 D. L. Mills, Surface Sci. 48 , 59 (1975)
26 H. Ibach and D. L. Mills, Electron Energy-Loss Spectroscopy and Surface Vibrations (Academic, New York, 1982)
27 B. N. J. Persson, Phys. Rev. Lett. 50 , 1089 (1983)
28 A. Selloni and R. Del Sole, Surface Sci., to be published
29 S. Selci, P. Chiaradia, F. Ciccacci, A. Cricenti, N. Sparvieri and G. Chiarotti, Phys. Rev. B 31 , 4096 (1985)
30 J. Northrup and M. L. Cohen, Phys. Rev. Lett. 49 , 1349 (1982)
31 P. Vogl, H. P. Hjalmarson and J. D. Dow, J. Phys. Chem. Sol. 44 , 365 (1983)
32 D. J. Chadi and R. Del Sole, J. Vac. Sci Technol. 21 , 319 (1982)
33 S. D'Addato, S. Nannarone and M. De Crescenzi, Surface Sci., to be published

Light Emission from Nominally Smooth and Statistically Rough Ag Tunnel Junctions

P. Dawson and D.G. Walmsley

Physics Department, University of Ulster, Coleraine, N. Ireland BT52 1SA

Light emission from both nominally smooth $Al-Al_2O_3-Ag$ tunnel junctions and statistically rough $CaF_2-Al-Al_2O_3-Ag$ tunnel junctions under high values of applied bias (>4V) is observed.

The nominally smooth Ag-junctions emit narrowband uv radiation which is sharply peaked at ∿3.8 eV. The peak emission energy actually displays a weak, but significant dependence on the angle of emission. This observation strongly suggests that the emission is primarily due to the decay of a radiative surface plasmon polariton (SPP) mode excited in the Ag electrode, and is not, as has been suggested, due to Ag volume plasmon-photon coupling. Also, the emission is strongly p-polarised, and displays a Hertz-dipole-like angular distribution; these observations are consistent with radiative SPP emission.

Both narrowband 3.8 eV emission and broadband, mostly visible emission, is observed from single, statistically rough Ag-junctions. Importantly, we find that the broadband emission cuts off at 3.5 eV even when the applied bias exceeds 4 V. This cut-off energy is higher than that reported elsewhere, and confirms that for Ag-junctions the bulk of the broadband output is indeed mediated by the fast SPP mode (which has a maximum energy of 3.5 eV) and not by the slow or junction SPP mode. Both components of the output are found to be most intense in the direction normal to the junction surface, and to be substantially unpolarised. For the broadband emission, these features are consistent with results of theoretical calculation; for the narrowband 3.8 eV emission they contrast sharply with the observed characteristics of the 3.8 eV emission from smooth Ag-junctions. It would thus appear that the radiative SPP mode is roughness scattered, in much the same manner as the intrinsically nonradiative fast SPP mode. Resolution between the narrowband and broadband components of the output is much superior in the tunnel junction arrangement than in experiments where high-energy (keV) electron beams impinge on rough Ag films. It is considered that the difference in resolution is due principally to the different nonradiative SPP excitation spectra in each case.

Observation of Various Electromagnetic Modes in a Nematic Liquid Crystal Using the ATR Method

K.R. Welford and J.R. Sambles

Department of Physics, University of Exeter, Exeter, Devon, U.K.

Attenuated Total Reflection (ATR) techniques are here used to couple radiation into a structure composed of a Nematic liquid crystal (BDH-E7) confined between two aligning layers of obliquely evaporated SiO_x. The liquid crystal when aligned is uniaxial in nature with refractive indices of 1.522 and 1.745 at a wavelength of 632.8 nm. The ordinary index (1.522) of liquid crystal is relatively high, and necessitates using high-index glass prisms (n = 1.800) as the semi-infinite medium on which the active silver layer is deposited.

A rapid evaporation (>10 nms^{-1}) is used to produce flat polycrystalline silver films of thickness ~45 nm on the prism. A layer of SiO_x is then evaporated onto a region of the silver film at an angle of 30° from the normal. A surface-plasmon-polariton (SPP) is excited at the Silver/Air interface, and the ATR spectra is monitored. By fitting the spectra to theory at two different wavelengths, accurate thickness and relative permittivity values for the Silver can be obtained. With this as a starting point, a second ATR spectra for the Silver-SiO_x-Air interface may be fitted to theory, giving then an accurate thickness and relative permittivity values for the SiO_x layer. A thin (~10μm) layer of E7 is now confined between this substrate and a glass backing plate coated with a transparent conductor (ITO) and are arranged to produce homogeneous and parallel alignment of the liquid crystal director. This structure is now investigated at 632.8 nm, using the ATR technique with different $\underline{E}$-fields applied across the liquid crystal layer. A realignment of the director parallel to the field is induced. The realignment occurs initially in the centre of the liquid crystal layer i.e. furthest from the aligning layers, and becomes more extensive as the $\underline{E}$-field is increased. At relatively low voltages (~5 v) the director becomes mostly aligned perpendicular to the Silver-SiO_x interfaces, although very close to the aligning layer the director remains orientated parallel to the Silver-SiO_x interface.

The plane of the incident light is chosen to be parallel to the plane of the director. Measuring the ATR spectra for this structure demonstrates the existence of a SPP mode at the Silver-SiO_x interface. As the applied field is increased from zero, so the director begins to realign with the field. The effective index of the liquid crystal layer for the transverse magnetic SPP begins to change from 1.522 to 1.745 in the centre region of the liquid crystal. At an applied field of 2.5 v the ATR spectra develops features which indicate the existence of guided waves in the structure. As the applied field is further increased, so the guiding region becomes broadened, reducing the coupling length between the SPP and the guiding region. The coupling strength to the guided modes is thus increased at larger fields, and the SPP gradually moves to higher angles of the incident beam. At voltages of the order of 20 v the SPP has all but disappeared leaving only the guided mode features. A detailed calculation using Frank-Oseen theory and multilayer Fresnel equations produce theoretical ATR spectra, which are in excellent agreement with the data, showing clearly for the first time how one may detect simultaneously the SPP resonance and guided modes.

Surface Reflectivity of GaAs(110)

F. Ciccacci, S. Selci, and G. Chiarotti

Dipartimento di Fisica, Università di Roma "Tor Vergata", Italy

P. Chiaradia, A.C. Felici, and C. Goletti

Istituto di Struttura della Materia del CNR, Frascati, Italy

Electronic transitions between surface states in semiconductors can be detected by optical techniques. In particular the Differential Reflectivity spectra, that is the relative change of the overall reflectivity between the clean and the oxidized surface, is directly related to the dielectric function of the surface layer. Data for the GaAs(110) cleaved surface show electronic transitions at around 3.0 eV. In Fig.1 (Rcl-Rox)/Rox vs photon energies is shown for two different oxygen exposures. At a very low coverage (corresponding to an exposure of 10^4 L) an oscillating structure is obtained (asterisks, dotted line). This structure is related to a bulk-like Franz-Keldysh effect induced by the inner elctric field in the space-charge region. In fact a small amount of defect states near the Fermi level induced by oxygen adsorption is enough to drastically modify the band bending. For higher coverages approaching one monolayer (corresponding to $2x10^6$ L) a much larger contribution due to surface transitions appears (open circles, full line in Fig.1).

Figure 2 shows the results obtained using polarized light for the light electric field parallel to the two axes of the surface unit cell. A small anisotropy in the spectra is observed. These measurements can be used as a probe of the symmetry properties of the surface states responsible for the transitions.

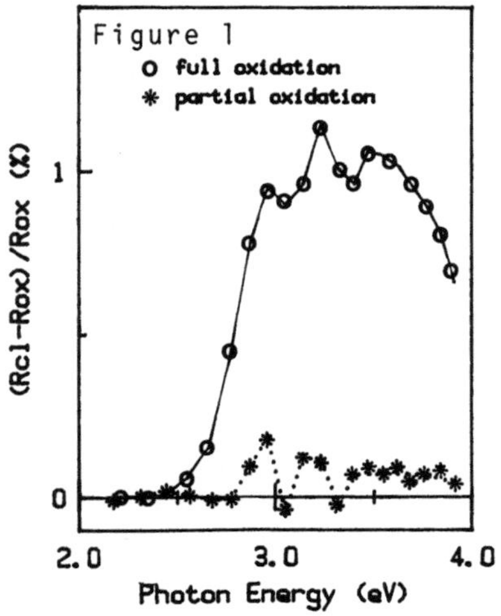

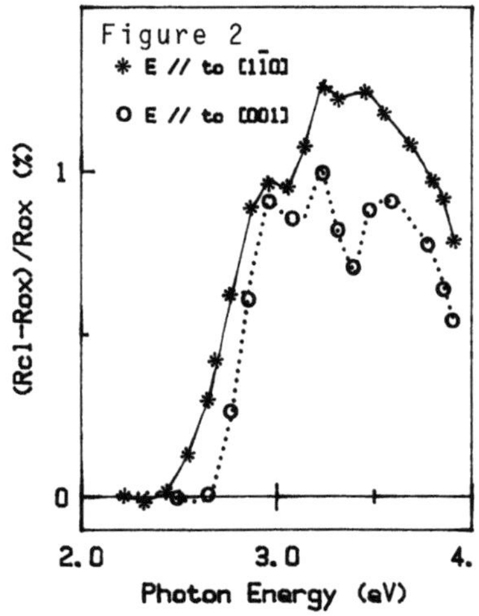

In situ Spectroscopic Studies of the Metal-Electrolyte Interface

G. Piazza

Istituto Chimico G. Ciamician, Università di Bologna, Via Selmi 2, I-40126 Bologna, Italy

Electrochemical interfaces are widely spread out in nature. In most of the spontaneous processes at a solid-liquid interface, a driving electric force generated by accidental heterogeneities acts at the solid surface coupled to non-uniform potential distribution. In an electrochemical system, the necessary physicochemical parameters (e.g., voltage or current, nature and composition of the electrolytic solution, etc.) can be empirically controlled in order to direct the course of a faradaic process. However, little is known about the influence of the nature of the metal side on the rate and mechanisms of these processes at a microscopic level. The need of ancillary spectroscopic techniques with molecular specificity and high sensitivity is steadly growing, particularly in connection with studies in electrocatalysis and energy conversion. A correct use of these techniques and a proper interpretation of the experimental results connot abstain from considering the possible perturbing effect produced by the strong interfacial electric field on the electronic and vibrational energy of the reacting species. This is a consequence of the Fermi-Thomas screening of the potential inside the metal (within ca. 0.5 Å) and its deployment in the solution side within a distance of a few Ångstroms. So that, even a potential of 1 V away from the potential of zero charge applied to the metal can produce a field of more than 10^7 V/cm across the first layer of the solvent (e.g., water) adsorbed onto the electrode. As a consequence, a strong Stark effect may be experienced not only by the electronic states of the species adsorbed at the electrode, but also by surface states of the metal which protrude outside its ideal crystallographic surface.

In this contribution, a brief account is given of the results obtained by the application of Electromodulated Reflectance Spectroscopy in VIS, UV and NIR for in situ studies of:

1- Surface reactions (e.g., various stages of oxide formation at polycrystalline platinum electrode by means of electrochemical-optical impedance measurements at various modulating potential frequencies);
2- Electrochromic and Stark effect on adsorbed dye molecules at Ag and Au polycrystalline electrodes (e.g., Triphenylmethane cation dye);
3- Empty surface states at single crystal Ag and Au electrodes of low Miller indices in relation to the tracing of the interfacial potential profile with and without specific adsorption of anions and their influence on the stability of the metal surface;
4- Optical bulk plasmon dispersion in Ag overlayer during its electrochemical deposition onto a Au(111) substrate.

Surface Enhanced Raman Scattering from Bare Silver Gratings of Various Groove Depths

J.L. Coutaz

LEMO, U.A. 833 du CNRS, Génie Physique, ENSIEG, B.P. 46, F-38402 Saint Martin d'Hères, France

Using RAMAN effect at a bare silver surface, we can excite a continuum of surface plasmons (SP) of STOKES frequency ω_{SP} varying from 0 to the upper limit value $\omega_P/\sqrt{2}$, where ω_P is the plasma frequency of silver [1,2]. The wavevector of each SP is described by the well-known relation :

$$k_{SP}(\omega_{SP}) = \frac{\omega_{SP}}{c}\left(\frac{\epsilon_0\ \epsilon(\omega_{SP})}{\epsilon_0 + \epsilon(\omega_{SP})}\right)^{1/2}$$

ω_{SP} is the frequency of each SP, $\epsilon(\omega_{SP})$ is the dielectric constant of silver and ϵ_0 the dielectric constant of the above medium (air). According to the usual RAMAN nonlinear polarization, this generation of SP is independant of the direction of the incident pump light.

If a grating is grooved on the silver, these STOKES SP can be diffracted towards the above medium, and so detected. Because of the dispersion involved in the longitudinal wavevector conservation law, the frequency of the diffracted SP depends on the angle of detection. We have experimentally verified this property using holographic gratings with 1800 grooves/mm and an 5145 Å wavelength incident laser light.

This nonlinear interaction can be enhanced by increasing the pump field in silver. This occurs when a SP at incident frequency is resonantly excited by use of the grating. As a result in this kind of experiment, two resonances are excited with the grating :

a) coupling of pump light into the silver.

b) diffraction of the continuum of SP at STOKES frequency towards the upper medium. The groove depth of the grating plays an important role in the efficiency of these resonances. As for some different other nonlinear optical effects at grating surface [3], there exists an optimum groove depth for which the STOKES signal is maximum.

We experimentally found [4] this optimization for the RAMAN effect at the bare silver grating described above. The incident angle is about 4° in order to excite a SP at pump frequency. The detector is always placed in the same direction to detect a constant diffracted frequency STOKES beam (ω_{SP} = 16500 cm^{-1}), the only varying parameter during the experiment being the groove depth. The maximum value of the diffracted intensity at STOKES frequency is measured for an optimum groove depth of 370 Å. Notice that :

a) the optimum modulation of the grating is very small : about 7 %.

b) no enhancement, as compared to the flat case, can be calculated, because the flat surface does not diffract any SP.

1. J.C. Tsang, J.R. Kirtley, T.N. Theis : Solid State Commun. 35 667 (1980)
2. A. Girlando, W. Knoll, M.R. Philpott : Solid State Commun. 38 895 (1981)
3. J.L. Coutaz et al. : to be published in Phys. Rev. B
4. J.L. Coutaz, R. Reinisch : to be published in Solid State Commun.

Brillouin Spectroscopy of Langmuir-Blodgett Films

R. Zanoni, C. Naselli, J. Bell, G. Stegeman, and C. Seaton

Optical Science Center and Arizona Research Laboratory,
University of Arizona, Tucson, AZ 85721, USA

S. Lindsey

Department of Physics, Arizona State University, Tempe, AZ 85287, USA

Using Brillouin scattering we have been able to measure the Brillouin Spectra of thin cadmium arachidate (CdA) films which were deposited by the Langmuir-Blodgett (L-B) technique onto aluminum film substrates. A 9-pass tandem Fabry-Perot interferometer [1] was used to frequency analyze light scattered by multiple, high frequency, thermally-excited surface acoustic waves guided by the CdA film.

Light from a frequency-stabilized argon ion laser, wavelength .5145 μm, was focused onto the sample surface at a 70° angle of incidence. Back-scattered light was collected and collimated for spectral analysis by the tandem Fabry-Perot. Representative Brillouin spectra are shown in Fig. 1. These spectra typically took 1 hour to accumulate while scanning the tandem Fabry-Perot at the rate of 1/sec/scan. Spectra were obtained for various incident field polarizations, in order to identify the acoustic modes.

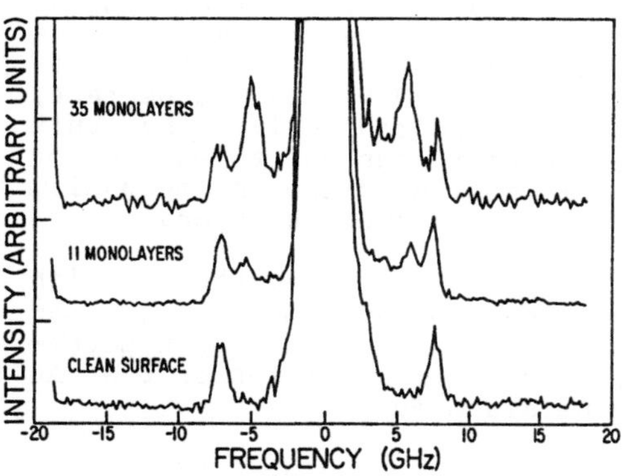

Fig. 1. Brillouin spectra of aluminum film surface coated with L-B film. The three spectra are a) uncoated surface, b) 11 monolayers, and c) 35 monolayers

The fact that distinct changes in the Brillouin spectrum occurred for as few as 11 monolayers (295 Å thick) is remarkable, and we expect that film-related Brillouin spectra can probably be obtained for 5 monolayers, or less. It is not possible to obtain accurate elastic constants from the current data without a detailed analysis which is currently underway. This work will be published in greater detail [2].

References

1. S. Lindsay, M.W. Anderson, and J.R. Sandercock, Rev. Sci. Instrum., 52 1478 (1981)
2. R. Zanoni, C. Naselli, J. Bell, G. Stegeman, S. Lindsay, and C. Seaton, (to be published in Thin Solid Films)

Elastic Surface Wave Anomalies Near the Cluster-to-Layer Transition of Au on NaCl Observed by Brillouin Scattering

B. Hillebrands, R. Mock, and G. Güntherodt

II. Physikalisches Institut+, Universität Köln, D-5000 Köln 41, F.R.G.

P.S. Bechthold and N. Herres

Institut für Festkörperforschung, KFA Jülich, D-5170 Jülich, F.R.G.

The cluster-to-layer transition of Au on NaCl (001) has been investigated by means of Brillouin scattering from elastic surface wave excitations /1/. The measurements were performed using a tandem Fabry-Perot interferometer. For the clean NaCl surface, both bulk modes (transverse (TA) and longitudinal (LA) acoustic modes) and surface modes (Rayleigh- and LA-mode) have been detected.

In the regime of Au clusters on NaCl a softening of the Rayleigh mode with increasing gold deposition is observed, which is attributed to the increase in surface mass density. A new surface acoustic mode, which tracks the Rayleigh mode, appears in the frequency range above the TA mode of NaCl. The physical nature of this mode may be due to a) modifications of the reflection coefficients of bulk phonons at the surface due to the enlarged surface mass density, b) coherent off-resonance internal cluster vibrations coupled via the substrate, resulting in a long-wavelength surface mode driven by bulk waves or c) additional inter-cluster forces due to charged or dipolar clusters. From the experimental evidence the mechanism b) seems most suited to describe the observed phenomena.

In the regime of Au layers on NaCl we observe the Rayleigh mode and the first and second order Sezawa modes. In the vicinity of the percolation threshold, which has been determined by electron microscopy, the inhomogeneities of the Au layer cause a softening of the Sezawa mode.

The frequency of the new surface acoustic mode crosses the TA mode frequency of NaCl at the percolation threshold, which can be identified this way.

/1/ B. Hillebrands, R. Mock, G. Güntherodt, P.S. Bechthold, N. Herres, to be published

---

+ Work supported by Deutsche Forschungsgemeinschaft, SFB 125

Part III

Wave Guides and Integrated Optics

Fundamentals of Waveguides

E. Garmire

Center for Laser Studies, University of Southern California,
Los Angeles, CA 90089, USA

1. Introduction

Many of the properties of dielectric waveguides can be discussed in a general way without solving any difficult transcendental equations or concentrating on specific cases. In this chapter we will consider many of the fundamental characteristics of propagation in waveguides which will be important in integrated optics. We will start from the simplest possible case, and progressively add more sophistication. It is anticipated that sufficient fundamentals are presented here to allow a basic understanding of all of integrated optics.

2. Mode Picture

For simplicity, consider the planar symmetric waveguide shown in Fig. 1. When the boundary conditions are symmetric, the basic properties of solutions of a linear second-order differential equation dictate that there will be two solutions: one which is symmetric and one which is anti-symmetric around the axis of symmetry. For the geometry of Fig. 1, the symmetric solution is

$$E = E\cos k_x x e^{i(k_z z - wt)}. \qquad (1)$$

This describes a cosine-like amplitude distribution, which propagates in the z direction with a propagation constant given by k_z. The anti-symmetric solution is given by the same expression with cos replaced by sin. This latter solution will be discussed later.

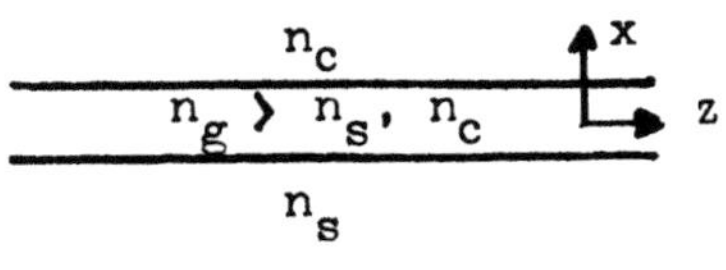

Fig. 1. Geometry for symmetric planar dielectric waveguide, with propagation along the z direction.

Equation (1) describes the field in a waveguide. We will find that there are a series of discrete solutions, depending on the refractive index and thickness of the waveguide. These solutions represent the modes of the waveguide. That this is the correct form for the modes can be seen by realizing that inside the waveguide, where the medium is uniform, the electro-magnetic field must obey the wave equation. Solutions to the wave equation in uniform media must be either cos/sin or exponential. Both forms are included in (1) when the k's are considered complex if necessary. The value of the k's for any particular solution may be found by matching boundary conditions at the boundaries of the waveguide. A formal solution will follow our discussion of some general properties of waveguides.

3. Ray Picture

In order to relate the mode picture and the ray picture of propagation in a waveguide, rewrite (1) as

$$E = (E/2)(e^{ik_x x} + e^{-ik_x x})e^{i(k_z z-wt)}$$

and define a new vector, shown in Fig. 2:

$$\vec{k}_\pm = \pm k_x \hat{1}_x + k_z \hat{1}_z. \qquad (2)$$

Thus the guided mode can be written as

$$E = (E/2)[e^{i(\vec{k}_+ \cdot \vec{r}-wt)} + e^{i(\vec{k}_- \cdot \vec{r}-wt)}]. \qquad (3)$$

This field describes the sum of two travelling waves, propagating at angles given by

$$\tan\theta = k_x/k_z \qquad (4)$$

as shown in Fig. 2. Ray optics provides a convenient method for analyzing waveguides, as long as it is remembered that the ray describes the normal to a plane phase front and not a direction of energy propagation.

Fig. 2. Geometry for ray optics model.

An equivalent ray optics description can be provided for the sine-like waveguide modes. In particular, the mode "angles" are also described by (4).

If the waveguide is not symmetric, the solutions to the waveguide will have a linear combination of cosine-like and sine-like solutions.

4. Dispersion

The most important characteristic of waveguides for most integrated optics applications is the fact that they have a dispersion which is different from that of plane waves. Dispersion is the relation between propagation vector and frequency. This can be determined from the wave equation:

$$\nabla^2 E - n_g^2 \varepsilon_o \mu_o \partial^2 E/\partial t^2 = 0.$$

Plug in the field from (3) to obtain

$$k_\pm^2 = n_g^2 \varepsilon_o \mu_o w^2. \qquad (5)$$

This equation says that the rays introduced above in (2) travel with the same propagation constant as a plane wave travels in bulk material of refractive index n_g, the same composition as the waveguide. That is

$$|k_\pm| = n_g w/c \equiv k_g.$$

The propagation constant in the mode picture can now be related to the transverse mode profile through (2), which becomes

$$k_z^2 = k_g^2 - k_x^2. \qquad (6)$$

We define an effective refractive index describing propagation of a mode down the waveguide as

$$n_{eff} \equiv k_z c/w = \sqrt{n_g^2 - (k_x/k_o)^2}, = n_g\sqrt{1-\theta^2}.$$

where we have defined the free space propagation constant as

$$k_o \equiv w/c = 2\pi/\lambda.$$

This effective refractive index defines the phase velocity of the mode in the waveguide:

$$v_p = w/k_z = c/n_{eff} = \frac{(c/n_g)}{\sqrt{1-(k_x/k_g)^2}} \sim \frac{c/n_g}{(1-\theta^2)^{1/2}}. \qquad (7)$$

The group velocity is given by the derivative of the frequency with respect to the wave-vector, and is found by squaring and differentiating (7), using $k_g = n_g w/c$:

$$v_g = dw/dk_z = (c/n_g)\sqrt{1-(k_x/k_g)^2} = (c/n_g)(1-\theta^2)^{1/2}. \qquad (8)$$

5. <u>Perfectly Confining Waveguide</u>

To see how the dispersion relations behave in a waveguide, we solve the simplest possible problem - the waveguide whose field goes to zero at the boundaries and is therefore "perfectly confining". This case will be valid for waveguides with very high dielectric discontinuities or for highly reflecting metallic walls. The geometry and field distributions are shown in Figure 3.

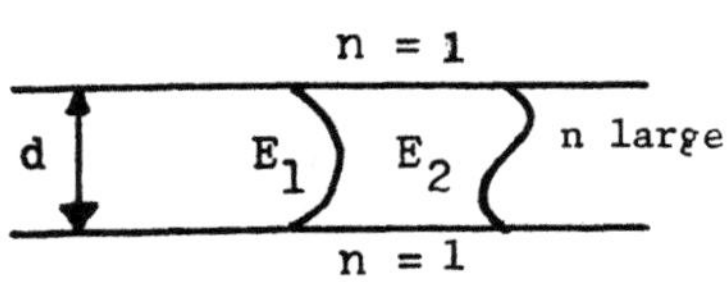

Fig. 3. Geometry for waveguide with perfectly confining walls

From the requirements that the fields go to zero at $|x| = d/2$,

$\cos(k_x d/2) = 0$, which imples $k_x d/2 = m\pi/2$, $m = 1,3,5,..$

$\sin(k_x d/2) = 0$, which imples $k_x d/2 = m\pi/2$, $m = 2,4,6,..$

This means that there is a complete set of possible transverse wave vectors given by

$$k_x = m\pi/d, \qquad m = 1,2,3,4...$$

where **odd** values of m represent **symmetric** solutions and **even** values of m represent the **anti-symmetric** solutions.

The effective refractive index for the mth mode of this guide is therefore given by

$$n_{eff} = n_g\sqrt{1-(m\lambda/n_g d)^2}. \tag{9}$$

The group velocity at which information propagates down the waveguide is

$$v_g = (c/n_g)\sqrt{1-(m\lambda/n_g d)^2}.$$

Consider that the wavelength is fixed, but that the waveguide can vary in thickness. The effective index as a function of waveguide thickness is shown in Fig. 4. Notice that as the waveguide becomes infinitely thick, the effective index asymptotically reaches the refractive index of the material in the waveguide. This can be seen from the form of (9). Notice also that there is a minimum waveguide thickness below which the refractive index becomes unphysical for each mode. This minimum determines the **waveguide cutoff condition**:

$$d \geq m\lambda/n_g. \qquad m = 1,2,3,4.... \tag{10}$$

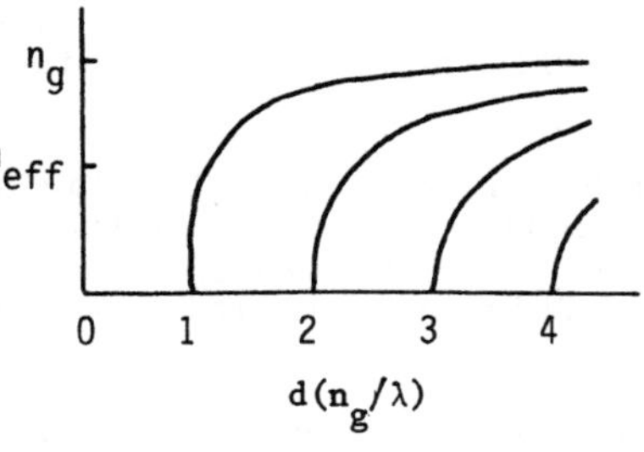

Fig. 4. Effective refractive index vs normalized guide thickness for waveguide with perfectly confining metal walls.

By locating the value of d on the graph, the number of modes and their effective index can be determined. The apparent possibility of a refractive index going to zero is an artifact of the "perfectly confining" approximation. It will be shown below that the **minimum** effective refractive index is that of the substrate.

We will now show that dielectric waveguides which are not "perfectly confined" have similiar properties, with the fields described as a series of modes characterized by their individual effective indices and cutoff thicknesses.

6. Symmetric planar dielectric waveguide

We now consider waveguides which are not "perfectly confining" as shown in Fig. 5. In this general case, the field penetrates into the surrounding cladding layers, where it has the form of a decaying exponential. Thus the general expression is

$$E = \begin{cases} Ae^{-p(x-d/2)} & x \geq d/2 \\ B\cos(hx) & -d/2 \leq x \leq d/2 \\ Ce^{p(x+d/2)} & x \leq -d/2. \end{cases}$$

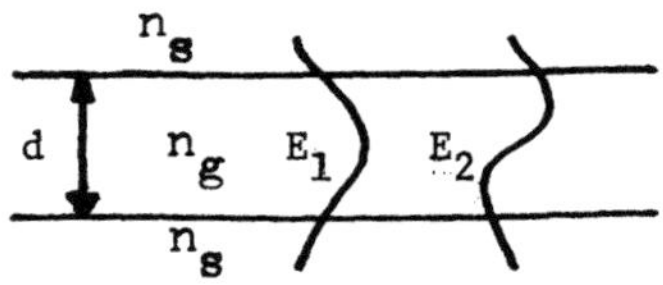

Fig. 5. Symmetric dielectric waveguide, with field penetration into the substrate. Modes 1 and 2 are shown.

From (6) the propagation constant can be seen to be

$$k_z^2 \equiv (n_g k_o)^2 - h^2. \qquad (11)$$

Similiarly,

$$k_z^2 = (n_s k_o)^2 + p^2. \qquad (12)$$

The last two equations can be rewritten as

$$h^2 d^2 + p^2 d^2 = (n_g^2 - n_s^2) k_o^2 d^2 \equiv V^2. \qquad (13)$$

V is the **V parameter**, which is very useful to describe waveguides, and can be thought of as a normalized waveguide thickness, for a given refractive index discontinuity and wavelength. V is a measure of how well the structure guides light.

From the knowledge of the V parameter we can determine a considerable amount about the waveguide without actually solving for the numerical solution. For example, (13) describes a circle in hd, pd space. This puts a limit on the maximum value of h or p. The two limits are:

1. Maximum value for p; this occurs when h = 0. From (11) this means that

$$k_{z,max} = n_g k_o \equiv k_g.$$

The maximum value for the propagation constant is the same as for a plane wave propagating in bulk material of the waveguide. This is the same as discussed in the "perfectly confining" waveguide, and corresponds to a mode which is pulled tight into the waveguide, with very little tail in the surrounding region.

2. Maximum value for h; this occurs when p = 0. This corresponds to the tail of the guided wave extending to infinity. That is, the waveguide is at cutoff. Notice from (12) that the propagation constant in this case is the same as that in the surrounding cladding layer:

$$k_{z,min} = n_s k_o \equiv k_s.$$

That the minimum value for the propagation constant is the same as for a plane wave propagating in the substrate or surrounding medium is not surprising, since in this case the tails of the guided wave extend to infinity. The light sees much more of the substrate than of the waveguide. This is a more physical limit than was obtained in the perfectly confining case. Also,

$$h_{max}d = \sqrt{(k_o d)^2 n_g^2 - n_s^2} \equiv V.$$

Thus the V parameter gives the maximum possible value for the transverse propagation constant in the guide. We will show that this will limit the number of modes which can exist in the waveguide.

7. Mode Size

A waveguide is generally characterized by its effective refractive index, which is determined by calculating k_z. The mode-profile is also important, and is determined by knowledge of the mode-number m and of the value for p. In particular, since p describes the exponential fall-off of the field outside the waveguide, the mode can be considered to have an effective thickness given by

$$t = d + 2/p.$$

In the case of maximum p, the mode thickness equals the waveguide thickness. The second case, p -> 0, corresponds to a mode thickness going to infinity. In a general case, the mode thickness will be somewhat larger than the waveguide thickness.

The complete solution for the field in a dielectric waveguide is determined by matching boundary conditions at $x = \pm d$. It is at this point that the polarization of the field in the waveguide must be considered, since the boundary conditions depend on polarization. Each of the orthogonal polarizations must be separately analyzed.

8. TE Polarization

Consider the electric field parallel to the plane of the waveguide. This is generally called the TE polarization. The boundary conditions require that the tangential electric field and the normal magnetic field are continuous. The normal magnetic field is related to the derivative of the tangential electric field, however. Thus the boundary conditions can be expressed as the requirement that Ey and ∂Ey/∂x are continuous. The field distribution can therefore be written as

$$Ey = \begin{cases} \cos(hd/2)e^{-p(x-d/2)} & x \geq d/2 \\ \cos(hx) & -d/2 \leq x \leq d/2 \\ \cos(hd/2)e^{p(x+d/2)} & x \leq -d/2. \end{cases} \quad (14)$$

The continuity of the derivative leads to the following **eigenvalue equation**:

$$pd = (hd)\tan(hd/2) \quad \text{even modes}$$
$$pd = -(hd)\cot(hd/2) \quad \text{odd modes} \quad (15)$$

These, coupled with (13), lead to transcendental equations which must be solved numerically. It is convenient to consider

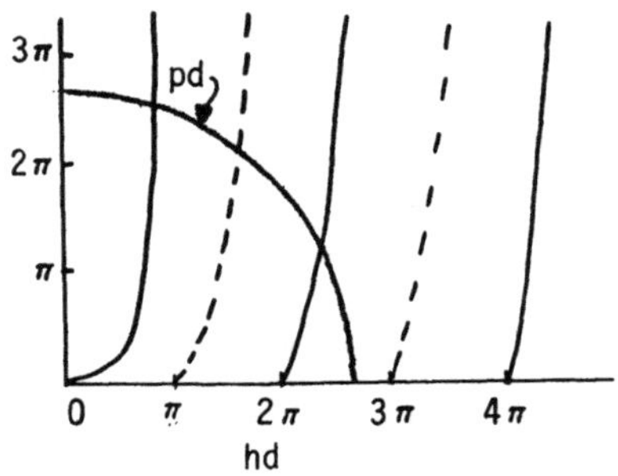

Fig. 6. Graphical solution of eigenvalue equation for modes of symmetric dielectric waveguide. Solid lines are even modes (eq. 15) and dashed lines are odd modes.

the simultaneous graphical solution of the two equations, as shown in Fig. 6. This describes pd, hd space, in which (13) describes a circle and (15) is also plotted. Intersections of these two graphs yield solutions for the particular waveguide. By studying this graphical solution we can see that the behavior is similiar to that described for the "perfectly confining waveguide". That is, there are discrete modes with different cutoff values of V (normalized waveguide thickness). The discrete cutoff is determined by the limits of the range of (hd)tan(hd/2) and (hd)cot(hd/2). In this case, however, the lowest order mode is always guided, no matter how small the waveguide thickness, since (hd)tan(hd/2) has a finite solution no matter how small hd. This expresses the well-known fact that a symmetric waveguide always contains at least one mode (and two polarizations).

The numerical solutions to dielectric planar waveguides give the effective index as a function of normalized waveguide thickness. The results are shown in Fig. 7, for varying degrees of asymmetry in the waveguide, as will be discussed below. From the form of the solutions it can be seen that the cosine and sine-like modes interleave. That is, the thinnest waveguides ($V < \pi$) have only a single symmetric cosine-like mode. However, as the size of the waveguide increases (V increases), it goes through the cutoff condition and begins to guide the second mode, which is anti-symmetric, with a sine-like field profile. This occurs when $V = \pi \equiv V_1$.

In general, the waveguide will contain m modes if $V \leq V_m$, where

$$V_m \equiv m\,\pi. \tag{16}$$

The transverse propagation constant must always be smaller than V/d. Thus

$$h_m \leq m\pi/d,$$

so that by (11) the longitudinal propagation constant is

$$k_{z,m} \geq \sqrt{(n_g k_o)^2 - (m\pi/d)^2}. \tag{17}$$

Equation (16) describes the cutoff condition for the waveguide. That is, the number of modes which are confined is given by

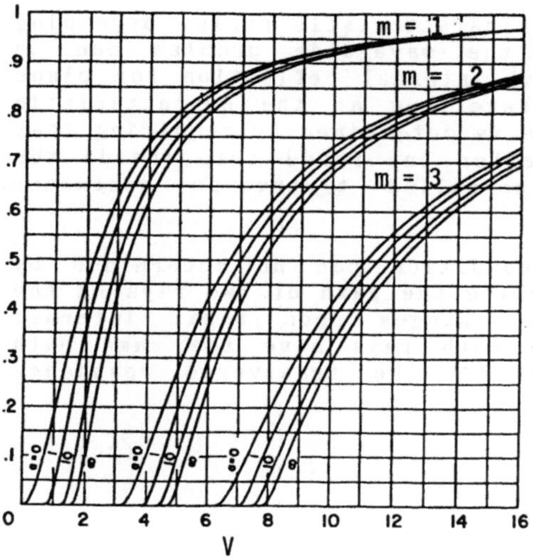

Fig. 7. Effective refractive index (normalized so that 1 = n_g and 0 = n_s) vs normalized waveguide thickness (V parameter). Curves with a = 0 are symmetric guides, while curves which are perfectly asymmetric have a = ∞.

$$V = k_o d\sqrt{n_g^2 - n_s^2} \leq V_m = m\pi.$$

Thus the total number of modes M is the largest integer given by:

$$M \leq (d/\lambda)\sqrt{n_g^2 - n_s^2}. \qquad (18)$$

Another way to write this condition is that the Mth order mode will be confined if the thickness of the waveguide is given by

$$d \geq \frac{M\lambda}{\sqrt{n_g^2 - n_s^2}}. \qquad (18)$$

The amount of penetration of the mode field into the substrate is determined by knowledge of p. In particular, from the limits to k_z, we may write

$$p^2 \geq (n_g^2 - n_s^2)k_o^2 - (m\pi/d)^2.$$

The effective thickness of the mode becomes

$$t \leq d + \frac{2}{\sqrt{(n_g^2 - n_s^2)k_o^2 - (m\pi/d)^2}}.$$

In terms of the V parameter, this is

$$t \leq d\left[1 + \frac{2}{\sqrt{V^2 - (m\pi)^2}}\right]. \qquad (19)$$

9. Alternative Description by Ray Optics

A completely equivalent picture to the above analysis can be determined by ray optics. We have shown that light can be considered as travelling at a characteristic angle, depending on the mode. Solution to the waveguide problem can be determined by considering total internal reflection for plane waves inside the waveguide travelling at the characteristic angle. The eigenvalue equation is determined by requiring that this plane wave be periodic after one round trip inside the waveguide. This is the so-called **transverse resonance condition**.

The transverse resonance condition can be determined by referring to Fig. 8. By following the path of two rays on the phase front, including phase changes upon total internal reflection, and requiring that both rays have the same path length within a multiple of 2π, the transverse resonance condition is determined:

$$2n_g k_o d\cos\theta - \phi_s - \phi_c = 2m\pi. \tag{20}$$

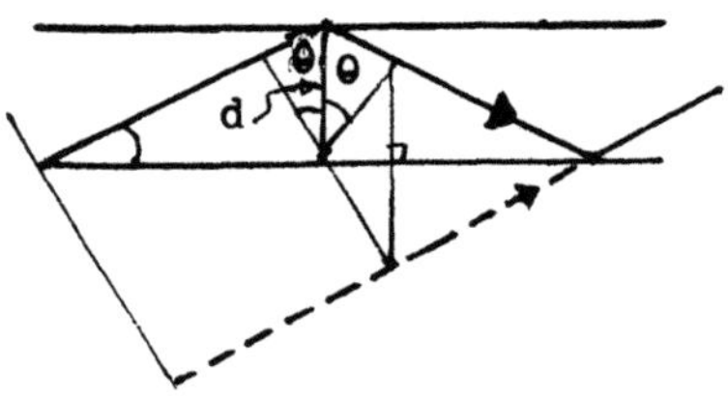

Fig. 8. Geometry for transverse resonance condition. Two path lengths must be the same to within an integral number of wavelengths.

The phase change upon total internal reflection is determined from the Fresnel reflection coefficients, and is given in any optics book as

$$\tan(\phi_{TE}/2) = \sqrt{(n_g\cos\theta)^2 - n_s^2}\,/n_g\sin\theta \tag{21}$$

The angle θ is here defined from the surface of the waveguide and not from the normal, as it is in most optics books. Simultaneous solution of (20) and (21) determines the eigenvalue equation. This may be done graphically as with (13) and (15). The results are identical to those obtained by the mode evaluation. The ray optics picture can be used interchangeably with the mode picture in any waveguide which is not nonlinear.

10. TM-Polarized Modes

The analysis to this point has been for TE polarized waves. That is, light polarized along the plane of the waveguide. Completely analogous solutions can be found for light polarized perpendicular to the plane of the waveguide. This is called the **TM polarization**. The simplest way to analyze this case is to redo the mode analysis leading to (14) and (15) for the **magnetic** field rather than the electric field, since it is this field which has no z component (transverse magnetic). In this case the boundary conditions are slightly different. In particular, the requirement is continuity of the dielectric constant times the electric field, which leads to a requirement that $n^2\,\partial B_x/\partial y$ must be continuous. Under these conditions, the eigenvalue equation is altered to have the form

$$pd = (n_s/n_g)^2(hd)\tan(hd/2) \qquad \text{even modes}$$

$$pd = -(n_s/n_g)^2(hd)\cot(hd/2) \qquad \text{odd modes.}$$

Notice that under the usual conditions, that the refractive index in the waveguide and surrounding layers have close to the same values, that the TM eigenvalue equations reduce to the TE eigenvalue equation. In isotropic material the TE and TM modes in planar waveguides are very nearly degenerate.

11. Asymmetric Waveguides

So far we have solved only for symmetric waveguides in which both surrounding layers have the same refractive index. When the top cladding layer is different from the bottom substrate layer, the same procedure which led to (14) and (15) is followed, but the results are algebraically more complicated. One practical limit is particularly simple, however. That is the case shown in Fig. 9, in which the refractive index discontinuity between the waveguide and the cladding layer is large while the refractive index discontinuity with the substrate is small.

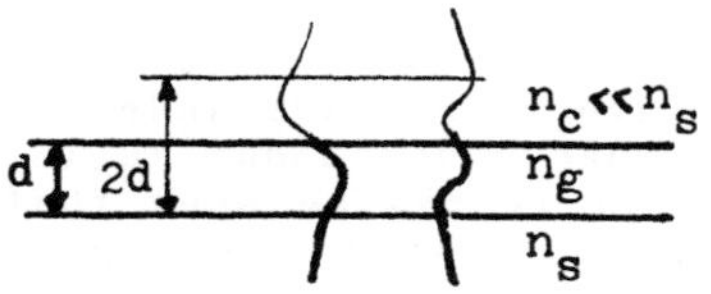

Fig. 9. Geometry and mode profiles for highly asymmetric waveguide. Light lines indicate equivalent symmetric waveguide of twice the thickness.

Notice from the figure that the modes in this case correspond to the anti-symmetric modes of a waveguide of twice the thickness. The previous analysis for a symmetric waveguide can be used, therefore, if only the odd modes are considered, and if d = 2a, where a is the thickness of the asymmetric waveguide. Note that the mode number must be adjusted accordingly, since only every other mode is present. That is, m' = m/2. Thus, the cutoff condition becomes

$$a \geq \frac{\lambda m'/4}{\sqrt{n_g^2 - n_s^2}} \qquad m' = 1,2,3,4... \qquad (22)$$

What is particularly important is that even the lowest order mode of the asymmetric waveguide has a cutoff thickness. This is fundamentally different from the symmetric waveguide, which always, in principle, confines at least one mode.

12. Rectangular Waveguides

For most integrated optics applications the waveguides are not planar, but are confined in two dimensions, as shown in Fig. 10. The exact solution to the fields in such a waveguide requires numerical methods. However, simple approximate methods are valid in most experimental cases.

The most useful approximate technique is the **effective index method**. This method is best described by referring to Fig. 11.

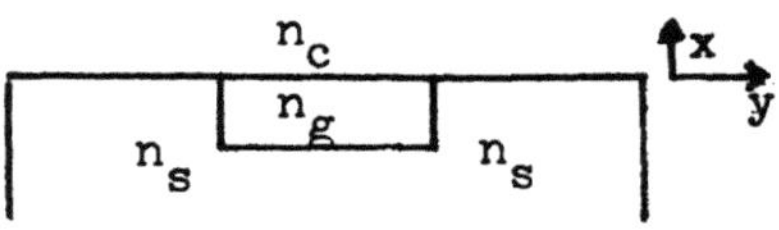

Fig. 10. Cross-sectional geometry for rectangular waveguide

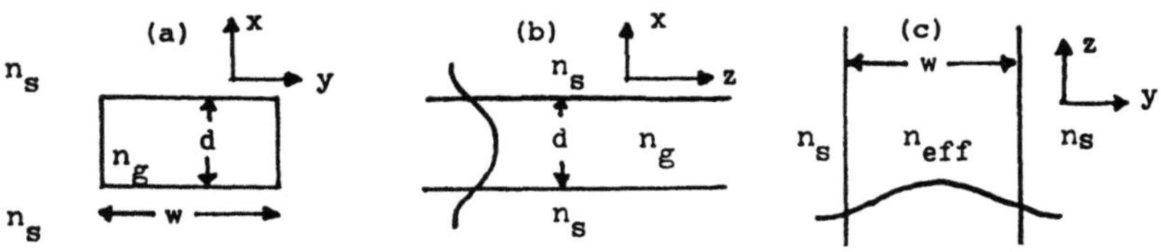

Fig. 11. Geometry to calculate rectangular guide using the effective index method.

The guiding properties in the x and y directions are considered separately, and in each dimension the waveguide is considered as effectively planar. Consider first guiding in the planar waveguide defined by refractive index variations in the x dimension, as previously calculated, and shown in Fig. 11b. The result of this analysis was that guiding resulted in an effective refractive index for light inside the waveguide. The effective index method uses this effective refractive index as the value to be used for the waveguide when confinement in the y dimension is calculated. From Fig. 11c it can be seen that we must solve for an effective planar waveguide in the y dimension with the original value of the index in the substrate but with the effective refractive index in the waveguide. This method gives very close to the correct result for rectangular waveguides.

The effective refractive index is given by referring to (17),

$$(n_{eff})^2 = (k_{z,1}/k_o)^2 = n_g^2 - (h_1/k_o)^2. \qquad (23)$$

The subscript 1 is added to indicate that this was calculated for the first planar waveguide. This expression is inserted as the refractive index in the waveguide of the second guide.

Consider the cutoff condition for the lowest order mode in a highly asymmetric waveguide. When the waveguide is rectangular it becomes

$$d \geq \frac{(\lambda/4)}{\sqrt{n_g^2 - (h_1/k_o)^2 - n_s^2}} . \qquad (24)$$

Notice that this increases the thickness of the guide needed to confine light.

13. Coupling into Waveguides

There are two fundamental methods for coupling into waveguides, both shown in Fig. 12. The first is to focus light into the end of the waveguide. This abrupt coupling, occurring at a fixed value of z, is called endfire coupling. This coupling is

Fig. 12. Geometry for coupling into waveguide

normally evaluated by calculating an overlap integral between the incident light field and the desired modes of the waveguide.

The second method is coupling distributed along the length of the guide. Prism, grating, tapered and directional couplers represent distributed coupling. These couplers are typically evaluated by coupled mode theory. Both coupling techniques require utilizing the fact that the modes of the waveguide form a complete set of orthogonal functions. That is, a waveguide mode E_m obeys the following relationship for all m and n:

$$\int E_m E_n^* \, dx = P\delta_{mn} \qquad (25)$$

The integral is taken over all values of x. The delta function is one when m = n and is zero for m not equal to n. P is a normalizing factor. This complete set of orthogonal functions makes it possible to describe any electric field in the waveguide in terms of these modes:

$$E(x) = \sum_{m=1}^{\infty} c_m E_m(x) \qquad (26)$$

where c_m is an amplitude factor describing how much of each mode is required to describe the field function E(x).

14. Radiation Modes

To fully describe any field E(x) it is necessary to include radiation modes as well as guided modes. These are modes of the waveguide whose value of p is imaginary. The spatial profile of these modes, therefore, are radiation-like rather than evanescent outside the waveguide. From (12) it can be seen that imaginary p means that radiation modes have an effective refractive index less than that of the substrate. This corresponds to a mode angle greater than the critical angle for total internal reflection. Radiation modes are not totally internally reflected, but are leaky, radiating energy to infinity. Incident light which is coupled into radiation modes represents a loss to guided wave coupling efficiency. The orthogonality and completeness of the waveguide modes are important properties which must be used when analyzing waveguide couplers. It is important to note that in general these properties are not valid in nonlinear waveguides.

15. Endfire Coupling

Equation 26, which results from the fact that the modes of the waveguide form a complete set, allows us to calculate endfire coupling. Consider that E(x) is the incident light beam. By performing the expansion in (26), we determine the amplitude factors c_m, which in turn determine the fraction coupled to each mode. Maximizing the power coupled into a given waveguide mode means matching the incident beam amplitude profile with that of the waveguide mode. This means maximizing the overlap integral between the incident light and the mode. In the expansion of (26), the mode coefficient is expressed in terms of the incident light field E(x) by

$$c_m = \frac{\int E^*(x)E_m(x)dx}{\int |E(x)|^2dx \int |E_m(x)|^2dx}. \qquad (27)$$

The fraction coupled into mode m is

$$f_m = |c_m|^2. \qquad (28)$$

Remember that the field of the mode often extends beyond the waveguide, so that the integration is over all space, with the field inside the waveguide expressed as cosine or sine, while the field outside the waveguide is exponential (guided waves) and oscillatory (radiation modes). In fact, the coupling is roughly maximized when the input beam has a width equal to the thickness of the guided mode, t, given by (19).

16. Distributed Coupling

Coupled mode theory is generally used to describe distributed coupling, and uses a modal description of the incident light as well as of the desired coupled mode. It is assumed that these modes can be well defined before the coupler is introduced, and that the interaction causes distributed coupling which occurs over many wavelengths. The slow exchange of light between the incident beam and the waveguide can be described by the **coupled mode equations**;

$$dc_m/dz + ik_{z,m}c_m(z) = \Sigma K_{mn}c_n(z). \qquad (29)$$

The **coupling coefficients** K_{mn} are calculated for each particular coupler by determining the power which drives the field of the guided mode. When the coupler alters the boundary conditions, this can be thought of as changing the dielectric constant in the altered region. This changed dielectric constant causes the incident field to induce a polarization which drives power into the waveguide. The coupling coefficient can be written as

$$K_{mn}(z) = iC\int [n(x,z)^2-n_o^2]E_m^*E_n dx \qquad (30)$$

where n(x,z) is the altered refractive index and n_o is its value in the unaltered structure. C is a normalizing constant.

From the form of (29) it can be seen that constructive power growth between modes m and n will occur only if $k_{z,n} = k_{z,m}$, or if K has a periodicity which will cancel the difference in the propagation constants. That is, if we may write

$$K_{mn}(z) = K_{mn}\exp(\pm iKz),$$

then growth of power into guided mode m from incident mode n occurs only if

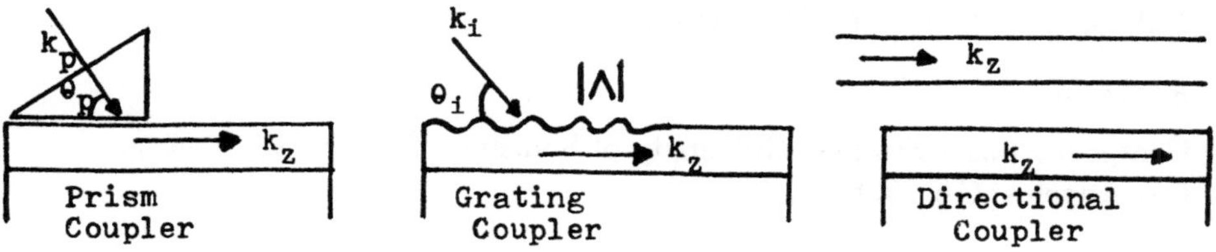

Fig. 13. Three geometries for coupling into waveguides

$$k_{z,m} = k_{z,n} \pm K. \tag{31}$$

Equation (31) is sufficient to understand the basic principles of all the distributed input couplers shown in Fig. 13.

The most common distributed coupler is the prism. In this case the coupler has no periodicity in the K direction, so that it is necessary that the incident wave and the guided wave have the same z components. If light incident at an angle is to couple into the waveguide and have the same z component as the guided mode, it is necessary for the light to be incident from a region of refractive index higher than the waveguide. This is done with a prism, as shown. Evanescent wave coupling between light in the prism and light in the guided mode obeys coupled mode theory,and the required input coupling angle is

$$k_p \cos\theta = k_{z,m} \quad \text{or} \quad n_p \cos\theta = n_{eff}.$$

Through coupled mode theory it can be seen that if light can couple into the waveguide, it can also couple out of the waveguide. To achieve maximum coupling efficiency, therefore, it is important to terminate the input coupler when maximum coupling efficiency is achieved. This is why a **prism** is usually used, as shown in the figure.

Another important input coupler is the **grating** coupler. In this case a corrugation is applied to the surface of the waveguide, so that the coupling coefficient has a periodicity. With the requirement of (31), it can be seen that the input angle should be

$$k_i \cos\theta_i = k_{z,m} \pm K$$

where k_i is the wave vector of the incident wave in medium of refractive index n_i. This condition on angle can be rewritten as

$$n_i \cos\theta_i = n_{eff} \pm K/k_g.$$

The last coupler of particular interest is the **directional** coupler, which is composed of two waveguides in close proximity. Light can couple between the two waveguides through the interaction of their evanescent waves. The condition for complete coupling is that the effective refractive indices of the modes in each of the waveguides are the same.

Integrated Optics Technology

E. Garmire

Center for Laser Studies, University of Southern California,
Los Angeles, CA 90089, USA

1. Introduction

Integrated Optics is the use of guided waves on a substrate to achieve optical circuits which can be used in a number of applications. An example of such an integrated optical device is the waveguide switch shown in Fig. 1. This waveguide device is used to switch light between fibers.

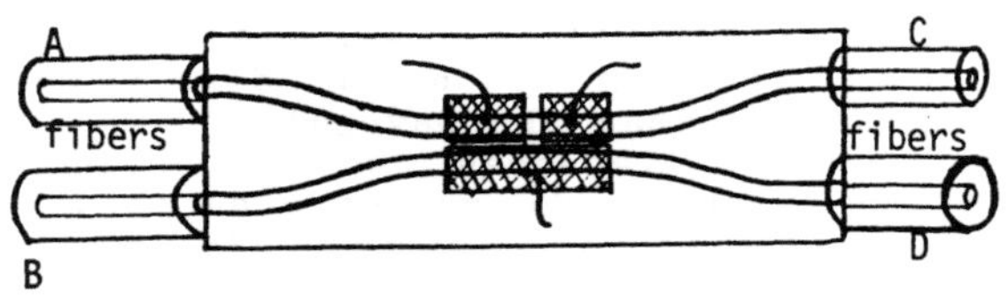

Fig. 1. Directional coupler switch between two fibers

Applications for integrated optics devices lie in the areas of optical communications, optical signal processing, sensors, high-speed electrical test equipment, in nonlinear optics and in passive taps and branches. The important technologies developed to date use electro-optic materials such as $LiNbO_3$, III-V semiconductors, which integrate lasers and detectors as well as electro-optic components, and passive materials such as glass and polymers.

In this chapter we will review applications and materials, indicating the current state-of-the-art in devices.

2. Applications

2.1. Communications

Figure 1 shows an integrated optics switch for fiber optic communications, which allows optical signals from channel A or B to be switched to either C or D.

The alternative to such an an optical switch in a fiber optic communication system is a combination of detector, electronic switch and lasers. If the system is wide-band, these must all be high-speed components. By contrast, however, wide-band communications can be transmitted in the optical domain through a slow integrated optics switch. That is, fast communication is possible, even if the switching is slow. Optical switches may be low-cost components for local area networks with multiple users.

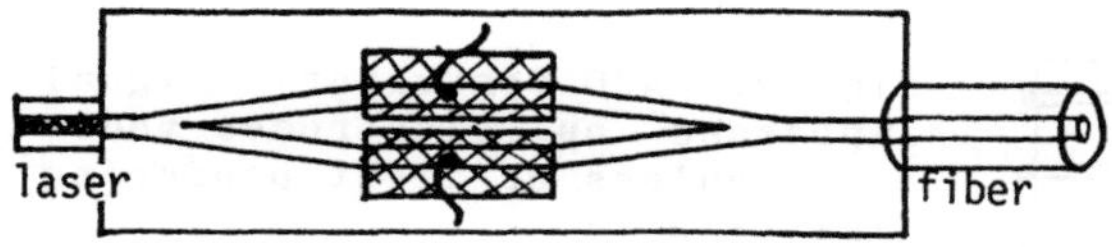

Fig. 2. High-speed interferometric modulator attached to laser

High-speed modulators, such as shown in Fig. 2, represent another important optical communication application. It has been found that direct modulation of lasers for communication purposes causes "chirp." That is, the frequency of the laser changes with time. This is a serious problem in long wavelength (1.55 μm) communications, since dispersion of the fiber means that a change in wavelength alters the speed of propagation of a pulse and causes errors in communication. It seems that it may be most practical to make good quality, stable, single-mode cw lasers and to use integrated optics to modulate them.

A third important application for integrated optics in optical communication systems is the multiplexing/demultiplexing function. An example of a possible integrated optics multiplexer/demultiplexer is shown in Fig. 3. In this case, a single fiber is used to transmit several wave-lengths, each modulated with a different message. The use of several gratings, each of different spacing, means that different wavelengths can be deflected into the same guided wave beam. These beams are then focussed by a luneberg lens onto the output facet, where a fiber is butt-coupled to the integrated optics chip.

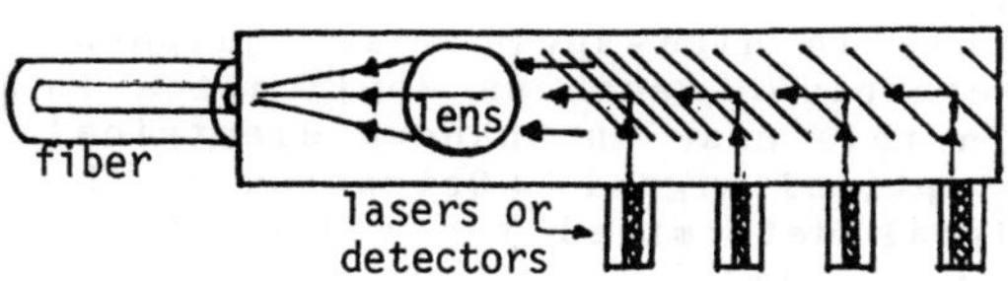

Fig. 3. Wavelength multiplexer/demultiplexer, which uses a chirped grating to separate wavelength.

2.2. Signal Processing

The use of integrated optics for signal processing has been one of the most promising applications. Commercial optical signal processors have long been available, and considerable effort has been underway to reduce these to integrated optics devices, with potential for high performance and low cost. In 1975 a major program was begun to develop an integrated optics rf spectrum analyzer, the fundamentals of which are shown in Fig. 4. This technology is now mature, although performance continues to be limited by cross-talk between channels. Indeed, integrated optics has not yet supplanted bulk acousto-optic rf spectrum analyzers in the commercial market.

A more successful application is the high-speed A/D converter, as shown in Fig. 5. The initial and final signals

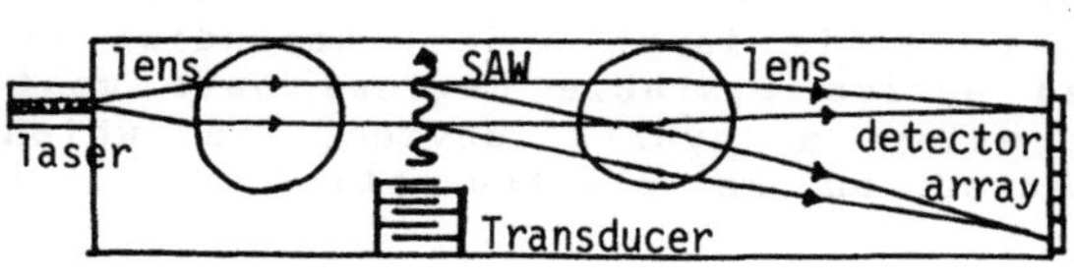

Fig. 4. Rf spectrum analyzer, which measures angle of light deflection from a surface acoustic wave

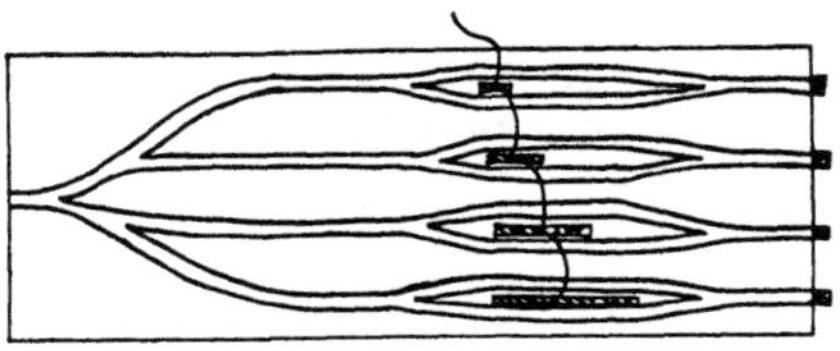

Fig. 5. A/D Converter. Signal impressed on electrodes modulates pulses of light producing digital bits

are electrical,and an electrical discriminator means that some degradation in the optical signal will not spoil the performance. The integrated optical device has performed A/D conversion faster than any available electronics,and indicates that there is a large future for specialized signal processing applications using integrated optics.

2.3. Optical Sensors

There has been a recent revolution in sensors, with the development of fiber optic devices which are able to measure temperature, pressure, electric field, magnetic field and rotation, by measuring their effect on the path length in an optical fiber coil. In some cases, less expensive and more reliable sensors may be available using integrated optics. Temperature-sensitive integrated optics devices have been demonstrated both by using a resonant ring and an interferometer with unequal arm length.

2.4. High Speed Electrical Test Equipment

One of the new and exciting applications for integrated optics is the possibility of their use in high-speed test equipment for electrical signals. An example of such an application is the electro-optic sampler. In this case an unknown electrical signal is used to modulate an optical signal. Detection of the modulation on the optical signal determines the value of the electrical signal as a function of time.

2.5. Nonlinear Optics

The use of integrated optics keeps light confined in a small cross-section for a long path length. In nonlinear optics this makes it posible to have efficient frequency conversion. In particular, 25% second harmonic from 0.83 μm to visible wavelengths has been demonstrated; this makes possible blue light from a GaAs diode.

Another nonlinear optical device of interest for a variety of applications from signal processing to aberration-correcting laser mirrors is degenerate four-wave mixing. Integrated optics degenerate four-wave mixing has been demonstrated. Guiding of both pump waves,as well as the probe and signal beams, allows degenerate four-wave mixing with rather low intensity.

Finally, nonlinear optics makes possible optical logic functions, such as bistable devices - that is, devices with two stable output states for a single input state, depending on past history. Integrated optics bistable devices have most conveniently been made using a hybrid device, in which electrical feedback is used on an electro-optic or other modulator.

2.6. Branches and Taps

Passive materials may be used for star couplers, branches and taps. When developed to be reproducible and inexpensive, they will replace the all-fiber taps which are currently used.

Although in almost all areas of application there are competitive technologies, such as high-speed electronics, or all-fiber systems, eventually the extraordinarily wide bandwidth capabilities and the low-cost batch-fabrication possibilities of integrated optics, as well as its ability to handle specialized signal processing functions rapidly, will allow this technology to surpass its competition.

3. Materials

An important question which is not yet fully answered is, what materials will form the basis of integrated optics technology? We will look at the status of the major technologies to date.

3.1. Lithium Niobate

3.1.1. Fabrication

It is generally considered that $LiNbO_3$ is the most mature integrated optics technology. The fabrication technique of choice is thermal diffusion of a thin titanium film which has been photolithographically processed to delineate the optical circuit. Other fabrication technologies, such as proton exchange (heating in benzoic acid), are not as mature, and are not generally used for fabricating complex devices.

Experiments have shown that the titanium diffusion technology is reproducible within a given laboratory, although the specific recipes differ from laboratory to laboratory. Typically, 800 Å of Ti is heated to 1000°C for a few hours in flowing O_2 which has been bubbled through water to saturate it with vapor. Researchers generally agree that neither substrate material nor the processing technology is the limit to reproducible performance. It appears that the mask tolerances are the most crucial aspect of achieving high performance. Even high quality e-beam written masks seem to be the cause of device performance which differs from design specifications.

The difficulty of reproducible waveguide fabrication can be seen from a simple analysis by referring to Figure 6. For small changes in waveguide width, the fractional change in propagation constant is proportional to the fractional change in waveguide width through

$$k_g^2 = k_z^2 + (m\lambda/W)^2; \qquad dk_z/k_z = (m\lambda/k_z W)^2$$

Assume $k_z \sim k_g = 2\pi/n\lambda$. For a path length of L, and ensuring that the optical path length difference in the two guides is less than π, the tolerance condition on the width is given by

$$dW/W = 2nW^2/\lambda L.$$

For L = 1 cm, λ = 1 μm, n ~ 2, W = 5 μm, we find dW/W < 0.01. This puts a tolerance of 500 Å on the width. Since such close tolerances are hard to achieve in masks or in photoresist processing, trimming after fabrication, or active compensation may be required to achieve high-performance devices.

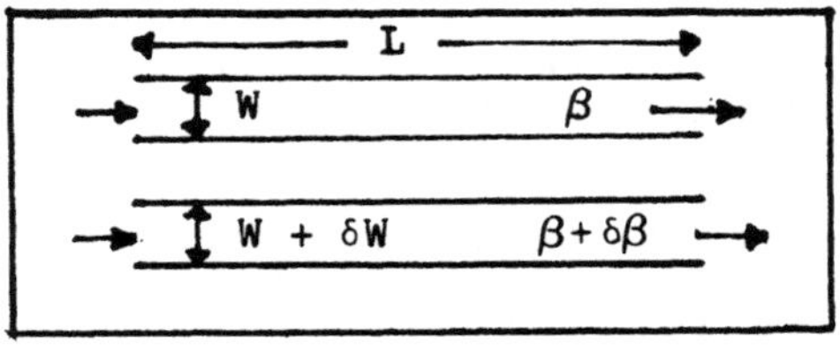

Fig. 6. Geometry of parallel waveguides for tolerance calculations.

3.1.2. Optical Damage

One of the outstanding problems discussed was optical damage (or photorefractive effect). The photorefractive effect is a change in refractive index with optical exposure which occurs in electro-optic insulating materials. The electric fields which cause the change in refractive index are due to the migration of carriers which are released by optical absorption. This exposure-caused change in refractive index causes a drift in the optical output as exposure is increased, and is termed optical damage when it is undesirable. The effect is well known in $LiNbO_3$ at shorter wavelengths, and has even been used to write semipermanent holograms. It is now apparent that this problem also exists at GaAs wavelengths, for there are several reports of changes in optical output with time at 0.8 µm at power levels less than 100 µwatts. However, apparently optical damage is not a problem at 1.3 µm, for no effects have been seen at this wavelength, even for tens of milliwatts of input optical power.

Several techniques have been considered to reduce the seriousness of optical damage in $LiNbO_3$ at 0.83 µm or shorter wavelengths. This includes using MgO doped $LiNbO_3$ which has recently been shown successful for frequency doubling. This material is slightly conductive, which removes optical damage by allowing charges to be mobile. However, preliminary reports indicate that after titanium diffusion, optical damage is almost as bad in this material as in undoped $LiNbO_3$. A technique of photo-fixing has been developed which involves flooding the illuminated sample with UV light, capturing the carriers outside the waveguide channels. This increases the optical damage resistance considerably. Finally, many systems designers expect to provide a means of actively adjusting for possible optical drift, particularly in light of the need to compensate for mask irregularities.

In summary, optical damage can be ignored at 1.3 µm, while at 0.83 µm there are several techniques to keep its deleterious effects under control.

3.1.3. Optical Loss

Typical numbers for the best fiber-channel waveguide-fiber insertion loss at 1.3 µm are around 1 dB, still somewhat higher than desirable for systems applications. The limitations to loss are: first, the necessity of mode-matching the channel waveguides with the fiber modes; second, channel waveguide propagation loss; third, loss at bends and junctions; and finally, optical loss introduced by electrodes. In order to maintain overall low insertion loss, each of these sources of loss must be reduced. Mode-matching is difficult, because the asymmetric $LiNbO_3$ mode profile does not match the circular fiber mode profile. Deeper titanium diffusion matches the modes better, but requires higher modulation voltages. Propagation loss is minimized by using high-quality substrates and by carefully designing buffering layers to isolate the optical fields from the metallic electrodes. Reduction of loss in bends and junctions requires proper mask design. Rapid progress is being made in this area with the recent reduction of overall loss from 6dB to 1dB.

This loss is sufficiently low that prototype devices may be used in experimental systems, but continuing research is required to reproducibly reach the overall 0.5 dB insertion loss which is desirable for most systems applications.

3.1.4. Electro-optic Switching and Modulation

Three basic device geometries appear to have the most promise for the purposes of electro-optic switching and modulation. These are the directional coupler switch, shown in Fig. 1, the Mach-Zehnder inteferometric modulator, shown in Fig. 2, and the intersecting waveguide switch, shown in Fig. 7. While, in principle, the interferometer may be configured to be a switch as well as a modulator, it is not the most convenient geometry except for a modulator. The advantage of the interferometric modulator is its inherent symmetry, which means that optical damage effects are less serious in this configuration than in the directional coupler, in which all power enters one arm.

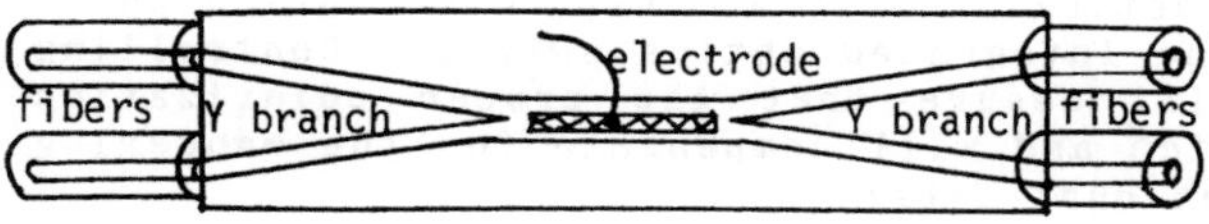

Fig. 7. Intersecting waveguide switch

In recent years the intersecting waveguide switch has begun to look particularly promising. This device can be seen as the logical extension of the directional coupler in the limit that the spacing between the waveguides goes to zero. There is no consensus at the present time as to how the intersecting waveguide switch compares to the directional coupler. A practical difference is its use of small-angle Y branches rather than curved waveguides, each of which may introduce loss. One of the apparent advantages of the intersecting waveguide switch is that it is typically shorter than directional couplers which have been reported to date. In addition, the smaller distance between the arms in the active region would suggest that lower switching voltages are possible; however, cross-talk may be more of a problem than in the directional coupler with the channels fully separated.

It is generally accepted that the technology of each of these devices is mature, and that they can be reproducibly fabricated; the specific design depends on the specific application. Bandwidth, cross-talk, insertion loss, buffer and electrode design, all depend on specific device design, and application and optimization requires a tradeoff of these concerns. This is one reason why general- purpose integrated optics switches and modulators are not commercially available at the present time.

There are a number of other useful electro-optic devices which have been demonstrated in $LiNbO_3$. An important one is the grating deflector/modulator shown in Fig. 8. This has been used for matrix multiplication by impressing one electrical signal on the incoming light beam and the second on the electrodes. The output is the matrix multiplication of these two signals.

Fig. 8. Electro-optic grating deflector using interdigital electrodes.

With fiber optical communications, polarization is generally not preserved, and unfortunately, none of the switches and modulators which are most convenient are polarization-independent. Geometries which ensure polarization-independence exist, but they typically require very high voltages. This is an outstanding research problem which has not been solved. Since it appears that practical integrated optics switches depend on polarization, their application will require polarization control. They may be used with polarization-preserving fiber. It is likely, however, to be more practical to put an integrated optics phase shifter in front of the integrated optics switch. Controlling this phase shifter will assure that the proper polarization always enters the switch and will compensate for the wandering polarization of conventional fiber.

3.1.5. <u>Research Still in Progress</u>

Currently under development are buffer layers, which do not degrade the electrical performance but provide optical isolation. The transparent electrode material indium tin oxide is an example of a new buffer layer which looks promising. Reliability and long-term degradation are advanced development studies which are just beginning. Reduction in fiber-to-fiber insertion loss to a reproducible 0.5 dB must be made before these switches will be practical for local area networks. High speed packaging is crucial to optimized performance and well-known microwave techniques are currently being applied to this problem.

Convenient, reliable and inexpensive fiber pigtailing is the most important development which must be completed if the systems designers are to use integrated optics switches. Furthermore, it is necessary to develop techniques to handle drifts in voltage, optical damage, and control of polarization. Finally, it is important to determine the limits to cross-talk for each of the devices. There must be discussion between device developers and systems designers, so that useful optimization can proceed, both on the device level and on the systems level.

3.2. III-V Semiconductors

3.2.1. Fabrication

The least developed of the integrated optics technologies are waveguides and switches in III-V semiconductors such as GaAlAs and GaInAsP. However, rapid progress in this area is being made because of the recent availability of MOCVD material. This provides relatively inexpensive, flat and reproducible layers. Examples of waveguides under construction are strip-loaded GaAlAs single-mode waveguides surrounded by cladding layers with an even larger aluminum concentration, as shown in Fig. 9. For low-loss propagation it is important to have the bandgap of the waveguide very much larger than the wavelength of observation. Under these circumstances, losses typically 1 dB/cm are observed. There has been an unconfirmed rumor of losses as low as 0.2 dB/cm.

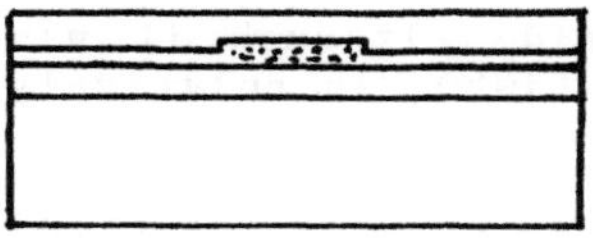

Fig. 9. Cross-sectional geometry for III-V integrated optics

3.2.2. Research Areas

The most immediate research in III-V semiconductor-integrated optics is the development of waveguides which mode-match to fibers, since insertion losses reported are still very large (~7 dB). In addition, fabricating low-loss curved waveguides remains a challenge. Directional couplers which utilize multiple parallel channels rather than curved waveguides have been demonstrated. Finally, electro-optic switching devices require further research and development, with no consensus to date as to the optimum geometry.

The prognosis for III-V integrated optics looks very good, particularly in the area of optical switching. This will provide optical components for monolithic opto-electronics, and will complement the lasers, detectors and electronic elements currently under development. Furthermore, integration of electronic components with electro-optic III-V switches should make possible monolithic ultra-high-speed optical switches. Finally, switches and modulators in III-V semiconductors may become competitive with $LiNbO_3$ because of their inherent small size and the ability to use semiconductor processes in their fabrication.

3.3. Glass

Several glass-passive components are currently available and may be competetive with fiber optics components.

Multimode waveguides, buried below the surface of the glass, have been developed, which mode-match very well to graded index fibers. The fabricational technique is ion exchange of thallium followed by a second sodium exchange to decrease the refractive index near the surface. These buried guides have been used to make star couplers with very small insertion loss.

Single mode waveguides have been made to produce stars and taps for fiber optics communication systems. The loss is typically about 0.1 dB/cm. There is interest in developing passive ring resonators for inertial rotation sensing, which would require losses much lower, on the order of 0.01 - 0.001 dB/cm. With current technology (potassium ion exchange in BK7 glass), the loss in glass waveguides is limited to the order of 0.1 dB/cm. Lower loss can be achieved by using special soda lime glass substrates which are prepared without fining agents.

Research areas which remain in the field of glass waveguides are the development of ultra-low loss waveguides, and pigtailing fibers to chips. The low cost of batch fabrication and potential reproducibility may make glass integrated optics passive components competitive with currently available fiber components.

4. Conclusions

Integrated Optics is a rapidly moving field, although it was begun as early as 1969. The technology in $LiNbO_3$ is relatively mature, but research in III-V semiconductors is rapidly progressing, with the availability of MOCVD material. Specialized signal processors have shown considerable potential for integrated optics devices, as have optical sensors. Commercial applications in the communications field have awaited single mode fiber technology, which has recently arrived. This has provided the incentive for rapid development of new integrated optics devices, although loss reduction remains an ongoing challenge. There is still room for innovation, as well as increasing materials development in the field of integrated optics.

Surface Wave Electrooptics and Acoustooptics

R.M. De La Rue

Department of Electronics and Electrical Engineering, The University, Glasgow G12 8QQ, Scotland, U.K.

1. Introduction

My objective in this series of lectures is to give an introduction to an area of research activity which is already well developed and which has very clear technological applicability. The linear electrooptic (Pockels) effect and the photoelastic effect together provide convenient 'passive' means for manipulating many, if not all, properties of guided light beams. In many cases, it is essential that the light be coherent and therefore, typically, obtained directly or indirectly from a laser.

Research in this area of 'integrated optics' has progressed on a foundation provided by a remarkably limited number of materials. The most notable of these is the ferroelectric single-crystal, lithium niobate, although its sister material, lithium tantalate, has played a minor part. The only serious alternatives are provided by the III-V compound semiconductors such as gallium arsenide and indium phosphide (and related ternary and quaternary compounds formed in epitaxial layers on substrates of these two materials). Gallium arsenide and indium phosphide are, of course, key optoelectronic materials for light emission and photodetection. Interest in electrooptic devices in these materials stems primarily from the objective of monolithic integration. In contrast, integrated optics based on lithium niobate, with its relatively large band gap, is necessarily a hybrid technology.

2. Basic Crystal Optics and the Electro-optic Effect

Single-crystal materials possessing a linear electrooptic effect must be anisotropic and must not possess a crystallographic centre of symmetry. The same requirement holds for piezoelectricity. Our present objective is to set up a minimal description of the tensorial nature of the electrooptic and elastooptic properties of a single-crystal medium. We use a Cartesian set of axes (x,y,z) or alternatively (1,2,3), which can be related in a well-defined manner to the true crystallographic axes. In lithium niobate, the z-axis (according to the IRE convention) is identical to the optical c-axis, while there are (because of the three-fold symmetry property about the z-axis) three crystallographic axes in the plane perpendicular to the z-axis, any one of which can be chosen as the y-axis, with the x-axis then automatically fixed.

The relationship between optical frequency electric flux density, D, and the electric field strength, E, is then written as:

$$D_i = \varepsilon_{ij} E_j \tag{1}$$

or, explicitly, as:

$$D_1\hat{x}_1 + D_2\hat{x}_2 + D_3\hat{x}_3 = (\varepsilon_{11}E_1 + \varepsilon_{12}E_2 + \varepsilon_{13}E_3)\hat{x}_1 + (\varepsilon_{21}E_1 + \varepsilon_{22}E_2 + \varepsilon_{23}E_3)\hat{x}_2 + (\varepsilon_{31}E_1 + \varepsilon_{32}E_2 + \varepsilon_{33}E_3)\hat{x}_3 \quad (2)$$

Eq.(2) follows from the compact form of Eq.(1) through the convention that a repeated index implies summation over the range of values of the index. Here, j extends from 1 to 3. In technological units, the electric flux density, electric field strength and dielectric permittivities are measured respectively in Coulomb. m^{-2}, Volt. m^{-1} and Farad. m^{-1}. It is often more convenient to use the relative permittivity components given by

$$\varepsilon_{ij} = \varepsilon_{ij}^{(r)} \cdot \varepsilon_o \quad (3)$$

where $\varepsilon_o = 8.854.10^{-12}$ Farad. m^{-1} is the free-space value.

ε_{ij} is a 3x3 tensor but this does not mean that nine different numbers are required to characterise it. Standard reference works (NYE [1], KAMINOW [2], MILEK and NEUBERGER [3]) show that a set of 'principal axes' (x_1, x_2, x_3) exist, for which the dielectric permittivity is purely diagonal, i.e. that only the three 'principal' dielectric constants $(\varepsilon_1, \varepsilon_2, \varepsilon_3)$ are required. It can also be shown that, in general, there are two optical waves with different plane polarisations and velocities which can propagate along a given direction. The velocities are given by: $v = C/n$ (4)
where C is the free-space velocity of light and n is the refractive index (a dimensionless quantity). The general-direction propagation properties of the medium are then summarised in the indicatrix relationship:

$$\frac{x_1^2}{n_1^2} + \frac{x_2^2}{n_2^2} + \frac{x_3^2}{n_3^2} = 1 \quad (5)$$

where $n_1 = \sqrt{\varepsilon_1}$, $n_2 = \sqrt{\varepsilon_2}$ and $n_3 = \sqrt{\varepsilon_3}$. The refractive index is <u>not</u> a proper tensorial property of the medium, but this is not a problem because a dielectric <u>impermeability</u> tensor B_{ij} can be defined, for which, in general:

$$B_{ij}x_ix_j = 1 \quad (6)$$

and where, using the Kronecker delta:

$$B_{ij}\varepsilon_{jk} = \delta_{ik} \quad (7)$$

and, along principal axes only, $B_{ii} = 1/\varepsilon_{ii}$. It can be shown that both B_{ij} and ε_{ij} are symmetric, i.e. that $B_{ij} = B_{ji}$. This symmetry property permits the reduced tensorial notation to be used, i.e. $B_{11} = B_1$, $B_{22} = B_2$, $B_{33} = B_3$, $B_{23} = B_{32} = B_4$, $B_{13} = B_{31} = B_5$, $B_{12} = B_{21} = B_6$. The indicatrix relation can then be re-written as:

$$B_1x_1^2 + B_2x_2^2 + B_3x_3^2 + 2B_4x_2x_3 + 2B_5x_1x_3 + 2B_6x_1x_2 = 1 \quad (8)$$

In many crystals and for lithium niobate, in particular, there are only two distinct values for the refractive index: ordinary index, n_o, for polarisation in a plane perpendicular to optical axis (C- or z-axis) and the extraordinary index, n_e, for polarisation along the optical axis.

The linear electro-optic effect and the photoelastic effect respectively describe the changes in the optical frequency dielectric properties of the medium when an additional electric field is present and when the medium is elastically deformed. The change in dielectric impermeability is expressed by:

$$\Delta B_{ij} = r_{ijk}E_k + p_{ijk\ell}s_{k\ell} \tag{9}$$

Before proceeding further, a word of warning - notation and nomenclature for the various physical properties and quantities in different source works are not the same. Here, r_{ijk} is the 3x3x3 electro-optic tensor, $p_{ijk\ell}$ is the 3x3x3x3 photoelastic tensor and $s_{k\ell}$ is the 3x3 elastic strain tensor. There are important reasons for showing the electrooptic and photoelastic effects together in Eq.(9). One reason is that the state of stress or strain of the medium, because it is ipso facto piezoelectric, influences the apparent value of the electro-optic coefficients. There are two sets of electro-optic coefficients r^S_{ij} (the constant strain, 'true', clamped, or high-frequency electrooptic coefficients) and r^T_{ij} (the constant stress, unclamped, free or static electrooptic coefficients). The relationship between the two sets of coefficients is:

$$r^T_{ij} = r^S_{ij} + \sum_k p_{ik}d_{jk} \tag{10}$$

Here the reduced notation has already been used, so that both the indices i and k run from 1 to 6, while j only runs from 1 to 3. The piezoelectric coefficients, d_{jk}, form (in the reduced notation) a 3x6 tensor. In lithium niobate, reported values for the coefficient r_{33} are $30.8.10^{-12}$ m/V at constant strain and $32.2.10^{-12}$ m/V at constant stress (MILEK and NEUBERGER [31]), while for r_{22} the difference is proportionately much larger:- between $3.4.10^{-12}$ m/V at constant strain and $6.8.10^{-12}$ m/V at constant stress.

The tensor for the electrooptic coefficients in lithium niobate, which is classified crystallographically as trigonal 3m, can now be set out:

$$\begin{pmatrix} 0 & -r_{22} & r_{13} \\ 0 & r_{22} & r_{13} \\ 0 & 0 & r_{33} \\ 0 & r_{42} & 0 \\ r_{42} & 0 & 0 \\ -r_{22} & 0 & 0 \end{pmatrix}$$

But, following PAPUCHON [4], it is useful to present this tensor in a slightly different form:

The small dots indicate zeros, while the links indicate coefficients of equal magnitude, and encirclement indicates that signs are opposite. Notice, (i) that only eight coefficients are non-zero and (ii) that there are only four different values. Physically, giving the tensor coefficients in the unreduced notation is better, because it is then easier to understand what components of field and polarization are being linked by the electrooptic effect.

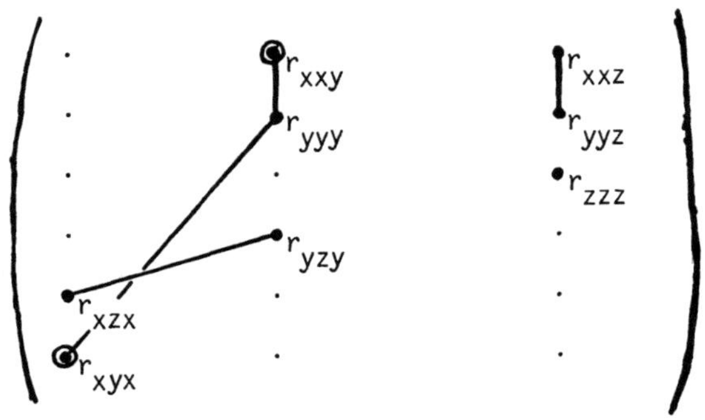

Because the different coefficients play key roles in different types of guided-wave device structure, it is worth looking at them individually. Firstly, consider r_{zzz} or, in reduced notation, r_{33}. This coefficient implies a change in the refractive index (i.e. in the propagation velocity) for an optical field polarized along the z-axis, produced by an applied electric field also along the z-axis, but without any change in the optical field polarization. The value of r_{33} in lithium niobate, $30.8.10^{-12}$ m/V is considered to be large! The coefficient r_{xxz} ($=r_{13}$) implies a change in the refractive index for an optical field polarized along the x-axis, produced, also, by an applied electric field along the z-axis and, again, without any change in the polarization. r_{13} is 10.10^{-12}. mV^{-1} and therefore requires three times the electric field strength (i.e. in practise approximately three times the voltage applied between electrodes). Both r_{13} and r_{33} have been extensively employed in guided-wave electrooptic devices because they can provide, within limits, pure phase modulation. r_{yyz} ($=r_{23}$) is identical in magnitude to r_{xxz} and applies in an analogous way for y-polarized light.

The coefficient r_{xzx} ($=r_{51}$) describes the change of z-polarized light to x-polarized light by an applied field along the x-direction, or vice versa. Likewize, r_{yzy} ($=r_{42}$) leads to changes of optical polarization between y and z and vice versa with applied electric field along the y-direction. The value of r_{42} and r_{51} is 28.10^{-12} mV^{-1}. r_{42} and r_{51} couple two differently polarized optical waves with substantially different velocities, i.e. with, approximately, the ordinary refractive index, $n_o = 2.28$ for the x- and y-polarisations, and the extraordinary refractive index, $n_e = 2.20$ for the z-polarisation (at $\lambda = 0.6328\mu m$). In order to obtain interaction between the two polarizations over a substantial distance, a special periodic electrode structure has to be used to provide the necessary difference wave vector for effective phase matching.

The remaining group of coefficients (r_{xxy} ($=r_{12}$), r_{yyy} ($=r_{22}$) and r_{xyx} ($=r_{61}$)) has the value $6.7.10^{-12}$ mV^{-1} (at constant stress), is arguably less useful and even, in some cases, may lead to distinct limitations on good device performance.

3. Guided-wave Electrooptic Devices: Phase Modulation

Consider the diagrams for the electro-optic device configurations shown in Fig. 1(a) and 1(b).

The increased-index stripe region immediately below the substrate surface is typically produced by high temperature (c. 1000C) diffusion of a thin ($\sim$ 300Å) layer of titanium into a lithium niobate substrate. This region is typically 5μm by 2μm in cross-section and several millimetres long, has only

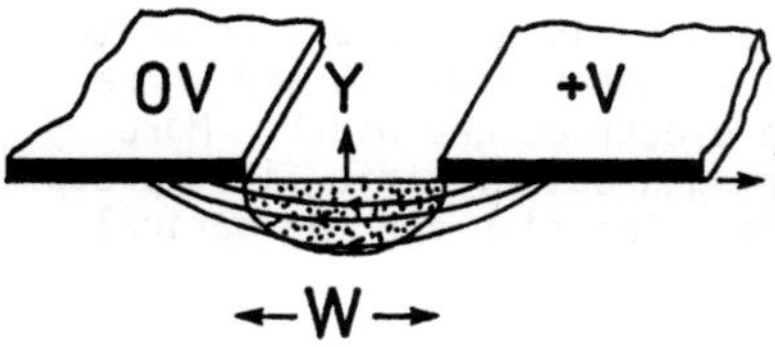

Fig. 1(a)

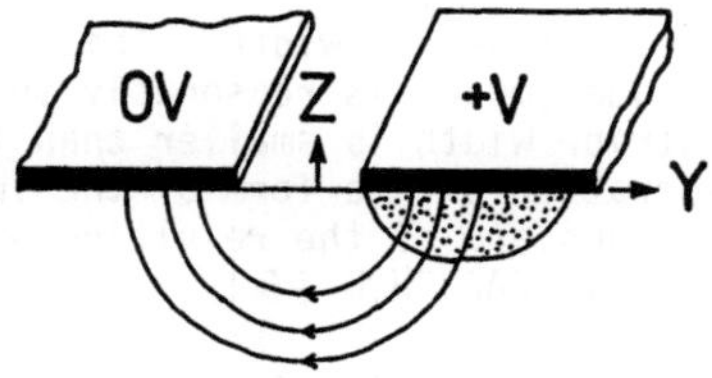

Fig. 1(b)

Electro-optic stripe waveguide device configurations on lithium niobate

a small increase in refractive index ($\sim$ 0.01) and forms an inhomogeneous single-mode waveguide. In Fig. 1(a), metal electrodes have been deposited and formed lithographically on either side of the stripe waveguide. With a voltage difference between the electrodes there will be an inhomogeneous electric field across the waveguide region. With the choice of crystal axes shown, a quasi-TE waveguide mode with polarisation along the z-axis will be affected predominantly by the z-component of the applied electric field via the r_{33} electrooptic coefficient.

Assuming that the situation can be simplified greatly, the change in dielectric impermeability can be written as

$$\Delta B_{33} = r_{33}E_3 \tag{11}$$

In terms of the refractive index, this becomes:

$$\Delta B_{33} = \Delta\left(\frac{1}{n_3^2}\right) = \frac{-2\Delta n_3}{n_3^3} = r_{33}E_3 \tag{12}$$

The change in refractive index is, therefore:

$$\Delta n_3 = -\frac{n_3^3}{2} r_{33}E_3 \tag{13}$$

Suppose that a voltage of 10V is applied between the electrodes. The maximum field strength between the electrodes is then approximately 2V/μm (or 2MV/m) for an electrode gap of 5μm. How long should the electrode length, L, be to obtain a phase change of, say, π radian? Using $n_3 = 2.2$, $r_{33} = 30.10^{-12}$ m/V and E = 2V/μm in Eq. (13) gives $|\Delta n_{33}| = 3.2.10^{-4}$. The resulting phase change is then:

$$\eta = 2\pi.\frac{L}{\lambda_0}.|\Delta n_{33}| \tag{14}$$

where λ_0 is the free space wavelength. Taking λ_0 = 0.6328μm and setting

$\eta = \pi$ leads to a value of L = 0.99mm.

KAMINOW et al [5] give a value of V/η = 0.32V/radian for a similar structure with the same crystallographic orientation, but with a length of L = 30mm and an electrode gap of 9.3μm. Scaling our structure gives a value of V/η = 0.195 V/radian. The discrepancy between these two values for V/η is due to our simple assumption about the effective value of the electric field strength. As KAMINOW et al [5] show, a more appropriate expression for the electric field strength is

$$E = \frac{V2}{\pi W} \tag{15}$$

where W is the electrode gap width. This expression is reasonably accurate where the waveguide (mode) is reasonably shallow compared with W and where the waveguide stripe width is smaller than the electrode gap width. More detailed consideration of the form of the inhomogeneous electric field produced below the electrodes, and of the resulting optimum location for the optical waveguide appears in PAPUCHON [6].

An important practical factor (and one which is closely related to other contributions in this school) is that significant losses are produced by the optical waveguide mode being in contact with the metal electrodes. These losses may be avoided either by making the electrode gap larger than the waveguide stripe or by isolating the substrate from the electrodes via a suitable low-index buffer layer. The penalty for using a large electrode gap width is that either the voltage requirement or the device length is increased. The buffer layer (typically about 0.2μm of silica grown by vapour phase deposition) must show both low optical propagation losses and very stable, insulating, electrical characteristics - slow drifts in the effective field strength in the waveguide region associated with charge movement under steady or slowly-varying applied voltage are a major practical problem of surface-wave electro-optic devices. The two fairly obvious choices for electrode metallisation, from both optical and electrical viewpoints, are aluminium and gold - with aluminium favoured because it adheres well and is much cheaper!

The configuration of Fig. 1(a) may also be used to produce a phase modulator for a (quasi-) TM polarised guided optical mode but the voltage requirement is increased by a factor of three because r_{13} is used rather than r_{33}. This electrode configuration is appropriate for both x-cut (with light propagating in the y-direction) and (as shown) y-cut lithium niobate - and likewise for lithium tantalate. For z-cut lithium niobate and tantalate the configuration of Fig. 1(b) is appropriate. Here, most of the effect of the electric field in the waveguide region is along the z-direction - so that r_{33} is used with TM-modes and r_{13} with TE-modes. For low optical propagation loss (particularly in the case of TM-modes), use of a buffer layer is now essential.

A pure phase modulator is, in itself, a useful device in specific systems applications (notably fibre optic sensors). Work first carried out at Glasgow University has exploited a simple conceptual extension of phase-modulation to provide a simple frequency shifter suitable for operation up to several MegaHertz (WONG et al [7]). This device is called (after an electronic ancestor) a 'serrodyne' device. Applying the linear saw-tooth voltage waveform shown in Fig. 2 leads, ideally, to a pure time frequency shift at the repetition frequency of the sawtooth since a linear change in phase is a constant frequency shift and a shift in phase of 2π is equivalent to zero-phase-shift. Performance of such a device is restricted by the degree of perfection attainable in the drive voltage waveform and also by any amplitude modulation (for which there are several possible causes). However, the simplicity of the device structure makes it attractive for some applications when compared with more sophisticated structures (HEISMANN [8].

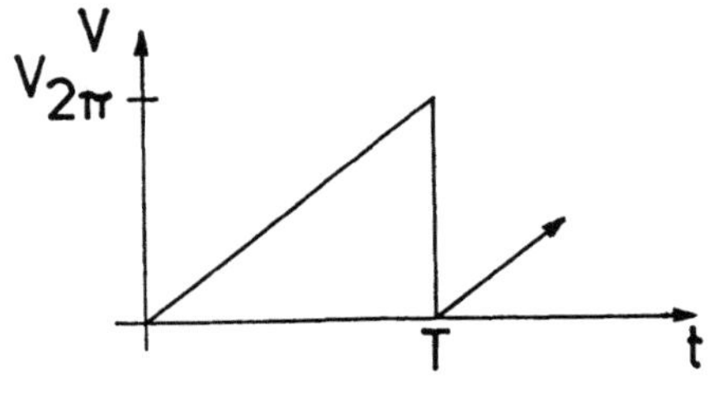

Fig. 2 Sawtooth drive voltage for serrodyne device

4. The Electro-optic Directional Coupler

At microwave frequencies, waveguide directional couplers are a standard piece of 'plumbing'. The equivalent optical waveguide structure, with electro-optic control, may well come to be regarded similarly as a standard item of integrated optics hardware. Figure 3 shows the favoured configuration, with two stripe optical waveguides placed sufficiently close together that light launched in one waveguide can be coupled into the other via the evanescent tails of the modes in each waveguide.

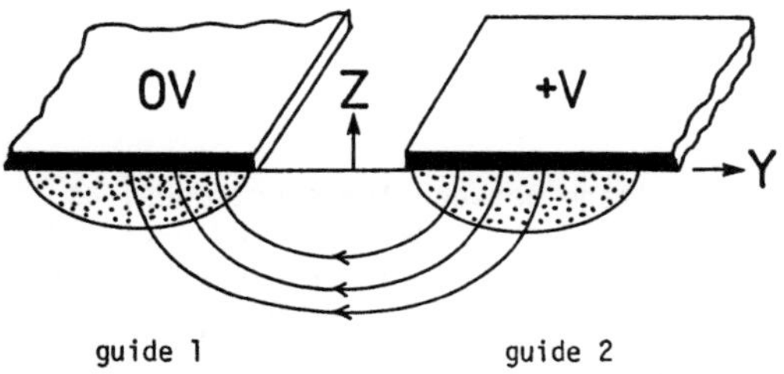

Fig. 3 Electro-optic directional coupler device configuration

Fig. 4 Weakly-coupled two-pendula system

A close analogy exists between the lossless optical waveguide directional coupler and the lossless mechanical system consisting of two pendula coupled by a weak spring, Fig. 4. If the pendula have identical length and mass and only one pendulum is set in motion, oscillation will gradually transfer across via the coupling spring until, momentarily, the first pendulum comes to a complete stop and the second pendulum is swinging with the same amplitude as the initial swing of the first pendulum. If this time to total transfer is called T, it is obvious that motion transfers back totally after 2T. The coupled pendulum system has two characteristic oscillation modes - with different frequencies:-

(a) with the two pendula swinging in exactly opposite directions and
(b) with the two pendula swinging in exactly the same direction.
Likewise, the directional coupler has two characteristic modes with different propagation constants (a) with the symmetric field distribution of Fig. 5(a) and (b) with the antisymmetric field distribution of Fig. 5(b).

Launching light into a single waveguide is equivalent to exciting both characteristic modes of the two-guide system with equal amplitude. The periodic transfer of the optical energy between the two guides then results from locally constructive or destructive interference of the two modes. When an

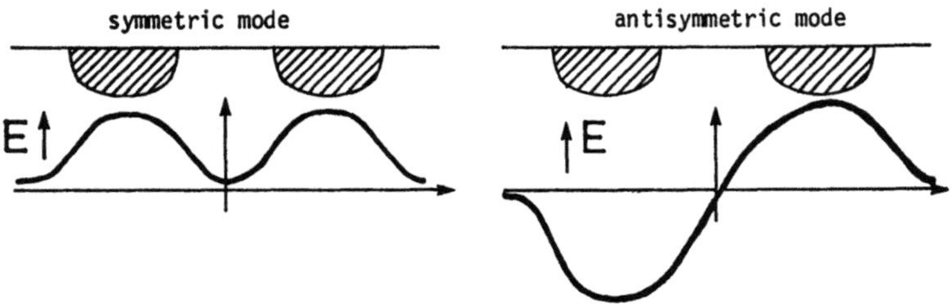

Fig. 5(a) Fig. 5(b)

Modes of a two-guide system (schematic).

electric field is produced by applying a voltage between the electrodes, as shown in Fig. 3, the refractive index is increased in one guide and decreased in the other (because the z-directed component of field is 'up' in one guide and 'down' in the other). In the mechanical analogy, this is equivalent to increasing the length of one pendulum and decreasing the other (while keeping the masses unchanged). Total energy transfer between the first and the second guides no longer takes place, because the individual guide modes are no longer synchronous. The situation can be analysed by using the coupled ordinary differential equations (SCHMIDT [9]):

$$\frac{dR}{dx} = j\delta.R - j\kappa.S \tag{10a}$$

$$\frac{dS}{dx} = - j\delta.S - j\kappa R \tag{10b}$$

Where R is the amplitude of the light at any point, x, along guide 1 (transverse spatial variation is not specified here) and S is the amplitude of the light at any point along guide 2. κ is the coupling constant of the guides with no voltage applied (the length for a complete transfer of power between equal guides is $L_c = \pi/2\kappa$). δ is one half of the (electro-optically produced) difference between waveguide propagation constants, i.e.

$$\delta = \Delta\beta/2 = \frac{\beta_R - \beta_S}{2} \quad . \tag{17}$$

We are interested in the particular case where $R = R_0$ and $S_0 = 0$, i.e. all the light is launched in guide 1. Differentiating and combining Eq. (16a) and (16b), gives

$$\frac{d^2R}{dx^2} + (\delta^2 + \kappa^2) R = 0 \tag{18}$$

This can be Laplace-transformed, with initial conditions included, to give:

$$\mathcal{R}(s) = \frac{sR_0 + j\delta R_0}{[s^2+(\delta^2+\kappa^2)]} \quad . \tag{19}$$

The inverse-transform of Eq. (19) is the required solution

$$R(x) = R_0 \ [\cos[\kappa x(1+(\delta/\kappa)^2)^{\frac{1}{2}}] + [j\delta/\kappa(1+(\delta/\kappa)^2)^{\frac{1}{2}}].\sin[\kappa x(1+(\delta/\kappa)^2)^{\frac{1}{2}}]]. \tag{20}$$

The power in guide 1 is given by RR* and it can readily be shown that in guide 2:

$$SS^* = \sin^2[\kappa x(1+(\delta/\kappa)^2)^{\frac{1}{2}}]R_oR_o^* \tag{21}$$

and that power conservation holds, i.e.:

$$RR^* + SS^* = 1 \quad . \tag{22}$$

In Eq. (20) and (21), the general position variable x has been used. The power division between the two guides at the output end of a device of length L_D is simply given by making the substitution $L_D = x$. Complete power transfer requires $\Delta\beta = 0$, i.e. exact phase-matching and also requires:

$$\kappa L_D = (2m_1 + 1)\pi/2 \qquad (m_1 = 0, 1, 2, \text{etc.}) \tag{23}$$

or, in terms of the previously defined coupling length L_c, the requirement is:

$$\frac{L_D}{L_c} = (2m_1 + 1) \quad . \tag{24}$$

When an electric field is applied between the electrodes of a device with L_D/L_c given by Eq. (24), complete transfer into guide 2 is impossible, but after maximum partial transfer, all of the input power can exit in guide 1, if the condition:

$$(\kappa L_D)^2 + (\delta . L_D)^2 = (m_2\pi)^2 \qquad (m_2 \text{ integer}) \tag{25}$$

is obeyed.

Essentially these concepts were exploited in the 'Cobra' electro - optic switching device of PAPUCHON et al [4] where the electrode configuration of Fig. 3 was used and (by trial and error) fabrication conditions established to make a device with coupling length equal to device length. Applying an appropriate voltage then introduced enough mismatch to satisfy Eq. (25) and therefore gave the required switch operation. For $L_D/L_c = 1$, it can be shown that $\Delta\beta = (12)^{\frac{1}{2}}\ \kappa$. The power-transfer situation, with and without mismatch, is shown schematically in Fig. 6.

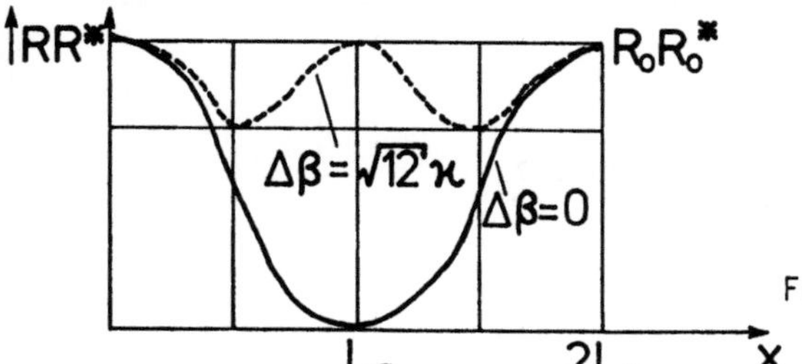

Fig.6 Power versus position in directional-coupler switch

The crucial problem in the directional-coupler just described is to obtain a device which is exactly one coupling length long. However, by using two sequential sets of electrodes of equal length, together with 'stepped $\Delta\beta$ reversal' it becomes possible to adjust voltages to obtain complete power transfer between guide 1 and guide 2 and then to change voltages so as to have all power back into guide 2. Application of this concept is examined in SCHMIDT [9]. Clearly, also, a device which is several coupling lengths long requires a smaller switching voltage. L_c can be considered to lie typically in the range 1mm to 5mm, with corresponding κ values in the range $1.5.10^{-3}\ m^{-1}$ to $3.10^{-4}m^{-1}$. The electro-optic effect is required to provide phase-mismatch comparable to these κ values. The reader is invited to estimate the required Δn and switching voltage for a hypothetical but realistic structure with L_C = 1mm, L_D = 5mm and an electrode gap width of 2µm.

5. Acousto-optic Interaction in the Bragg Regime

Standard treatments (UCHIDA and NIIZEKI [10], KLEIN and COOK [11]) of the acousto-optic interaction between a well-collimated light beam and a single-frequency acoustic beam of width L also involve coupled-mode theory. If L is large enough so that the condition

$$Q = \frac{K^2 L}{k_o} = \frac{2\pi\lambda_o L}{\Lambda^2} >> 1 \qquad (26)$$

is satisfied, then interaction takes place in the Bragg regime and interaction is strongest for incidence at the Bragg-angle

$$\theta_B = K/2k_o = \lambda_o/2\Lambda \qquad (27)$$

For this situation, a single first diffraction order is produced, having its frequency either up-shifted or down-shifted by an amount equal to the acoustic frequency depending on alignment. The acoustic wave forms a thick travelling phase-grating with its periodic refractive index modulation produced by the alternate regions of expansion and compression associated with the acoustic wave. In terms of the complex electric field amplitudes of the zeroth and first diffraction orders, $E_0(x)$ and $E_1(x)$, the coupled equations for a general angle of incidence near to the Bragg angle may be written as:

$$\frac{dE_o}{dx} = + \frac{\beta_1}{2} E_1 \qquad (28a)$$

and

$$\frac{dE_1}{dx} = -j2\beta_2 E_1 - \frac{\beta_1}{2} E_o \; . \qquad (28b)$$

Here β_1 is the maximum of the oscillatory phase modulation, with $\beta_1 = \Delta nk$, where k is the free-space optical propagation constant and β_2 is a measure of the misalignment given by:

$$\beta_2 = \frac{K(\theta_i - \theta_B)}{2} \qquad (29)$$

The interaction configuration is shown in Fig. 7.
K is the acoustic propagation constant.

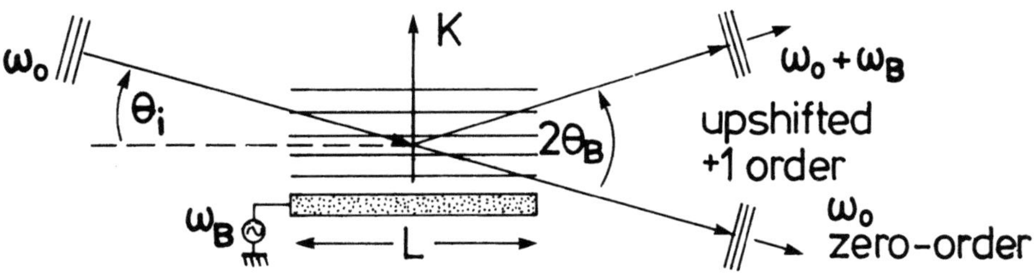

Fig. 7 Acousto-optic interaction in the Bragg-regime

Eq. (28a) and Eq. (28b) can be combined to give:

$$\frac{d^2E_o}{dx^2} + 2j\beta_2 \frac{dE_o}{dx} + (\frac{\beta_1}{2}) E_o = 0 \quad . \tag{30}$$

Again, Laplace transform techniques may be used to give:

$$\mathcal{E}_o(s) = \frac{sE_o(0) + 2j\beta_2 E_o(0)}{s^2 + j2\beta_2 s + (\beta_1/2)^2} = \frac{E_o(0)(s + j\beta_2 + j\beta_2)}{(s + j\beta_2)^2 + \beta_2^2 + (\beta_1/2)^2} \tag{31}$$

where the initial conditions are $E_o(x)|_{x=0} = E_o(0)$ and $E_1(0) = 0$

Inverse transformation then gives:

$$E_o(x) = E_o(0)\exp(-j\beta_2 x) \begin{bmatrix} \cos[(\beta_1/2)(1+(2\beta_2/\beta_1)^2)^{\frac{1}{2}}x] \\ +(j\beta_2/(\beta_1/2)(1+(2\beta_2/\beta_1)^2)^{\frac{1}{2}}.\sin[(\beta_1/2)(1+(2\beta_2/\beta_1)^2)^{\frac{1}{2}}x] \end{bmatrix} \tag{32a}$$

The equation for $E_1(x)$ is:

$$E_1(x) = \frac{E_o(0)\exp(-j\beta_2 x)}{(1+(2\beta_2/\beta_1)^2)^{\frac{1}{2}}} \cdot \sin[(\beta_1/2)(1+(2\beta_2/\beta_1)^2)^{\frac{1}{2}}x] \tag{32b}$$

Power conservation for this lossless interaction is then given by:

$$E_o(x)E_o^*(x) + E_1(x)E_1^*(x) = |E_o(0)|^2 \tag{33}$$

Apart from notational differences, Eq. (32a) and Eq. (32b) are identical to Eq. (12) of UCHIDA and NIIZEKI [10]. It is also apparent that Eq. (32a) and Eq. (32b) have a strong formal resemblance to Eq. (20) and Eq. (21) obtained previously for the electro-optic directional coupler. The factor $\beta_1/2$ depends on the acoustic wave amplitude,and is a measure of the acousto-optic interaction strength directly analogous to the coupling coefficient κ in the directional coupler device. The angular misalignment factor β_2 plays the same role in the acousto-optic interaction as the mismatch factor δ in the electro-optic directional coupler. The acousto-optic interaction in the Bragg-regime transfers power periodically with position across the acoustic beam (if the beam is wide enough). Only if the angle of incidence equals the Bragg angle can the zero-order optical beam be completely depleted,and the output power goes into the first-order beam.

The formal similarity between plane-wave acousto-optic interaction and electro-optic directional coupler operation is obviously an intellectually interesting result. However, there are also many practical differences between acousto-optic and electro-optic devices. In acousto-optic devices, the acoustic waves are generated via the piezoelectric effect by an oscillatory electrical source, whereas in the electro-optic devices the voltage can be at any frequency from d.c. up to an RC time-constant determined cut-off frequency. The fact that the acousto-optically diffracted first order beam has a pure frequency shift may be useful in heterodyne processing systems. In the simple case considered here, there is no frequency shift inherently produced in the electrooptic device.

6. Acousto-optic Interaction: basics

The acousto-optic interaction was introduced rather abruptly in section 5 so that the analogy between coupled-mode theory treatments of electro-optic directional couplers and acousto-optic plane wave interaction could be emphasized. Considering Bragg-regime interaction at exactly the Bragg angle gives, at x=L, from Eq. (32b):

$$\frac{E_1 E_1^*}{|E_0(0)|^2} = \frac{I_1}{I_0} = \sin^2\left(\frac{\beta_1 L}{2}\right) \qquad (34),$$

since $\beta_2 = 0$.

For engineering purposes, particularly in bulk-wave acousto-optic interaction, it is useful to be able to relate the phase modulation $\beta_1 L$ to the acoustic power input level, using simple relationships which ignore complexities associated with anisotropy. We then have:

$$\beta_1 = \frac{pn^3}{2}.s.\frac{2\pi}{\lambda} = \frac{pn^3}{2}.\left(\frac{2P_s}{\rho v_s^3}\right)^{\frac{1}{2}}.\left(\frac{2\pi}{\lambda}\right) = (M_2 P_s/2)^{\frac{1}{2}}.(2\pi/\lambda) \qquad (35)$$

where p is the relevant photoelastic coefficient and s the relevant strain term, λ is the free-space wavelength, P_s is the acoustic power density in Watt/m^2 and v_s is the acoustic wave velocity. M_2 is the figure of merit of the material, given by:

$$M_2 = \frac{n^6 p^2}{\rho v_s^3} \qquad (36)$$

All else being equal (and it isn't), a large value of M_2 is desirable, so that materials with high refractive index and/or slow acoustic wave velocity are indicated. The high index of semiconductors indicates their use at wavelengths in the infra-red, e.g. germanium for acousto-optic modulation at the 10.6μm wavelength of CO_2 lasers; while the very slow shear wave velocity of single-crystal paratellurite (TeO_2) propagating along the [100] direction has led to its use in some applications in signal-processing at communications frequencies around 50MHz. Other figures of merit M_1, M_3 and M_4 have been defined [12], but M_2 is most relevant (with an appropriate generalisation) to the case of surface acoustic wave-guided optical wave interaction, with which we shall be particularly concerned.

Substituting from Eq. (35) into Eq. (34) gives the diffraction efficiency versus acoustic power relationship:

$$(I_1/I_0) = \sin^2[(\pi/\lambda).(M_2.P_s/2)^{\frac{1}{2}}.L] \qquad (37)$$

7. Acousto-optic Signal-processing

Acousto-optic interaction provides an effective means of applying one-dimensional information at fairly large realtime bandwidths (hundreds of MegaHertz) across coherent light beams with widths as large as several centimetres [13,14]. One of the key operations is frequency spectrum analysis [15], where the variation of the angle between the zeroth and first diffraction orders with the electrical (and, therefore, acoustical) input frequency is exploited. The Fourier-transformation properties of a lens are used to convert angular changes into position changes in the focal plane, which in turn can be extracted as electrical information, in the time domain, via a scanned photodetector array.

It is not the objective of these lectures to explore signal-processing in detail. However, one fundamental point will be made - concerning linearity. The relationship between fractional diffracted optical power and acoustic (i.e. implicitly electrical) drive power level, Eq. (37), is inherently non-linear. For signal-processing it is generally desirable to have a linear relationship, and this is readily obtained by operating at low drive-power levels so that the small-angle approximation gives:

$$(I_1/I_0) = (\pi/\lambda)^2 (M_2 P_s/2).L^2 \ , \qquad (38)$$

a relationship which becomes more accurate as efficiency decreases. It might be thought that we are showing an excessive concern with linearity, since the non-linearity could in principle be compensated. However, non-linearity leads, when signals at several different frequencies are present, to intermodulation effects and spurious signals [16]. For signals restricted to an octave bandwidth range, or less, third-order intermodulation is of greatest concern.

8. Surface Acoustic Waves: their Generation and Interaction with Guided Optical Waves

The interdigital transducer (IDT) method for launching and detecting surface acoustic waves has become by far the most widely used and technologically important. A recently published work [17] sets out, in considerable detail, how surface acoustic waves, predominantly on piezoelectric substrates such as single-crystal lithium niobate and quartz, provide the basis of a powerful and moderately flexible high-frequency signal processing technology in which the IDT structure plays a central role. It is worth remarking that this surface acoustic wave signal processing technology provides direct competition for acousto-optic signal processing technologies [18], having important advantages in the relatively simple and precise lithographic processes required for device fabrication.

The configuration of an IDT structure, typically fabricated in a thin aluminium film on a piezoelectric substrate, is shown in Fig. 8(a), while Fig. 8(b) shows a schematic cross-section of the structure with the associated electric field distribution at some instant.

When an oscillatory voltage is applied to the transducer structure at a frequency for which the surface acoustic wavelength matches the periodicity (i.e. the centre-to-centre distance between alternate fingers), efficient excitation of surface acoustic waves occurs, through the piezoelectric

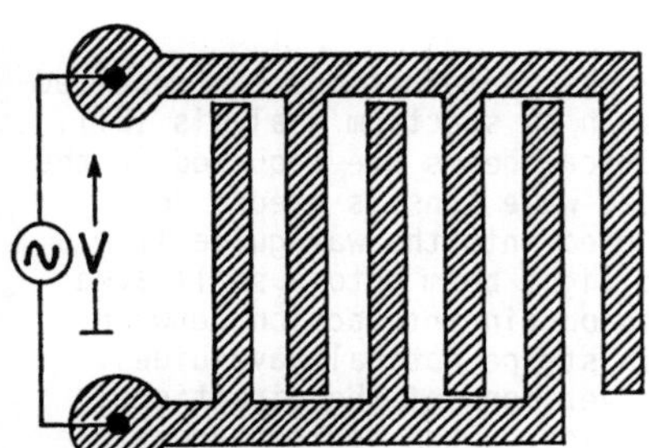

Fig. 8(a) Plan view of IDT electrode structure

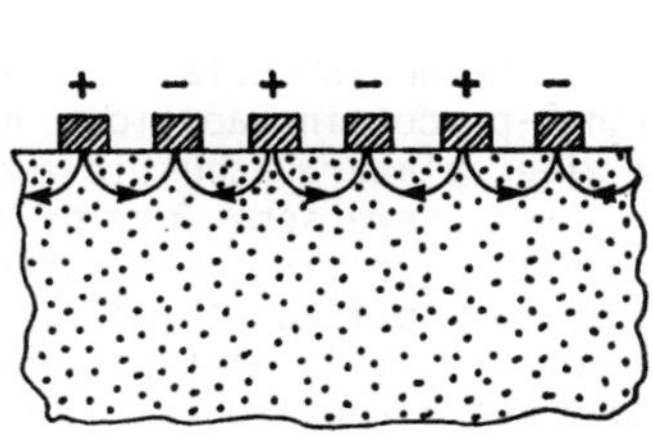

Fig. 8(b) Cross-section of electrode structure

effect. The structure is bi-directional and so has a minimum insertion loss of 3dB. The same structure acts equally well (being passive and reciprocal) as a receiver for surface acoustic waves so that, for instance, two transducers placed accurately several centimetres apart on a single substrate provide a delay on the order of a few microseconds and with a minimum insertion loss of 6dB. In the same time, an electromagnetic wave travels a distance measured in kilometres - and this is, historically, the initial reason for using acoustic wave devices in radar systems. At the cost of additional complexity, the bi-directionality can be removed and the delay line insertion loss can be significantly reduced. The surface acoustic wave IDT is an acoustic array antenna and so, with appropriate transformations, impedance concepts can be applied, enabling the design of transducer structures which are properly matched to the external electrical system [19].

With the advent of surface acoustic wave and guided optical wave technologies, a logical step was to investigate acousto-optic devices using the interaction of the two types of surface wave. The attractive possibilities following from an efficient interaction were soon demonstrated, once a satisfactory optical waveguide fabrication technology was formed for lithium niobate [20]. Further work [21,22] then showed the importance of the indirect piezoelectro-optic contribution in giving efficient interaction. Lithium niobate has therefore emerged as by far the strongest candidate for choice of base material because of its combination of strong piezoelectricity, very low acoustic propagation losses(particularly when expressed in dB/microsecond) and very adequate figure of merit when enhanced by the indirect interaction. As with other guided-wave interactions, the overlap integral approach is useful in calculating the interaction between complex field distributions and Eq. (39) below represents, by way of example, the direct photoelastic contribution to a mean p-coefficient which could be used in Eq. (35) or Eq. (36):

$$\tilde{p}_{Ao} = \frac{\Lambda}{2\pi\delta_o} \frac{\int_{-\infty}^{o} [p_{32}s_2(x_2)+p_{33}s_3(x_2)]U_m^2(x_2)dx_2}{\int_{-\infty}^{o} |U_m(x_2)|^2 \, dx_2} \qquad (39)$$

In Eq. (39) δ_o is the amplitude of the surface ripple, p_{32} and p_{33} are photoelastic coefficients, $s_2(x_2)$ and $s_3(x_2)$ are surface acoustic wave strain components and $U_m(x_2)$ is the field strengh of the m'th TE optical mode. Acoustic wave propagation along the crystallographic z-axis on y-cut $LiNbO_3$ is assumed. Λ is the acoustic wavelength.

9. Acousto-optic Stripe Waveguide Devices

Most of the interest shown in surface acousto-optic interactions has resulted from possible signal-processing applications such as spectrum analysis [21], where relatively wide, planar, acoustic and optical beams are required in the interaction region and where some form of guided-wave lens is needed to collimate the strongly diverging beam butt-coupled into the waveguide from a semiconductor laser and to focus the deflected light beam onto a small area photodetector. However, it is also possible to obtain interaction between plane acoustic wave beams and light confined in stripe optical waveguides. Kersten [22] has reviewed work on acoustooptic versions of the directional-coupler, Mach-Zehnder interferometer and x-junction. Such devices are close analogues of electro-optic stripe waveguide devices - but have less flexibility in practical operation. Nevertheless, acousto-optic stripe waveguide devices could be of great value in applications (e.g. fibre-optic sensors) where a periodic phase-shift or a pure frequency-shift is required. For such applications low device-to-fibre coupling losses are required, but are difficult to achieve, e.g., in an x-junction geometry compatible with efficient acousto-optic interaction and low cross-talk.

10. Waveguide In-plane Scattering

In-plane scattering is one of the key factors which restricts dynamic range in devices such as integrated optical spectrum analysers. Several papers have now appeared in which in-plane scattering is related to waveguide fabrication parameters and to microanalytical studies of the formation process [23-28].

11. Thermo-optic Effects

The prism-coupler technique commonly used in basic experiments on optical waveguides is sufficiently sensitive to variations in waveguide effective index that thermooptically-induced index changes can be observed quite readily. Such changes have been observed, during acousto-optic interaction, both on proton-exchanged and on titanium-diffused lithium niobate waveguides [24,29]. Because the functional dependence on input power of surface temperature and index change is the same, within measurement error, identification of the observed effects as of thermo-optic origin is clear-cut, but our measurements indicate approximately $(\text{Power})^{0.8}$ dependence in titanium-diffused guides and $(\text{Power})^{1.3}$ dependence in proton-exchanged guides. This difference is, presumably, an indication of differences in the nature of the heat-flow away from the transducer region, where heat is generated by conduction losses.

12. Intermodulation in Surface-wave Acousto-optic Devices

The importance of intermodulation has already been mentioned in section 7. The necessary non-linearity may arise from drive amplifier non-linearity, purely acoustic non-linearity or, in the limit, acousto-optic nonlinearity. Singh and De La Rue [30] have investigated intermodulation in a surface acousto-optic device and described precautions taken to eliminate drive-amplifier non-linearity effects in measurements at high-drive power levels. Nevertheless, further work is needed to provide clear identification of the primary source of non-linearity and to establish the importance of intermodulation for device performance under realistic operating conditions.

13. Proton-exchanged Electro-optic and Acousto-optic Waveguide Devices

Within a short time after its discovery, the proton-exchange process had been exploited for a number of different devices. At the same time, problems of stability also emerged but, despite these problems, interest in this process continues to grow [28,31]. Microanalytic studies arising from a collaborative program initiated at the 1981 Erice summer school have played a decisive role in relating hydrogen concentration profiles (through a resonant nuclear interaction [32]) to lattice distortion and to refractive index profiles. A recent publication [33] describes an investigation of the equivalent process for deuterated ion sources, with results which appear to show that, although deuterated benzoic acid produces a similar refractive index increase to normal benzoic acid, the resulting deuterium concentration profile is radically different from the index profile - and therefore has a different relationship to index from that for proton-exchange with normal benzoic acid.

In the case of electro-optic devices, the first published demonstration [7] showed that the electro-optic effect was comparable with that for titanium-diffused lithium niobate, although not stable under a d.c. applied voltage. A subsequent paper [34], for a slightly different structure involving a buffer layer, interpreted results as showing a greatly reduced electro-optic coefficient. A very recent paper [27] finds that the electro-optic coefficient in a stripe-guide device made from a 1% dilute lithium benzoate in benzoic acid melt is substantially maintained. Work on proton-exchanged acousto-optic devices [24,35] has indicated reduced piezoelectricity, increased acoustic propagation losses and a consequently reduced acousto-optic interaction efficiency. Proton-exchange under fairly extreme conditions [36,37] has been used in a polarization-rotating device and to obtain a substantial reduction in acoustic wave velocity. Clearly more work will be needed to achieve practical proton-exchanged devices.

14. Electro-optic Waveguide Devices in III-V Semiconductor Materials

The cubic crystal symmetry properties of III-V semiconductors such as InP and GaAs give the electrooptic tensor:

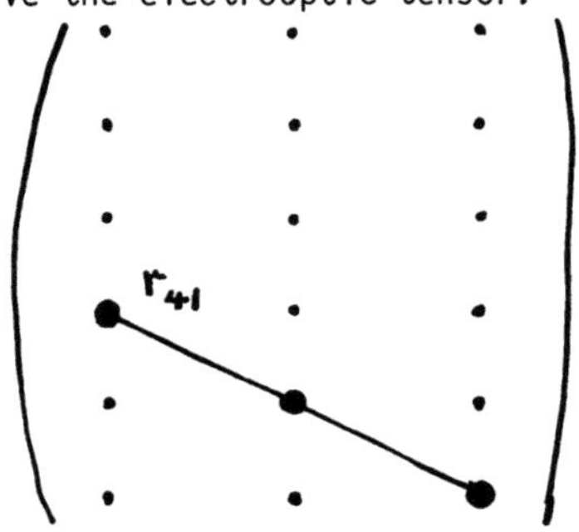

Only the three identically valued coefficients, r_{yzx}, r_{zxy} and r_{xyz}, are non-zero. The presence of substantial lattice strain such as might result from epitaxial growth, could remove the cubic symmetry and give a more complicated electrooptic tensor.

Exploitation of the electro-optic effect in an actual device requires an orientation rotated with respect to the principal axes used to define the electro-optic tensor above. As an example, rotating to a situation where light propagates along the (110) direction with polarisation along the ($\bar{1}$10) direction enables phase-modulation to be obtained with an applied electric field along the (001) direction [38]. Campbell et al [39] used essentially this configuration in a guided-wave device where amplitude modulation was obtained by using the electro-optic effect to induce a channel guide in the region below a reverse-biased Schottky contact. p-n homojunctions [40] or heterojunctions may also be used to apply electric field in stripe-waveguide devices. Both p-n junctions and Schottky contacts are able to give a situation where most of the applied potential drop occurs across the waveguide region, even if a back-contact some hundreds of microns away is used. In consequence, the applied electric field is used more effectively than is typical for devices on lithium niobate where potential drops more uniformly and where all electrodes are necessarily on the top surface of the structure.

Finally, the reader's attention is drawn to three papers by Suzuki and Tada [41-43]. These describe measurements on bulk samples of InP and GaAs of both the elasto-optic and electro-optic properties and show, in detail, how piezoelectricity and photoconductivity are allowed for in measurements of electro-optic coefficients. Although the electro-optic coefficients are much smaller in III-V materials ($r_{41}^T = -1.32.10^{-12}$ m.V^{-1} [42]) the higher index (n=3.29 at λ = 1.06μm) partially redresses the balance in comparison with lithium niobate, because it appears as n^3 in the phase-shift expression (Eq. (13) and (14). One possible reason for interest in electro-optic devices on III-V group materials is the fact that electromagnetic wave velocities are similar at microwave frequencies (in the range, say, 3GHz to 30GHz) and at optical frequencies, so that velocity-matched travelling wave interaction along a suitable electrode structure can be achieved over a long propagation distance [9]. Despite this promise, limitations of material quality have so far prevented success, whereas wide bandwidth devices have been realised on lithium niobate, despite the considerable velocity mismatch [44].

15. Future directions and conclusions

The author's view is that, in the short run, lithium niobate will provide the base material for practical electro-optic and acousto-optic guided wave optical devices. III-V group materials must, however, provide much of the base material for optoelectronic systems, more generally, and the explosion of interest in engineered lattices, i.e. multiple-quantum-well material, may well be the key to much larger electro-optic effects in III-V semiconductors as well as other novel device effects. Their greater versatility and their close match to single-mode fibres favour electro-optic devices on lithium niobate over acousto-optic devices. Although titanium diffusion seems likely to remain the standard waveguide formation technology, alternatives such as proton-exchange are well worth intensive investigation.

References

1. J.F. Nye: Physical Properties of Crystals (Oxford University Press, London, 1957)
2. I.P. Kaminow: An Introduction to Electrooptic Devices (Academic, New York, 1974).
3. J.T. Milek and M. Neuberger: Linear Electrooptic Modulator Materials (IFI/Plenum, New York, 1972).
4. M. Papuchon et al: Appl. Phys. Lett. 27,289 (1975).
5. Kaminow et al: Appl. Phys. Lett. 27, 555 (1975).
6. M. Papuchon: "Integrated Optical Modulation and Switching", in 'New Directions in Guided Wave and Coherent Optics', ed. D.B. Ostrowsky and E. Spitz, (Martinus Nijhoff, The Hague, 1984) pp 371-404.
7. K.K. Wong et al: Opt. Lett. 7, 546 (1982).
8. F. Heismann: Proc. SPIE 517, 206 (1984).
9. R.V. Schmidt: "Integrated Optics Switches and Modulators", in 'Integrated Optics-Physics and Applications' ed. S. Martellucci and A.N. Chester (Plenum, New York, 1983), pp 181-210.
10. N. Uchida and N. Niizeki: Proc. IEEE 61, 1073 (1973).
11. W.R. Klein and B.D. Cook: IEEE Trans. SU-14, 123 (1967).
12. I.C. Chang: IEEE Trans. SU-23, 2 (1976).
13. W.T. Rhodes: Proc IEEE 69, 65 (1981). See other papers in the January special issue.
14. D. Casasent: Electro-optical Systems Design, 39 (June 1981) and 41 (September 1981).
15. T.M. Turpin: Proc. IEEE 69, 79 (1981).
16. D.L. Hecht: IEEE Trans. SU-24, 7 (1977).
17. D.P. Morgan: Surface-Wave Devices for Signal Processing (Elsevier, Amsterdam 1985).
18. R.M. De La Rue: "Integrated Acousto-Optics: the Technology and the Competition", see ref. 6, pp.531-544.

19. W.R. Smith et al: IEEE Trans. MTT-17, 865 (1969).
20. R.V. Schmidt and I.P. Kaminow: IEEE J. QE-11, 57 (1975).
21. R.M. De La Rue: "Integrated Optical Spectrum Analysers using Geodesic Lenses on Lithium Niobate", see ref. 9, pp.283-296.
22. R. Th. Kersten: Proc. SPIE 517, 258 (1984).
23. C. Canali et al: Proc. SPIE 517, 119 (1984).
24. A. Dawar et al: ibid, 64 (1984).
25. M.N. Armenise et al: Proc. IEE (UK) 131 pt.H, 295 (1984).
26. J. Singh and R.M. De La Rue: IEEE Journal LT-3, 67 (1985).
27. K.K. Wong et al: IOOC/ECOC, Venice, (1985).
28. S.M. Al-Shukri et al: Proc. SPIE 578, paper 01 (1985).
29. J. Singh and R.M. De La Rue: J. Appl. Phys., to be published.
30. J. Singh and R.M. De La Rue: submitted for publication to Proc. IEE (UK) section J, Optoelectronics.
31. Papers by a number of authors in Proc. SPIE 578, (1985).
32. C. Canali et al: Proc. SPIE 460, paper 09 (1984).
33. C.E. Rice et al: J. Appl. Phys. 57, 4437 (1985).
34. R.A. Becker: Appl. Phys. Lett. 43, 131 (1983).
35. R.L. Davis: Proc. SPIE 517, 14 (1984).
36. V. Hinkov et al: 3rd ECIO, Berlin, 169 (1985) (Published by Springer).
37. V. Hinkov et al: submitted to Appl. Phys. (A) (1985).
38. M. Izutsu: Glasgow University internal memorandum.
39. J.C. Campbell et al: Appl. Phys. Lett. 26, 640 (1975).
40. A. Carenco et al: Appl. Phys. Lett. 40, 653 (1982).
41. N. Suzuki and K. Tada: Jap. Jour. App. Phys. 23, 1011 (1984).
42. N. Suzuki and K. Tada: ibid, 291 (1984).
43. N. Suzuki and K. Tada: ibid 22, 411 (1983).
44. T. Sueta and M. Izutsu: J. Opt. Commun. 3, 52 (1982).

Reflection and Transmission of Guided Waves at the Boundary Between Two Dissimilar Waveguides

T.P. Shen, R.F. Wallis, and A.A. Maradudin

Department of Physics, University of California, Irvine, CA 92717, USA

G.I. Stegeman

Optical Sciences Center and Arizona Research Laboratories, University of Arizona, Tucson, AZ 85721, USA

The reflection and transmission of a thin film guided wave incident at arbitrary angles onto the boundary between two dissimilar waveguides are problems of continuing interest in guided wave technology. The normal modes of individual waveguide with shorting planes are found and the point-matching method is used to satisfy the electromagnetic boundary conditions across the transverse interface between two waveguides.

The calculation has been done for two waveguides of the same thickness ($h = 0.45\lambda$) and the material parameters $n_c = 1.0$, $n_s = 1.46$, $n_f = 1.7$ (waveguide A), and $n_c = 1.46$, $n_s = 1.5$, $n_f = 2.0$ (waveguide B), respectively. A TE_o guided wave incident from A onto the boundary with B is shown in Fig. 1. There is an analog to the Brewster angle phenomenon near the incident angle being 50°.

This research was sponsored by the Army Research Office under contracts DAAG-29-85-K-0026 and DAAG-29-85-K-0025.

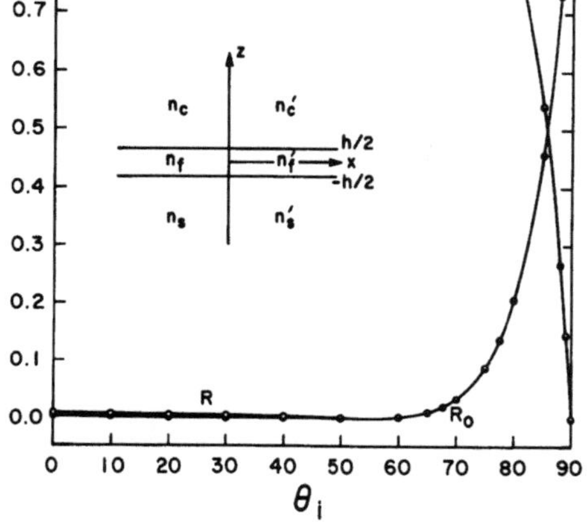

Fig. 1. The reflectivities and transmissivities versus the angle of incidence for TE_o guided wave incidence from A onto B. R and T are the total reflectivity and transmissivity, respectively. R_o (T_o) and R_1 (T_1) are the reflectivities (transmissivities) into TE_o and TE_1 guided waves.

Part IV

Nonlinear Surface Excitations and Guided Waves

Nonlinear Surface Polariton Interactions: Surface Enhanced Nonlinear Optical Effects

R. Reinisch

LEMO, U.A. 833 du CNRS, Génie Physique, ENSIEG, B.P. 46,
F-38402 Saint Martin d'Hères, France

M. Neviere

Laboratoire d'Optique Electromagnétique, U.A. 843 du CNRS,
Faculté des Sciences et Techniques, Centre de Saint-Jérome,
F-13397 Marseille Cedex 13, France

1. Introduction

Surface enhanced nonlinear optical (SENLO) phenomena have attracted a great deal of interest [1-27]. By SENLO effects, we mean nonlinear optical (NLO) effects whose efficiency is enhanced as compared to the case where the nonlinear (NL) interaction involves the NL medium alone. We shall be mainly concerned with surface enhanced Raman scattering (SERS) (including Raman excitation of surface polariton [6] (S Pol)), surface enhanced second harmonic generation (SESHG), surface enhanced Kerr effect (SEKE) which leads to a lowering of the threshold of bistable optical devices. As will be seen, all these "SE" phenomena have in common an enhancement which, at least partly, has a linear or NL electromagnetic (EM) component.

It is the aim of this chapter to study, on theoretical and experimental basis, the EM contribution to the enhancement of these SENLO phenomena.

For reasons which will become clear hereafter, it is convenient to consi der separately SERS and SESHG on the one hand and SEKE on the other hand.

SERS and SESHG have the following common feature : they both require a medium with negative permittivity and whose entrance face is rough. This medium is usually a metal, but this is not always the case. For example, Ushioda et al [6] have observed enhanced Raman excitation of S Pol propagating along a rough air-GaP interface.

In the case of SESHG the NL medium is constituted by the rough metallic medium itself, whereas for SERS, the NL medium consists of organic molecules adsorbed on the rough metallic surface.

It is usually believed that this enhancement is related to two mechanisms :

* a chemical one (for SERS only) , due to interactions between metal and molecules

* an EM one (for SERS and SESHG) , linked with resonant excitation of surface EM waves (EM resonances) at the pump frequencies and/or signal frequency. These resonances involve nonlocal surface waves or local ones[7,11].

We only consider the contributions of the EM resonant excitation of nonlocal surface waves (such as : surface plasmons (SP), S Pol, guided waves (GW)) to the enhancement of NLO phenomena. Instead of considering rough surfaces,we shall deal with gratings i.e. periodic structures. In order to be specific,we focus on NL interactions of the type : $(\omega_1,\omega_2) \rightarrow \omega_3 = (\omega_1+\omega_2)$ in a grating ruled on a NL material i.e. a NL grating. Hence the name of diffraction in NLO given to this study. By the way, we see that these SENLO effects are nothing but a special case of diffraction in NLO, namely when EM resonances take place.

Before going further on, some comments are in order.

The existing theories of diffraction all have in common the following features : they deal with the linear diffraction of light. In these studies, a plane wave with frequency ω is incident on a grating ruled on a linear material (in the optical sense) and the interest is in the diffracted field at the same frequency ω . A rigorous study of this problem, i.e., a study which does not consider the groove depth of the grating as a perturbation, was developed in the last ten years by several authors [13,28,29] (for a review, see Ref. 28). The rigorous feature of the formalism developed by the authors ref. 28 allowed them to explain, among other things, what is usually termed as Wood anomalies of gratings. They showed that these anomalies, exhibited with a TM incident wave, are closely related to the resonant excitation of normal modes (which correspond to SP or S Pol) of the interface obtained by allowing the groove depth of the grating to go to zero.

The authors of Ref. 28 also showed that, in the case of coated gratings, other anomalies may exist which are associated with the existence of different normal modes (corresponding to GW) from those considered above. These anomalies occur not only with TM incident waves but also with TE ones. When the grating, instead of being ruled on a linear optical material, is impressed on a NL one, several questions may be asked : what about

i) the direction of propagation of the unbounded (propagating) NL diffracted order ?

ii) the efficiencies of the diffracted orders at the new frequencies but also at the pump frequency ?

iii) the anomalies described above ? It must be emphasized that the answer to this last question is of great importance. Indeed, as already noticed, SERS and SESHG are due, to a large extent, to the resonant excitation of SP, i.e. of one of the normal modes cited above.

Obviously, the enhanced feature of the SENLO phenomena is polarization dependent. In addition, we remind the reader that the rigorous studies of diffraction in linear optics [14,15,28] shows the existence of an optimum value of of the groove depth of the grating for which the intensity of the bounded (evanescent) diffracted field is the greatest. It is expected that a similar effect occurs in NLO.

We have developed an EM theory of diffraction in NLO which :

* accounts for the vectorial character of the NL processes we are interested in.

* does not consider the groove depth of the NL grating as a perturbative parameter.

This theory is valid for any groove shape. It not only applies to NL metallic gratings (SESHG) but also to coated gratings whatever the nature, linear or NL, of each layer may be. This analysis allows dealing with the two fundamental cases of polarization (TE,TM) of the incident pump beams and also of the signal. To our knowledge, such a theory has never been performed. The only other rigorous theoretical study dealing with this kind of phenomena (SESHG) is that published by Farias and Maradudin[20]. But, as emphasized by the authors themselves, the expansion of the second harmonic (SH) field they use in the groove depth region is not always valid.

Let us now consider SEKE. Devices constituded by Kerr (i.e. NL) media may exhibit a bistable optical (BO) behaviour[31,32]. This kind of BO devices (BOD's) are referred to as intrinsic ones,in contrast with hybrid BOD's[31,33] where a detector converts the output optical signal in an electrical one which is fed back to an internal electro-optic element.

The advantage of hybrid BOD's is that the induced nonlinearities are much larger than the optical ones produced by Kerr effect. But their response time is limited by the electrical feedback.

Intrinsic BOD's are attractive because, since no electrical feedback is needed, they allow the lowest possible switching time. The price to be paid is a high level of optical power (of the order of 10 GW/cm^2), since optical NL coefficients are small.

A way to overcome this difficulty is to increase the EM field inside the NL material. This can be achieved by using the resonant excitation of SP. BOD's using this kind of EM resonance have been demonstrated in refs.[23,24] where SP were excited at the interface between silver and a NL medium in the ATR configuration :

* the authors of ref. 23 use Fresnel equations and therefore are concerned with a quasi-linear theoretical study of the case where the nonlinearity comes from Kerr effect. Their calculation shows that a reduction of 25 in the incident power is possible when compared to the case of a single NL interface involving no SP [32]. Although lowered, the minimum switching intensity still remains at a high level, about 3 10^8 W/cm^2. However, the properties of TM NLGW are difficult to investigate[34].

* In ref 24, the NL medium is CS_2 and the bistable effect arises from the temperature dependence of its refractive index. The corresponding experimental results show that optical bistability is achieved with a rather low incident power : 10^4 W/cm^2. But temperature effects are slow.

We consider BOD's based on the use of an EM resonance different from the SP, namely the GW resonance. Indeed, it has been shown elsewhere [18,19] that the GW resonance leads to a much greater increase of the EM field inside the NL medium than the SP resonance does. The GW propagates along waveguides including one or several Kerr media. Thus we deal with NL waveguides. Such guiding structures have attracted a great deal of interest[35] : they allow the propagation of NL guided modes (GM) which have been studied in refs [35]. But the authors of ref[35] do not take the excitation process into account : they are only concerned with the homogeneous NL problem.

For sake of simplicity, we first study the case where these NLGM are excited using a prism coupler (OTTO geometry) and show that this NL ATR device exhibits OB[25,26]. Emphasis is placed on the link between resonant excitation of NLGM and OB. We then consider the case of grating excitation of NLGM[27]. Such a device, which has a BO behaviour too, is suitable for NL integrated optics. Only the TE case is investigated. We show that the use of NLGM leads to a lowering of the threshold of OB. This justifies the name of SEKE given to this effect.

Before going further on, the case of SESHG deserves a comment. As already mentioned, SESHG takes place in metals which are centrosymmetric. Consequently, no SH signal should be observed. In fact this is not so because there is a breaking of the inversion symmetry at the metal surface. As a result, the study of SESHG requires the knowledge of the NL polarization and boundary conditions at the metal surface.

2. Surface enhanced Raman scattering

As is now well known,the Raman effect of organic molecules adsorbed on roughened metals is enhanced by many orders of magnitude as compared to the case where no metal is present.

We are interested in the study of the nonlocal SP contribution to SERS from molecules adsorbed on metallic gratings. Several theoretical papers [12,13,17] have been recently published on this subject. In this section, the EM study of SERS is achieved in the framework of the Rigorous Electromagnetic Theory of Linear Diffraction (RETLD) [17,28].

2.1. Brief outline of the RETLD

We consider the geometry fig 1 : a TM polarized plane wave with frequency ω is diffracted by a metallic grating (periodicity d, groove depth δ , angle of incidence θ).

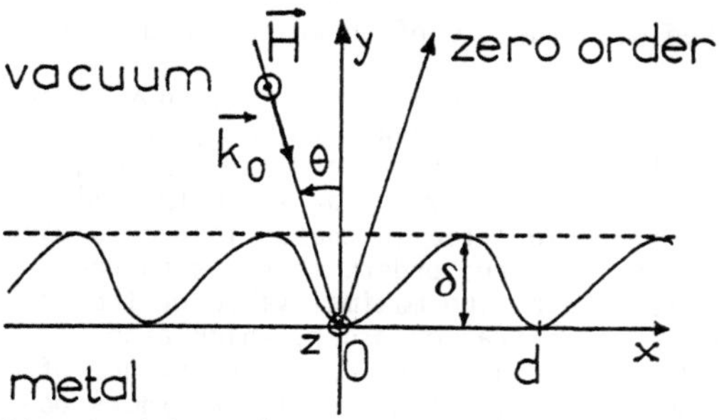

Fig. 1. Schematic representation of a grating and notations.

The RETLD allows the investigation of any periodic profile even in the case where the grating is coated by one or several dielectric or metallic layers. The RETLD we use is a differential method [28] in which a Fourier expansion of the EM field is done in the three regions : $y > \delta$ (medium 1 : permittivity ϵ_1 real positive), $y < 0$ (medium 2 : permittivity ϵ_2 complex) and $0 < y < \delta$ (medium 3). It is shown in ref 28 that the following expansions hold for the z component of the magnetic field.

* In region 1 : $y \geqslant \delta$

$$H(x,y) = e^{j(-\alpha y + \gamma x)} + \sum_{n=-\infty}^{+\infty} B_n e^{j(\alpha_{1,n} y + \gamma_n x)} \tag{1a}$$

with

$$\gamma = k_1 \sin\theta \quad (k_1^2 c^2 / \omega^2 = \epsilon_1, \quad \lambda = 2\pi c/\omega) \tag{1b}$$

$$\alpha^2 + \gamma^2 = k_1^2$$

$$\alpha_{1,n}^2 + \gamma_n^2 = k_1^2 \tag{1c}$$

$$\gamma_n = \gamma + n(2\pi/d) \tag{1d}$$

* In region 2: $y < 0$

$$H(x,y) = \sum_n T_n e^{j(-\alpha_{2,n} y + \gamma_n x)} \tag{2a}$$

with

$$\alpha_{2,n}^2 + \gamma_n^2 = k_2^2 \quad (k_2^2 c^2 / \omega^2 = \epsilon_2) \tag{2b}$$

* In region 3 : $0 \leqslant y < \delta$

$$H(x,y) = \sum_n H_n(y) e^{j \gamma_n x} \tag{3}$$

In Eqs. (1a) and (2a), B_n and T_n are the Rayleigh coefficients, which depend on θ and δ . It must be noticed that a diffracted order n is evanescent above the grating if $\gamma_n > k_1$. As shown in Ref 28 the Fourier coefficients $H_n(y)$ cannot be expressed in an analytical way. They appear to be the solution of an infinite set of coupled first order differential equations. The boundary conditions written at $y = 0$ and $y = \delta$, linked with this infinite set, allow the determination of the B_n, T_n, and $H_n(y)$, and thus of the EM field everywhere in space : above and below the grating but also in the modulated region. This is achieved with the use of an algorithm described in Ref. 28.

This theory leads to the following important result : at a given λ, there is a specific value θ_c of θ for which the intensity $|B_0|^2$ of the specularly reflected beam is minimal. This minimal value of $|B_0|^2$ strongly depends on δ. There is even [28] a value δ_c of δ for which this minimal value of $|B_0|^2$ is zero. Provided the λ/d ratio is chosen in such a way that there is only one propagating order, the zero order, there is no reflected beam. When this phenomenon takes place, the maximal value of the intensity of one of the reflected evanescent diffracted order, which also strongly depends on δ, is itself maximal. In fact, due to the existence of the losses, the greatest maximum of the intensity of one of these orders occurs for values δ_{opt} of δ and θ_{opt} of θ, which are not strictly equal to δ_c and θ_c respectively.

Figs. 2 are plotted for a Ag grating (groove profile : sinusoidal hereafter, λ = 5000 Å, groove frequency : 2400 mm^{-1}). The shape of these curves indicates the existence of a resonance phenomenon. In addition, figs 2a, 2b clearly exhibit the existence of the optimization process. We get :

δ_{opt} = 250 Å and δ_c = 210 Å

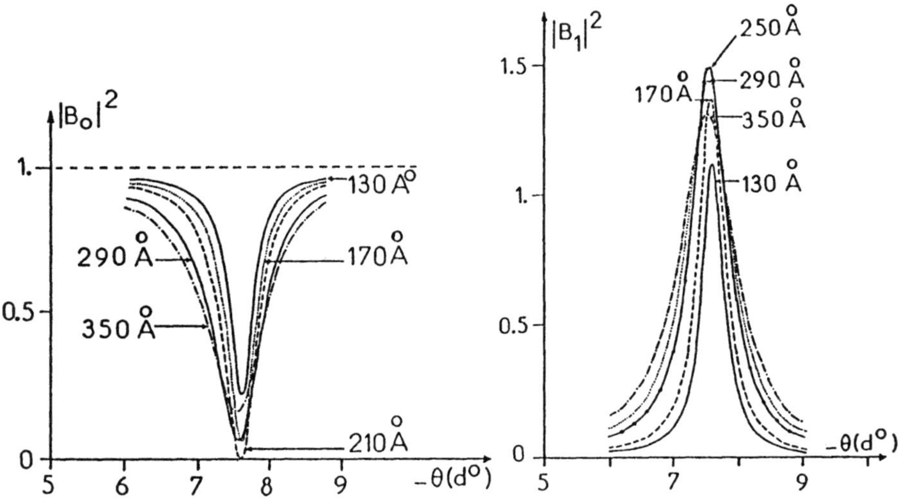

Fig. 2. (a) Evolution of $|B_0|^2$ as a function of the angle of incidence for various groove depths from 130 to 350 Å for a silver sinusoidal grating (2400 mm^{-1} groove frequency ; λ = 0,5 μm) ; (b) Same as (a) but $|B_1|^2/100$ is plotted instead of $|B_0|^2$.

Considering SERS, it is the EM field above the grating which acts as a pump. Thus the greatest efficiency is obtained when $\delta = \delta_{opt}$.

With these values of λ and d, the n = +1 diffracted order is evanescent above the grating. This diffracted order exhibits SP resonance when θ_{opt} = -7.58° and δ_{opt} = 250 Å. One then gets the following value for the greatest maximum of $|B_1|^2$: $|B_1|^2_{max\ max}$ = 149.3 MKSA and $|B_0|^2 \simeq 0$. Thus a very shallow modulation of the metallic surface leads to dramatic modifications in the reflected field since, in the absence of modulation (no diffraction), one would have obtained $|B_0|^2 \simeq 0.9$. This phenomenon is closely related to the resonant excitation of SP [15,28] or, equivalently, to the existence of Wood anomalies of gratings [30]. It is this SP resonance which gives rise to the EM contribution to SERS. Since in SERS organic molecules are adsorbed on the metal surface, we have to consider first coated gratings.

2.2. Influence of a thin dielectric coating

As already pointed out in Sec. 2.1., the RETLD still applies when a layer is deposited onto the grating. Therefore, using the RETLD we find that, when a thin (50-Å-thick) pyridine film is deposited onto the silver grating of Sec.2.1. (λ = 0.5 μm, and we still focus on the n = 1 diffracted order)

θ_{opt} = -6.4° (instead of -7.58° in the case of a bare silver grating),

δ_{opt} = 210 Å and and $|B_1|^2_{max\ max}$ = 140.4 MKSA.

This value is almost equal to that corresponding to the bare silver grating. It must be noticed that coating the grating leads to a shift, toward the low values of $|\theta|$, of the $|B_1|^2$ peak position or, equivalently, of the position of the zero of $|B_0|^2$. This is in agreement with experimental results [36].

In order to define the enhancement factor F, we consider a medium (same dielectric as that deposited on the metallic grating) with a smooth entrance face. A TM polarized plane wave is incident (same incidence and wavelength as for the grating case) on this medium. Let $|T|^2$ denote the intensity of the transmitted beam. F is equal to the ratio of the total intensity of the reflected diffracted field corresponding to the coated metallic grating to $|T|^2$. Two remarks need to be made : (a) It is the transmitted field which acts as a pump when one deals with Raman effect in the smooth dielectric medium. This explains why the denominator of F is $|T|^2$. (b) In the computation of the total intensity I of the reflected diffracted field we take into account 15 diffracted orders (from n = -7 to n = +7). Taking 1.5 for the refractive index of pyridine we find : $|T|^2$ = 1.437 $\longrightarrow$ F = 97.5 at λ = 0.5 μm.

2.3 SP contribution to SERS. Discussion

The SP resonance contribution to SERS is obtained as follows : 1) the incident pump beam is diffracted. 2) When $\theta = \theta_{opt}$, through coupling with the SP, one of the evanescent diffracted orders is resonantly excited. This results in an enhancement, F, which can be optimized ($\delta = \delta_{opt}$), of the whole reflected diffracted field. 3) This reflected field acts as a pump for the Raman effect from organic molecules adsorbed onto metallic gratings. Since the pump intensity is enhanced through SP resonance, so is the intensity of the Raman scattering. Points (1)-(3) also apply to the Stokes beam. Its frequency is only slightly different from that of the pump beam. From the numerical rigorous results obtained using the method outlined in section 2.1, it is then possible to get an approximate value of the enhancement factor, E, of SERS since E is nearly propertional to F^2. Table 1 gives the enhancement factor E for SERS derived through this simple formula, for the metals Au, Cu, Al, and Ag at the two wavelengths 0.5 μm and 1.06 μm.

Concerning Table 1, the following comments are in order:

* λ= 5000 Å. Except silver, no optimal value of the groove depth exists in the submicroscopic domain ($0 < \delta < 500$ Å) and, as a consequence, no strong SP resonance is expected to take place in this range of groove depth. We then took for these three gratings the groove depth corresponding to the optimal value of the bare silver grating illuminated at λ = 0.5 μm i.e., δ = 210 Å. Each of these gratings is coated by a 50-Å-thick pyridine layer. The angle θ is chosen such that $|B_1|^2$ is maximal.

* λ = 1.06 μm. Contrary to what occurs at λ= 0.5 μm, there is, in the submicroscopic domain, a value of δ_{opt}, not only for the Ag grating but also for the Au and Cu ones. For the Al grating, there is no value of δ_{opt} in this range of groove depth ; we thus took the value of δ which gives

TABLE 1. Enhancement factor E for SERS corresponding to Au, Cu, Al, and Ag gratings illuminated at wavelengths 0.5 and 1.06 μm.

λ μm	0.5			1.06		
d Å	4167			6174		
	refractive index	δ_{opt} Å	E	refractive index	δ_{opt} Å	E
metal						
Ag	0.05 + j2.87	210	9506	0.16 + j7.2	204	41493
Au	0.84 + j1.84	no δ_{opt}	3.6	0.22 + j6.36	276	13971
Cu	0.88 + j2.42	no δ_{opt}	8.8	0.21 + j6.56	260	17742
Al	0.67 + j5.58	no δ_{opt}	835	1.56 + i9.75	no δ_{opt}	1004

the smallest value for the minimum of $|B_0|^2$. For this wavelength, the numerical values of the refractive indices of these metals were extrapolated from the values known at $\lambda < 1$ μm.

In order to point out how optimizing the value of δ is important, we have computed, for a Ag grating, E at $\lambda = 1.06$ μm for d = 6174 Å and δ = 500 Å $\neq \delta_{opt}$. The angle θ is chosen so that $|B_1|^2$ is maximal. We get E = 9550. The comparison with the corresponding value of E quoted in Table 1 shows the importance of the optimization process.

The study of the field map in the vicinity of the silver grating, inside and outside the pyridine film, shows the following features :

a) Computations show that it is only at distance $y \simeq$ 2400 Å from the top of the groove that the field amplitude is divided by 1.778, which means that all the atomic layers of pyridine participate in the enhancement.

b) The field amplitude at y = const, considered as a function of x, is almost uniform along a grating period. The difference between the maximum and minimum is only about 20 %. Thus the field enhancement occurs on the whole corrugated surface, and not only at isolated locations.

To summarize, the SP resonance contribution to SERS is produced by submicroscopic modulations (<500 Å) of the metal surface. Strong enhancements are obtained provided the optimization is achieved. E is almost independent of the dielectric thickness and structure, since we have verified that a change in the value of the refractive index of pyridine has only little, or no influence, on the value of $|B_1|^2$. On the other hand, enhancement strongly depends both on the nature of the metal and on the wavelength. All these facts are consistent with experimental results [1, 4].

According to the data reported in Table 1, it is seen that the SP resonance contribution to SERS is not negligible at all, since the EM enhancement is of the order of 10^4. Such a conclusion is different from that of the authors of ref 13 who find an enhancement of modest value. This difference comes from the fact that Mills and Weber [13] use numerical values for which $\lambda/d = 0.6$ and $\theta = 45°$. Thus there are 3 propagating diffracted orders. As shown in ref 28, when several diffracted orders propagate (i.e. $\lambda/d < 1$), the optimized efficiency of the SP resonance is decreased as compared to the case where the only propagating diffracted order is the specular one ($\lambda/d > 1$).

Consequently, the SP resonance contribution to SERS is important provided the following conditions are fulfilled : i) the couple (d, λ) is such that $\lambda/d > 1$, ii) the optimization is achieved and iii) the metallic medium is Ag.

3. Surface Enhanced Raman Excitation of Surface Polaritons

In this section, we analyse the situation where the EM resonance takes place at frequency[16] $\omega_3 = \omega_1 - \omega_2$ ($\omega_{1,2}$: respectively frequency of the laser and stokes beams). This corresponds to the experiment performed by Ushioda et al[6] where S Pol are excited by Raman effect along air - GaP interfaces of increasing roughness.

3.1 Theoretical Study

As in section 2, we consider periodic rough surfaces (i.e. gratings) instead of statistically rough ones(fig.3).

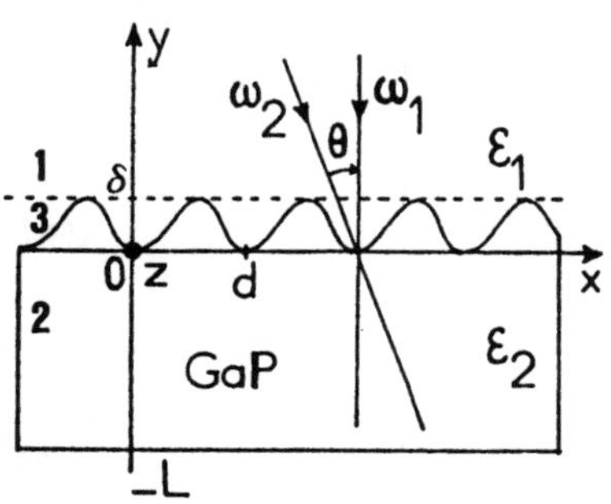

Fig.3 Scattering geometry. A grating is impressed on a GaP medium of thickness L

The surface polaritons are excited by Raman effect using the two-beam method where two laser beams (wavevectors $\vec{k}_2$ (ω_ν), $\nu = 1,2$, in the NL medium) are simultaneously sent on the NL medium. Their frequency difference, ω_3 is equal to that of the expected S Pol. Such a scattering geometry has often been used in the past [37] to get an efficient excitation of S Pol. This configuration simplifies the analysis without modifying the conclusions of the present study. A grating (groove depth δ , periodicity d) is impressed on the NL medium of thickness L.

The profile of the grating, given by the equation $y = f(x)$, divides the space into two parts : above the grating, the medium is a linear material with relative permittivity ϵ_1 which is real and frequency independent. Under the grating, the medium is nonlinear and has a relative permittivity $\epsilon_2(\omega_3)$. $\epsilon_2(\omega_3)$ is a complex function of ω_3 whereas $\epsilon_2(\omega_1)$ and $\epsilon_2(\omega_2)$ are real positive since at frequencies ω_1 and ω_2 the nonlinear medium is lossless.

Since we are interested in surface polaritons, the frequency ω_3 is chosen in such a way that ω_3 belongs to the absorption band of the nonlinear medium : thus $\mathrm{Re}[\epsilon_2(\omega_3)] < 0$. [Re(...) means real part of ...]. Throughout this paper, the diffraction of the pump beams is neglected. This is justified by the fact that their wavelengths (of the order of 0.5 μm) is much lower than the periodicity, d, of the grating. In addition, no EM resonance can occur at ω_ν ($\nu = 1,2$) because $\epsilon(\omega_\nu)$ is assumed to be real positive.

The Raman interaction between the two pump beams at frequencies ω_1 and ω_2 leads to an EM field at frequency ω_3 which is diffracted by the grating. The calculation of Raman intensity first requires the knowledge of the ω_3 diffracted EM field. This field is studied as follows. Maxwell equations read :

$$\vec{\nabla}\times\vec{E} = j\omega_3\mu_0\vec{H} \qquad (4a)$$

$$\vec{\nabla}\times\vec{H} = -j\omega_3\vec{D} \qquad (4b)$$

with $$\vec{D} = \epsilon_0\epsilon(x,y)\vec{E} + \vec{P}^{NL} \qquad (4c)$$

where $\epsilon(x,y) = \epsilon_1$ if $y \geqslant f(x)$ and $\epsilon(x,y) = \epsilon_2$ if $y < f(x)$

$\vec{P}^{NL}$ being the nonlinear polarization at frequency ω_3. Putting the value of $\vec{D}$ in the second Maxwell's equation leads to :

$$\vec{\nabla}\times\vec{H} = -j\omega_3\epsilon_0\epsilon(x,y)\vec{E} - j\omega_3\vec{P}^{NL} \qquad (4d)$$

Since we are dealing with surface polaritons which, as is well known, are TM, we focus hereafter on the TM electromagnetic field which is characterized by the z component of vector $\vec{H}$: $\vec{H}(x,y) = H(x,y)\vec{e}_z$ where $\vec{e}_z$ is the unit vector of the z axis.

Let us now derive the propagation equation satisfied by H(x,y).

From (4b), we get :

$$\vec{E} = -\frac{1}{j\omega_3\epsilon_0\epsilon(x,y)}(\vec{\nabla}H\times\vec{e}_z) - \frac{\vec{P}^{NL}}{\epsilon_0\epsilon(x,y)} \qquad (4e)$$

and (4a) implies :

$$\vec{\nabla}.[\frac{1}{k^2}\vec{\nabla}H] + H = j\omega_3[\vec{\nabla}\times(\frac{\vec{P}^{NL}}{k^2})]_z \qquad (5)$$

Where $k^2(x,y)c^2/\omega_3^2 = \epsilon(x,y)$ and $[\vec{\nabla}\times(\ldots)]_z$ denotes the z component of the curl operator.

It must be emphasized that a) (5) is valid for any region : not only in the homogeneous regions 1 and 2 but also in the inhomogeneous medium ($0 < y < \delta$) where ϵ and the susceptibility tensor are x and y dependent, b) since Maxwell equations are valid in the sense of generalized functions or distributions [28,38], (5) is also valid in the sense of distribution. Thus, (5) automatically takes into account the boundary conditions (BC) on the grating surface. The solution of (5) allows dealing with the diffraction of the EM field at frequency ω_3.

Let $\vec{E}(\omega_\nu) = \vec{e}_\nu E_\nu(\vec{r})e^{j\vec{k}(\omega_\nu).\vec{r}}$ be the electric field in regions 2 and 3 at ω_ν ($\nu = 1, 2$). Equation (5) together with the propagation equations at frequencies ω_1 and ω_2 constitute a set of coupled nonlinear partial differential equations which cannot be solved. The solution of this set of equations is achieved by a linearization well adapted to the two-beam methods in which the two incident beams must be symmetrically considered [39]. Indeed, during the interaction, the higher-frequency beam (ω_1) is depleted while the lower frequency one (ω_2) is amplified. Thus, we assume that the product $E_1(r)E_2^*(r)$ remains constant.

At this stage of the calculations, it is worth noting that the difference with the problems analysed in ref 28 is the existence of a source term in media 2 and 3 i.e. the nonlinear polarization (and subsequently the fact that there exists no incoming plane wave at ω_3).

Therefore, (5) is solved using a method [16] directly derived from that developed in ref 28. We then know the ω_3 EM field everywhere in space : not only outside the modulated region (media 1 and 2) but also inside it :

* In region 1, its expression is given by the second term of (1a) with :

$$\gamma_n = k_{2,x}(\omega_1) - k_{2,x}(\omega_2) + n(2\pi/d) \qquad (6)$$

Above the grating, the field at ω_3 is a superposition of an infinite number of plane waves with amplitude B_n.

* In the modulated region the expression of the ω_3 EM field is given by (3)

* Below the grating (region 2) : in addition to the diffracted plane waves with amplitude T_n (see (2a)), there exists a wave whose expression is :

$$A^{NL} e^{j\vec{K}.\vec{r}}$$

with : $\vec{K} = \vec{k}_2(\omega_1) - \vec{k}_2(\omega_2)$

$$A^{NL} = -\omega_3 \frac{(\vec{K}x\vec{P}^{NL}(\omega_3))_z}{k_2^2(\omega_3) - K^2} \qquad (7)$$

$$P_h^{NL}(\omega_3,\vec{r}) = \chi_{h,i,j}(\omega_3)e_{1,i}e_{2,j}E_1(\omega_1,\vec{r})E_2^*(\omega_2,\vec{r})$$

(*): complex conjugate

$[\chi(\omega_3)]$: NL susceptibility tensor at ω_3

The wavevector of the wave of amplitude A^{NL} is that of the NL polarization. This wave is, therefore, a driven one, contrary to the diffracted field (with amplitudes B_n and T_n) which are free waves. Notice that the amplitude A^{NL} of the driven wave, whose expression is given by (7), is independent of the groove depth δ, whereas B_n and T_n are expected to depend on δ. Since K_x and K_y are always real, this additional driven wave, whose existence is specific for a nonlinear process, always propagates in the bulk and will therefore never be a surface wave. The surface waves, if any, must be looked for in the diffracted field. They correspond to diffracted orders n for which $\alpha_{1,n}^2 < 0$. According to (1c) and (6), their existence depends on the relative values of d, ω_3 and θ.

The H_n, B_n and T_n are derived using the method of ref 16.

3.2 Raman intensity and comparison with the results of Ushioda at al.[6]

Let us first notice that Ushioda et al.[6] did not measure the intensity of the surface polaritons but that of the Stokes beam (the beam of ω_2 frequency with our notation). Thus, we must first determine the expression of the relative amplification of the ω_2 frequency beam between the planes $y = \delta$ and $y = -L$.

3.2.1 Expression of the power variation of the ω_2 frequency beam

The starting point of this calculation is the equation obeyed by the electric field at frequency ω_2 (Ref. 40) :

$$-2j\vec{k}_2(\omega_2).\vec{\nabla}E_2(\vec{r}) = \frac{\omega_2^2}{c^2}\vec{e}_2.\vec{P}^{NL}(\omega_2,\vec{r})e^{-j\vec{k}_2(\omega_2).\vec{r}} \qquad (8)$$

In (8), $\vec{P}^{NL}(\omega_2,\vec{r})$ is the nonlinear polarization at frequency ω_2 (Ref. 40) :

$$P_j^{NL} = \chi_{j,i,h}(\omega_2)E_i(\omega_1)E_h^*(\omega_3) + \chi_{j,i,k,l}E_i(\omega_1)E_k^*(\omega_1)E_l(\omega_2) \qquad (9)$$

where $[\chi(\omega_2 = \omega_1 - \omega_3)]$ is the nonlinear susceptibility tensor at frequency ω_2. $\chi_{j,i,k,l}$ is an element of the Raman tensor whose expression will be given hereafter in the case of GaP, and $E_h(\omega_3,\vec{r})$ are the components of the electric field at frequency ω_3 along the x, y, z axis whose expressions, as explained in section 3.1., includes a driven term in region 2.

From (8), we derive the equation fulfilled by the intensity $|E_2(\vec{r})|^2$ of the pump beam with frequency ω_2 :

$$\vec{k}_2(\omega_2).\vec{\nabla}|E_2(\vec{r})|^2=\frac{\omega_2^2}{c^2}\mathrm{Im}[\vec{e}_2.\vec{P}^{NL*}(\omega_2,\vec{r})e^{j\vec{k}_2(\omega_2).\vec{r}}E_2(\vec{r})] \quad (10)$$

The rigorous solution of (10) is difficult to obtain since, as a first step, it implies getting the expression of the intensity variation of the ω_2 frequency beam in the modulated region, i.e. from $y=\delta$ to $y = 0$, $|E_2(y=\delta, x)|^2$ being equal to the incident intensity at frequency ω_2. This is the difficult part of the integration of (10). Indeed, we have to carry out a numerical integration of this equation.since there is no explicit expression for the EM field at frequency ω_2 in the grating region. The second step corresponds to the integration of (10) in the bulk, i.e., from $y = 0$ to $y = -L$. As will be seen hereafter, this integration can be performed analytically.

Since the thickness L is much greater than the groove depth δ (for example, in ref 6, L = 15 μm whereas δ = 0.05 μm or 0.3 μm), the intensity variation of the ω_2 frequency beam in the modulated region is negligible compared to that in the bulk. Therefore, we shall integrate (10) from $y = 0$ to $y = -L$ assuming $|E_2(y=0,x)|^2 = |E_2(y=\delta,x)|^2$.

The power variation of the ω_2 frequency beam is a function which rapidly varies with x as does the grating surface. But the experimentalists are not interested in these microvariations, they are only concerned with the average value, with respect to x, of the power variation Δ of the ω_2 frequency beam. According to the periodicity of the grating profile, Δ does not depend on x. Thus, from (4e) and the expression of the magnetic field at ω_3 in region 2 together with the result that $|E_1(\vec{r})\, E_2(\vec{r})|^2$ is constant (according to the linearization done in Sec. 3.1.) we get, after a straightforward but tedious calculation , the following expression for Δ :

$$\Delta=\frac{|E_2(y=-L)|^2-|E_2(y=0)|^2}{|E_1(y=0)E_2(y=0)|^2} \quad (11)$$

$$\Delta=\frac{\omega_2^2}{k_{2,y}(\omega_2)c^2}[\mathrm{Re}(\chi_{j,i,h}(\omega_2)\sum_n a_{n,h,k,l}\frac{e^{j(\alpha_{2,n}+K_y)L}-1}{\alpha_{2,n}+K_y})$$

$$-\mathrm{Im}(\chi_{j,i,h}(\omega_2)b_{h,k,l}+\chi_{j,i,k,l})L]e_{1,i}e_{1,k}e_{2,j}e_{2,l} \quad (12)$$

In 12 the quantities $a_{n,h,k,l}$ arise from the existence of the free diffracted field, whereas $b_{h,k,l}$ come from the driven wave.

Inspection of 12 shows that not only all the diffracted waves at freqnency ω_3 but also the driven wave contribute to the power variation of the ω_2 frequency beam. The diffracted field leads to the term Re(...) whereas the driven one leads to Im(...). Therefore, it is seen that Re(...) is the only term, occurring in the expression of Δ, which depends on the groove depth.

3.3.3 Numerical results

We consider a sinusoidal grating with groove spacing d = 2.5 μm impressed on GaP. The permittivity and nonlinear susceptibility of GaP are given by [41] :

$$\epsilon_2(\omega_3) = \epsilon_\infty + \frac{S\omega_0^2}{\omega_0^2 - \omega_3^2 - j\omega_3\xi}$$

ϵ_∞, ω_0, ξ, and S, being, respectively, the high-frequency dielectric constant, transverse frequency, damping constant, and oscillator strength of GaP. The numerical values of which are [41, 42] :

$\epsilon_\infty = 9.09$, $\omega_0 = 367.3$ cm^{-1}, $\xi = 0.0035\omega_0$, S = 1.86

$$\chi_{j,i,h} = d_E\left[1 + \frac{C\omega_0^2}{\omega_0^2 - \omega_3^2 - j\omega_3\xi}\right] \quad (13)$$

where [41,42] : $d_E = 9\ 10^{-22}$ (MKSA), C = -0.53.

For ω_1, we take the same value as that used by the authors of Ref. 6 i.e., $\omega_1 = 19436$ cm^{-1}. According to (1c) and (6) and to the numerical values of d and ω_3, all the $n \geqslant 1$ diffracted orders are evanescent. In the following, we select the couple (ω_3, θ) to excite resonantly the n = 1 diffracted order. It is then consistent to consider that this order, which corresponds to a surface wave, represents the surface polaritons whose amplitude, for $y < 0$, is T_1. For the sake of simplicity, we assume that the ω_1 and ω_2 frequency beams are, respectively, polarized along the x and z axis. Thus [41,42] the nonlinear polarization at frequency ω_3 has only a nonzero component along the y axis :

$$P_y^{NL} = \chi_{y,x,z}(\omega_3)E_1E_2e_{1,x}e_{2,z}$$

were $\chi_{y,x,z}(\omega_3)$ is given by (13).

We then get [40] the following expression for the Raman tensor :

$$\chi_{j,i,k,l} = \chi_{z,x,x,z} = \frac{d_E^2C^2\omega_0^2}{S(\omega_0^2 - \omega_3^2 - j\omega_3\xi)}$$

where use has been made of the Armstrong, Bloembergen, Ducuing, Pershan symmetry relation [40].

The sample thickness L has the same value as that used by the authors of ref 6: L = 15 μm. We have plotted on Fig. 4 the quantity Δ computed at constant θ varying ω_3. The corresponding results of ref 6 (see Fig. 2 of Ref 6) were indeed obtained under these conditions. These curves have been drawn for the two groove depths used by the authors of Ref 6 : $\delta = 0.05$ μm and $\delta = 0.3$ μm.

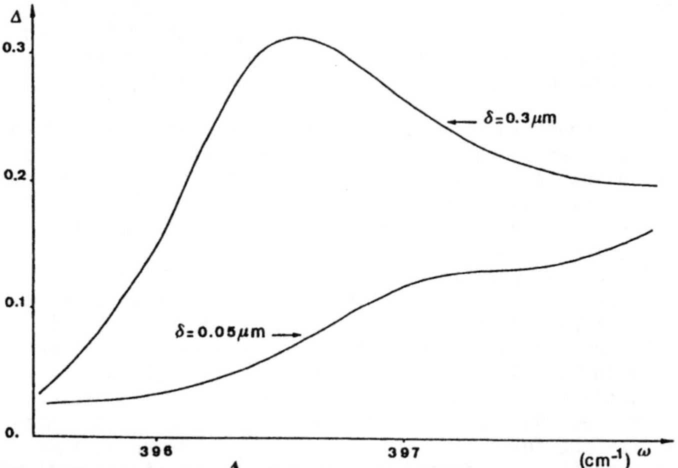

Fig.4 Increase of Δ with the groove depth of a GaP grating. d = 2.5 μm, $\theta = -3°$

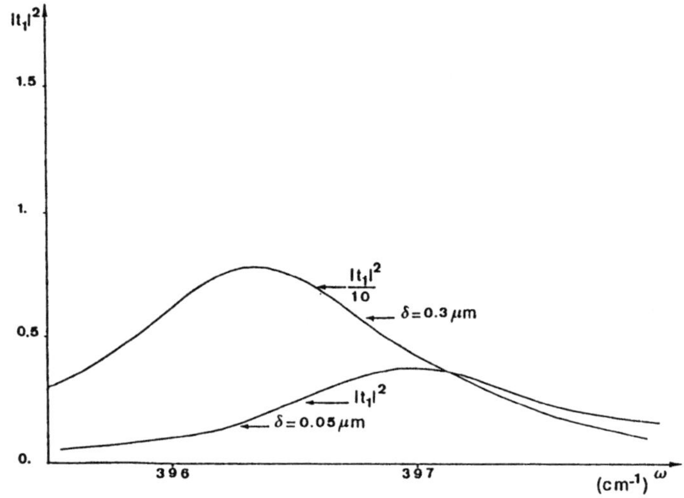

Fig.5 Same as fig.4, except that $|t_1|^2$ is plotted instead of Δ. Note that for δ = 0.3 μm, $|t_1|^2/10$ is plotted instead of $|t_1|^2$

Figure 4 agrees well with Fig. 2 of ref 6 since an increase of δ from 0.05 to 0.3 μm leads to (a) an enhancement of Δ equal to 2.8. This value is in agreement with that measured by Ushioda et al. (b) a very slight shift, toward the low-frequency range, of the Δ peak position. It is worth noting that the two curves shown in Fig. 4 and those plotted in Fig. 2 of ref 6 have the same characteristic asymmetric shape. Moreover, from the $|t_1|^2$ curves shown in Fig. 5, we see that the δ dependence of the Δ curves is much weaker thant that of the $|t_1|^2$ curves ($|t_1|^2 = |T_1/E_1E_2|^2$)

To understand these results, we have to remember that, as already pointed out, not only the diffracted waves, but also the driven waves contribute to Δ . The terms Re(...) and Im(...) which account, respectively, for each of these two contributions, do not behave in the same way as L is increased. Indeed, since the term Im(...) is proportional to L, the contribution of the driven wave to Δ increases linearly with L, whereas that of the diffracted field tends toward a limiting value as L is increased. Besides, numerical calculations have shown that Im(...), which does not depend on δ , grows monotonously with ω_3.

For a given value of L equal to 15 μm, computer calculations have shown that, in the frequency range where Re(...) takes its peak value, Im(...) is of the same order of magnitude as Re(...). This shows that, in the experiment reported in ref 6, the contribution of the driven wave is not negligible at all. This will be the case whenever L is greater than $1/\mathrm{Im}(\alpha_2)$. In ref 6, for ω_3 = 397 cm^{-1}, $1/\mathrm{Im}(\alpha_2)$ = 2 μm whereas L = 15 μm.

To summarize, Δ is the superposition of two functions which, with the numerical values used in ref 6, have the same order of magnitude as soon as the contribution of the diffracted field to Δ becomes important. Im(...) grows monotonously with ω_3 and is independent on δ whereas Re(...) behaves nearly as reported in Fig. 5 and is δ dependent. This explains the fact that Δ is less dependent on δ than $|t_1|^2$ and the asymmetric shape of the Δ curves. This proves the importance of the contribution of the driven wave. Let us emphasize the following important point: in Raman excitation of S Pol, Δ will exhibit a δ dependence only if the nonlinear medium is not too thick as compared with $1/\mathrm{Im}(\alpha_2)$: this result is a direct consequence of the existence of a driven wave at the S Pol frequency.

In conclusion, considering a grating impressed on a nonlinear medium, an increase of the groove depth leads to an enhancement of the power variation of the ω_2 frequency beam and a slight shift, toward the low-frequency range, of the Δ peak position. These two results agree well with the corresponding measurements made by Ushioda et al. This agreement is all the more noteworthy in that our calculation is concerned with surface polaritons propagating along a grating, whereas Ushioda et al.'s experiment was done on a statistically rough surface. From our results, we may infer that the statistical feature of the interface used by the authors of ref 6 does not play a major role, the important parameter being the groove depth.

4. Theory of diffraction in NLO

As already pointed out in section 1, SESHG takes place in metallic media. Thus we first consider the problem of SHG at metallic surfaces : nonlinear polarization and BC at the SH frequency.

One may wonder about the interest of such a study since an expression of the nonlinear polarization in metals has already been obtained by Bloembergen et al. [43]. The reason is the following : Sipe et al [44,45] have considered the nonlinear polarization derived by Bloembergen at al and have argued that its expression ,$\vec{P}^{NL}$, is "inherently ambiguous" [44] across S because one has to perform the product of a step function by a Dirac - δ function. This has prompted Sipe et al. to develop a hydrodynamic theory of second harmonic generation (SHG) in metals including the quantum pressure [44,45]. They find an expression of the NL polarization which is no longer ambiguous. This expression includes a surface term (which behaves as a Dirac - δ function) which must be placed outside the bulk metal.

We shall show using the theory of distribution [46-48] as a tool, that contrary to Sipe's claim, the expression of $\vec{P}^{NL}$ is in fact well defined. This is achieved by modelizing the entrance face of the metal by a transition region whose thickness is allowed to vanish. Our calculation is characterized by the fact that the corresponding limits are rigorously determined.

4.1. SHG at metal surface

4.1.1. Origin of the ambiguity pointed out by Sipe et al. [44].

According to the work of Bloembergen et al. [43], an EM field at circular frequency ω_1 ,incident on the surface S of a metallic medium, induces a NL polarization at 2 ω_1 which can be written as :

$$\vec{P}^{NL}(2\,\omega_1) = \gamma\vec{\nabla}[\vec{E}(\omega_1).\vec{E}(\omega_1)] + \beta\vec{E}(\omega_1)\vec{\nabla}.\vec{E}(\omega_1) \qquad (14)$$

with[44,47,48]: $\gamma = (\beta/4)(1-\epsilon_2(\omega_1))$ (15)

$\epsilon_2(\omega_1)$: permittivity of the metal at the pump circular frequency ω_1 ;

$\vec{E}(\omega_1)$: electric field at the pump circular frequency ω_1 in the metal.

Since $\epsilon_2(\omega_1)$ is space dependent, γ depends on the space coordinates in the metal and vanishes in vacuum, while β is a constant [44,47,48] (whose expression is not useful for our purpose).

As is well known [28,38], Maxwell equations are valid in the sense of distribution. Therefore (14), written in the sense of distributions, must have a meaning. This must be true, in particular, if we deal with an interface S separating a nonlinear metal from vacuum. It is shown in refs [47,48] that (14) becomes, in the sense of distribution [28,38,46] :

$$\vec{P}^{NL}(2\omega_1) = \gamma\left\{\vec{\nabla}[\vec{E}(\omega_1).\vec{E}(\omega_1)]\right\} + \gamma\ \vec{n}J[\vec{E}(\omega_1).\vec{E}(\omega_1)]\delta_S$$

$$+ \beta\ \vec{E}(\omega_1)\vec{n}.J[\vec{E}(\omega_1)]\ \delta_S \qquad (16a)$$

$\vec{n}$: unit vector of the normal to S oriented from metal to vacuum.

δ_S : distribution located on S and defined by $<\delta_S, \phi> = \iint_S \phi dS$ where ϕ is an infinitely differentiable function with compact support [28,46]. From a non-rigorous, but more intuitive, point of view, the two-dimensional distribution δ_S is nothing but the classical Dirac- δ function $\delta(n)$ where n represents the abscissa along the normal to S.

J(.....) : Jump of (.....) across S in the direction of $\vec{n}$.

{....} : operator written in the usual sense of function [28].

Use has been done of the fact that $\vec{\nabla}.\vec{E}(\omega_1) = 0$.

It is seen from (16 a) that $\vec{P}^{NL}$ includes a volume term (first quantity of (16a)) and two surface terms. Thus :

$$\vec{P}^{NL} = \vec{P}_V^{NL} + \vec{P}_S^{NL}\ \delta_S \qquad (16b)$$

The difficulty mentioned above comes from the surface part of $\vec{P}^{NL}$. As already pointed out, γ , occurring in (16) is not a constant : γ varies from its bulk value γ_b in the metal to zero in vacuum. Thus, γ varies as a step function (denoted Γ_S) :

$$\Gamma_S = \begin{cases} 0 \text{ above } S \\ \gamma_b \text{ below } S \end{cases}$$

Therefore, in (16) we are led to make the product of a discontinuous function Γ_S by a distribution δ_S. This product has no meaning [47,48]. The same conclusion applies to the quantity $\vec{E}(\omega_1)\ \delta_S$ since $\vec{E}(\omega_1)$ is discontinuous across S. Thus the two last terms of (16a) have no meaning and, as a consequence, the expression (14), of $\vec{P}^{NL}(2\ \omega_1)$, put under the form (16a), is ambiguous. This difficulty has already been pointed out by Sipe et al.[44] with similar arguments. Actually, such a calculation is not correct.

In order to understand where the error takes place in the above reasoning, one has to remember that the entrance face S of the metallic medium may be modelized by the limit, when $\eta \to 0$, of a continuous transition region, with thickness η , between the metal and vacuum.

Equations (16) are derived from [14] by assuming that the limit, when $\eta \to 0$, of the two terms $\gamma\ \vec{\nabla}[\vec{E}(\omega_1).\vec{E}(\omega_1)]$ and $\beta\ \vec{E}(\omega_1)\vec{\nabla}.\vec{E}(\omega_1)$ (occurring in (14)) is equal to :

i) $\lim\limits_{\eta\to 0}\gamma\ \lim\limits_{\eta\to 0}\vec{\nabla}[\vec{E}(\omega_1).\vec{E}(\omega_1)]$ for the first quantity

ii) $\lim\limits_{\eta\to 0}\vec{E}(\omega_1)\lim\limits_{\eta\to 0}\ \vec{\nabla}.\vec{E}(\omega_1)$ for the second quantity

This is generally incorrect,and the right way to perform the calculation is to look for the two limits :

$$\vec{L}_1 = \lim_{\eta\to 0}[\ \gamma\ \vec{\nabla}(\vec{E}(\omega_1).\vec{E}(\omega_1))]$$

$$\vec{L}_2 = \lim_{\eta\to 0}[\vec{E}(\omega_1)\vec{\nabla}.\vec{E}(\omega_1)]$$

Therefore when stating that (14) is ambiguous, one implicitly assumes that the limit, when $\eta \to 0$, of a product of two terms is the product of the limit of each term. In our case, the product of the limit of each term does not exist, and one has to look for the limits $\vec{L}_1$ and $\vec{L}_2$.

The determination of $\vec{L}_1$ and $\vec{L}_2$ is done in refs [47,48] (using an orthogonal curvilinear coordinates system u, v, w ; the equation of S being v(x,y,z) = 0) and will not be repeated here. We then get :

$$\vec{P}_S^{NL}\,\delta_S = \vec{E}_t \beta E_n^-[\epsilon_2(\omega_1) - 1]\,\delta_S$$

$$+\,\vec{n}(3/4)\,\beta\,E_n^{-2}[\epsilon_2(\omega_1) - 1][(1/3)\,\epsilon_2(\omega_1) + 1]\,\delta_S \qquad (17)$$

E_n^- : normal component of the pump electric field just below S.

$\vec{E}_t$: tangential component of $\vec{E}(\omega_1)$

It is worth noticing that :
i) the expression (14) of $\vec{P}^{NL}$ is well defined provided its surface term is correctly calculated : this surface part includes no ambiguities.
ii) the surface NL polarization is located <u>at</u> the entrance face S of the metal and not outside it.
iii) the surface NL polarization not only has a component parallel to S but also a component perpendicular to S.
iiii) the expression of $\vec{P}_S^{NL}\,\delta_S$ is <u>independent</u> of the profile of the permittivity in the transition region.

4.1.2. <u>Boundary conditions at the SH frequency on the metal surface</u>

Maxwell equations are given by (4a,4b) with $\omega_3 = 2\,\omega_1$; $\vec{P}^{NL}$ is deduced from (16),(17).

$$\epsilon(2\omega_1,v) = \begin{cases} 1 & \text{if } v>0 \\ \epsilon_2(2\omega_1) & \text{if } v<0 \end{cases}$$

Maxwell's equations being valid in the sense of distributions [28,38], let us rewrite (4b) in the sense of distributions :

$$\left\{\vec{\nabla}\times\vec{H}(2\omega_1)\right\} + [\vec{n}\times J(\vec{H}(2\omega_1))]\,\delta_S = -j2\omega_1\epsilon_0\,\epsilon(2\omega_1,v)\vec{E}(2\omega_1)$$

$$-\,j2\omega_1\vec{P}_V^{NL}(2\omega_1) - j2\omega_1\vec{P}_S^{NL}(2\omega_1)\,\delta_S \qquad (18)$$

Before going further on, let us consider (18). It involves classical terms and distributive terms (including δ_S). It is well known [28,46] that (18) implies the equality of the functions present at the left and right-hand sides and that the same thing holds for the distributive parts.

At this stage of the reasoning, we make the following important remark : the distribution $[\vec{n}\times J(\vec{H}(2\omega_1))]\,\delta_S$ has only a tangential component, whereas (see (17)) $\vec{P}_S^{NL}(2\omega_1)\delta_S$ not only has a tangential component but also a component perpendicular to S. Thus, in the right-hand side of (18), $\vec{E}(2\omega_1)$ cannot be a usual function. It must also have a distributive part which has, at least, a component perpendicular to S. Therefore, the electric field in (18), which represents the total electric field, is the sum of two terms, a classical (volume) part $\vec{E}_V$ and a distributive part $\vec{E}_S(2\omega_1)\,\delta_S$.

$$\vec{E}(2\omega_1) = \vec{E}_V(2\omega_1) + \vec{E}_S(2\omega_1)\,\delta_S \qquad (19)$$

where $\vec{E}_S(2\omega_1)$ has at least a component normal to S.
It is shown in refs (47,48) that $\vec{E}_S(2\omega_1)$ is perpendicular to S and that the following BC at $2\omega_1$ hold :

$$\vec{n}\times J[\vec{H}(2\omega_1)] = -j2\omega_1\vec{n}\times[\vec{P}_S^{NL}(2\omega_1)\times\vec{n}] \qquad (20a)$$

$$J[\vec{E}_{v,t}(2\omega_1)] = B\vec{\nabla}_t E_n^{-2}(\omega_1) \qquad (20b)$$

$$B = -\frac{\beta}{3\epsilon_0}\left[\frac{4\epsilon_2^2(\omega_1)}{3}\mathrm{Log}\left(\frac{\epsilon_2(\omega_1)}{\epsilon_2(2\omega_1)}\right)+(\epsilon_2(\omega_1)-1)(3-\epsilon_2(\omega_1))\right] \qquad (20c)$$

The $\vec{\nabla}_t$ operator is the two-dimensional gradient taken in a plane tangent to S.

Equation (20a) is already known but it is not so for (20 b). This equation shows that the tangential component of the electric field at $2\omega_1$ is no longer continuous when a surface NL polarization exists having a component perpendicular to S.

To summarize : the expression (14) of $\vec{P}^{NL}$ is mathematically well defined even on S. The demonstration [47,48] is valid whatever the profile of $\epsilon(v)$ in the transition region may be. $\vec{P}^{NL}$ is a sum of a volume term and of a surface one. The latter describes a surface NL polarization which is located at the surface S. The perpendicular component of $\vec{P}_S^{NL}\ \delta_S$ gives rise to surface electric field at $2\omega_1$ also located on S. The BC for the EM field at $2\omega_1$ are derived. As a new result, we obtain the expression of the discontinuity of the tangential component of the electric field at $2\omega_1$. This discontinuity comes from the existence of the surface EM field at the SH frequency.

4.2. Diffraction in NLO

For sake of simplicity, we only consider the following situations :
i) SHG in a bare silver grating. ii) $(\omega_1,\omega_2) \rightarrow \omega_3=\omega_1+\omega_2 \quad (\omega_1\neq\omega_2)$ in a metallic or dielectric grating coated by one or several dielectric layers. In this case, the nonlinearity comes from the dielectrics, not from the metal.

The scattering geometry is depicted in fig. 6

Two beams with frequencies ω_1 and ω_2 $(\omega_1>\omega_2)$ are incident on a grating ruled on a dielectric or metallic nonlinear medium and coated by one or several dielectric layers constituted by linear or nonlinear materials (Fig. 6). Above the modulated structure, space is filled with a linear material whose permittivity is real and positive. $\theta_\nu(\nu = 1,2)$ denotes the angle of

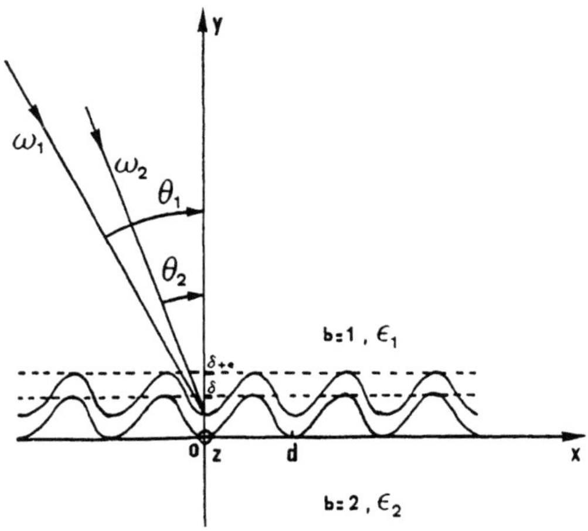

Fig. 6. Scattering geometry when the grating (periodicity d, groove depth δ) is coated by a single dielectric layer of thickness e. Index b characterizes the medium above or below the modulated region.

$$\epsilon(\omega_\nu,x,y)=\begin{cases}\epsilon_2(\omega_\nu) & y<0\\ \epsilon_1(\omega_\nu) & y>\delta+e\end{cases}$$

$$\nu = 1,2$$

incidence of each of the pump beams. The case of a nonlinear medium whose entrance face is constituted by a bare grating is dealt with by making e=0.

We assume $\partial/\partial z=0$. Thus the solutions at frequency ω_ν (ν=1,2,3) are either TM or TE polarized. In the following, the concept of associated smooth structure will be useful : It is the system derived from that of Fig. 6 by allowing every modulation depth to tend to zero.

Before going in the detail of the calculation, it is worth saying a few words about the physics involved in the phenomenon of diffraction in NLO : the ω_1 and ω_2 frequency beams are diffracted by the NL grating (bare or coated). The interaction between all the diffracted orders (propagating or evanescent) at ω_1 and ω_2 gives rise, in the NL media (whatever their location in fig. 6 may be) to a NL polarization at ω_3. This NL polarization radiates an EM field at ω_3 which, in turn, is diffracted by the grating : some of the reflected diffracted orders at ω_3 are propagating, all the others are evanescent (bounded to the NL grating). The observed signal at frequency ω_3 is to be found in the propagating diffracted orders.

The method used to study the NL diffraction is a direct transposition of the above description : First, we obtain the expression of the electric field of the diffracted beams at each frequency ω_ν (ν=1,2) below and inside the modulated region. Then, we determine the expression of the nonlinear polarization at frequency ω_3 below the modulated region but also inside it. Finally, we generalize the theory [16] developed in section 3 to the case where the diffraction of the pump beams is taken into account and where the signal at frequency ω_3 may be TE polarized. According to the periodic feature of the system fig. 6, it is convenient to expand all the pertinent quantities in Fourier series.

We make the usual undepleted pump approximation for the two pump beams. Thus, Maxwell's equations (4a,4b,4c) written at frequencies ω_ν (ν =1,2) with $\vec{P}^{NL}$ = 0 shows that (a) the two pump beams are linearly diffracted by the grating ; (b) their diffraction takes places independently of that of the signal at frequency ω_3 . According to point (a), the expression of $\vec{E}(\omega_\nu,\vec{r})$ (ν=1,2) can be obtained by using the RETLD [28].

4.2.1. Expression of the Electric Field at ω_1, ω_2

Whatever the polarization of the incident beams may be, the expression of the diffracted electric field at ω_1 and ω_2 can be written as[19] :

$$\vec{E}(\omega_1,\vec{r}) = \sum_l \vec{u}_{l,1} e^{j(-\alpha_{2,l,1}y + \gamma_{l,1}x)} \qquad y<0 \qquad (21a)$$

$$\vec{E}(\omega_2,\vec{r}) = \sum_m \vec{u}_{m,2} e^{j(-\alpha_{2,m,2}y + \gamma_{m,2}x)} \qquad y<0 \qquad (21b)$$

$$\vec{E}(\omega_1,\vec{r}) = \sum_l \vec{E}_{l,1}(y) e^{j\gamma_{l,1}x} \qquad 0<y<\delta+e \qquad (21c)$$

$$\vec{E}(\omega_2,\vec{r}) = \sum_m \vec{E}_{m,2}(y) e^{j\gamma_{m,2}x} \qquad 0<y<\delta+e \qquad (21d)$$

with:

$$\alpha^2_{2,l,\nu} + \gamma^2_{l,\nu} = k_2^2(\omega_\nu) \qquad (\nu=1,2)$$

$$\gamma_{l,1} = k_1(\omega_1)\sin\theta_1 + l\sigma \qquad (22a)$$

$$\gamma_{m,2} = k_1(\omega_2)\sin\theta_2 + m\sigma \qquad (22b)$$

where $\sigma = 2\pi/d$ and $k_b^2(\omega_\nu) = \epsilon_b(\omega_\nu)(\omega_\nu^2/c^2)$ (b=1,2 ; ν=1,2)

In (21) the components of $\vec{u}_{l,1}$, $\vec{u}_{m,2}$ are determined with the aid of computer programs described in ref. 28. Let us emphasize that $\vec{u}_{l,1}$, $\vec{u}_{m,2}$, $\vec{E}_{l,1}$, $\vec{E}_{m,2}$ depend on δ , but contrary to $\vec{E}_{l,1}$ and $\vec{E}_{m,2}$, $\vec{u}_{l,1}$ and $\vec{u}_{m,2}$ are independent on y. We are now in a position to get the expression of the nonlinear polarization at frequency ω_3.

4.2.2 Nonlinear polarization below the modulated region ($y < 0$)

a) Dielectric. The NL polarization at frequency ω_3 is written as[42]

$$P_h^{NL} = \chi_{h,i,j}(\omega_3)E_i(\omega_1,\vec{r})E_j(\omega_2,\vec{r}) \tag{23}$$

[$\chi(\omega_3)$] is the nonlinear susceptibility tensor of the medium on which the grating is ruled.
According to (22),(21a) and (21b)

$$\vec{p}^{NL} = \sum_{l,p} \vec{p}_{l,p-l} e^{j\vec{\kappa}_{l,p}\vec{r}} \tag{24}$$

with

$$p = l + m$$

$$\kappa_{l,p,x} = \gamma_{l,1} + \gamma_{p-l,2} \tag{25}$$

$$\kappa_{l,p,y} = -(\alpha_{2,l,1} + \alpha_{2,p-l,2})$$

$$p_{l,p-l,h} = \chi_{h,i,j}(\omega_3)u_{l,1,i}u_{p-l,2,j} \tag{26}$$

Let us introduce

$$\vec{K} = \vec{\kappa}_{0,0} \tag{27}$$

which would be the wave vector of the nonlinear polarization if the two pump beams were not diffracted.
Equation (23) may be rewritten as

$$\vec{p}^{NL} = e^{j\vec{K}.\vec{r}} \sum_p \vec{P}_p^{NL} e^{jp\sigma x} \tag{28a}$$

$$\vec{P}_p^{NL} = \sum_l \vec{p}_{l,p-l} e^{j(\kappa_{l,p,y}-K_y)y} \tag{28b}$$

b) Metals. Below the modulated region, $\vec{p}^{NL}$ reduces to the first term of (16a). The nonlinear polarization at frequency ω_3 can still be put under the form given by (28). Equations (22), (21a) and (21b) yield :

$$\vec{p}_{l,p-l} = j\gamma(\vec{u}_{l,1}\,\vec{u}_{p-l,2})\,\vec{\kappa}_{l,p} \tag{29}$$

4.2.3. NL polarization inside the modulated region

a) Dielectric. Equation (23) still applies. But now :

$$\chi_{h,i,j}(\omega_3) = \chi_{h,i,j}(\omega_3,x,y) = \sum_q \chi_q(y)e^{jq\sigma x} \tag{30a}$$

Use of (21c), (21d), (22) and (30a) leads to an expression of $\vec{p}^{NL}$ in the modulated region of the type of (28a) provided the following substitution is made : $p \rightarrow n = p + q$ and

$$P_{h}^{NL} = \sum_{p} \chi_{h,i,j,n-p}(y) V_{p,i,j}(y) e^{-jK_y y} \tag{30b}$$

with

$$V_{p,i,j}(y) = \sum_{l} E_{l,1,i}(y) E_{p-l,2,j}(y) \tag{30c}$$

b) Metal. The NL polarization is given by (16),(17). For reasons which will become clear hereafter, the Fourier series expansion of $\vec{P}^{NL}$ is not useful.

Let us notice that all the diffracted orders, regardless of their nature, contribute to the NL polarization. But [15,16,28] the efficiency of each contribution depends on the fact that a given diffracted order may or may not lead to a resonant excitation of one of the guided normal modes of the associated smooth structure.

4.2.4 NL diffraction of the ω_3 EM field

The starting point is Maxwell equations (4a, 4b, 4c) written at frequency ω_3. According to the hypothesis $\partial/\partial z = 0$, the solutions at frequency ω_3 are either TM or TE polarized. Let us first consider the TM case.

4.2.4.1 TM case

The propagation equation is given by (5). As already pointed out in section 3, this equation is valid in the sense of distributions, i.e. also in region $0 \leqslant y < \delta + e$. In the case of NL dielectrics, there is no problem and (5) can be used. But when dealing with SHG at metal gratings, we have to be cautious. Indeed, the results obtained in section 4.1 can be summarized as follows : i) existence of a surface NL polarisation, ii) discontinuity of the tangential component of the magnetic and electric field at $2\omega_1$ and iii) in addition, we remind the reader that $\rho = 1/k_2^2(\omega_3)$ is discontinuous on S.

Therefore, the operators $\vec{\nabla}.$ and $\vec{\nabla}x$ (occurring in r) have no meaning when written in the sense of distribution [28,46] since they require calculating quantities of the type $\Gamma_S\ \delta_S$.

In fact, use of (4e), (16), (17) and (20a) (also see refs 47, 48) shows that (5) can be rewritten under a well-defined form. Namely :

$$\vec{\nabla}.[\rho\{\vec{\nabla}H\}] + H = j\omega_3 \vec{e}_z.\vec{A}\,\delta_S \tag{31}$$

with (B is defined by (20c)):

$$\vec{A} = \vec{n} x J(\rho \vec{P}_V^{NL}) + \vec{W}_t$$

$$\vec{W}_t = -\vec{n} x \vec{\nabla}_t[(Bc^2/\omega_3^2) E_n^{-2}(\omega_1)]$$

It is convenient to split (31) into two first-order coupled differential equations by letting $K = \rho\{\vec{\nabla}_y H\}$. The solution of the resulting system is obtained by expanding H (equation (3)) and $\vec{A}\ \delta_S$ (see ref 28) in Fourier series. We then get the following expression for γ_n :

$$\gamma_n = K_x + n\sigma \tag{32}$$

Writing the BC (continuity of the tangential components of $\vec{E}(2\omega_1)$ and $\vec{H}(2\omega_1)$ at y = 0 and y = δ + e yields a TM boundary value problem (BVP) whose unknowns are the Fourier coefficients of H in the modulated region.

This requires the knowledge of the EM field at ω_3 outside the modulated region. The corresponding expressions have already been given in section 3.1. The only modification concerns the driven wave, which is now a superposition of an infinite number of elementary driven waves. This result comes from the fact that the NL polarization, being diffracted by the NL grating, is a superposition of an infinite number of elementary NL polarizations(28b).

4.2.4.2 TE case

Now $\vec{E}(\omega_3,\vec{r})$ is directed along the z axis. This case is much simpler to handle than the TM one. According to (16), (17) there is no TE SHG in metals. Thus we only have to consider dielectric NL media. The solution is given in ref 19 and will not be rewritten here. We have to solve a TE BVP whose unknowns are the Fourier coefficients of $E(\omega_3)$ in the modulated region.

Sophisticated computer programs lead to the numerical solution of these TM and TE BVP. This allows the determination of the Fourier coefficients of $H(\omega_3)$ (TM) or $E(\omega_3)$ (TE) in the region $0 < y < \delta + e$. Once this is done, it is an easy matter to get the Rayleigh coefficients B_n and T_n (see equations (1a),(2a)) of the ω_3 EM field above and below the modulated region.

4.2.5 Results and discussion

4.2.5.1 Direction of propagation of the nonlinear diffracted orders at frequency ω_3

The diffracted field contains propagating and evanescent waves. The direction of propagation of the radiated waves can be deduced from (1c) and (32) for the medium $y > \delta + e$ and from (2b) and (32) for the medium $y < 0$. We get the following result :

$$[n_b(\lambda_3)/\lambda_3]\sin\psi_{b,n} = [n_1(\lambda_1)/\lambda_1]\sin\theta_1$$

$$+ [n_1(\lambda_2)/\lambda_2]\sin\theta_2 + (n/d) \qquad (33)$$

with $n = 0 \pm 1, \pm 2, \ldots$, $n_b(\lambda_\nu)(b = 1,2;\ \nu = 1,2,3)$ is the refractive index of medium b at wavelength $\lambda_\nu(\lambda_\nu = 2\pi c/\omega_\nu)$, and $\Psi_{b,n}$ is the diffraction angle of the diffracted order n at wavelength λ_3 in medium b.

Equation (33) is very important, since it represents the nonlinear grating equation. This equation allows determining which diffracted orders n (at ω_3) are radiated in medium b (b = 1,2) (by $|\sin\psi_{b,n}| < 1$). For these orders, their direction of propagation is given, in medium b, by $\Psi_{b,n}$.

4.2.5.2 Numerical and experimental results.

The results derived from the analysis presented in the previous sections are compared with those obtained in an experiment of SESHG at silver gratings of constant periodicity, d = 5556 Å, and various groove depths [22] (pump wavelength : 1.06 μm, $\lambda_2 = 0.53$ μm)

Fig 7 is a plot of the observed and calculated SH peak intensity for various groove depths δ for the two propagating orders at $2\omega_1$: the 0 and - 1 ones. The value of θ_1 is chosen in such a way that there is a SP resonance for the + 1 diffracted order at ω_1: $\delta_{opt}(\omega_1) = 110$Å.

Let us make the following comments : i) as predicted theoretically, there exists an optimum value of δ for which the SH peak intensity is the greatest, ii) the measured value of δ_{opt} depends on the number, n, of the SH diffracted order : $\delta_{opt,0} = 250$ Å, $\delta_{opt,-1} = 300$ Å, iii) the optimum enhancement of the SH intensity, for the 0 and - 1 diffracted orders, is about 15

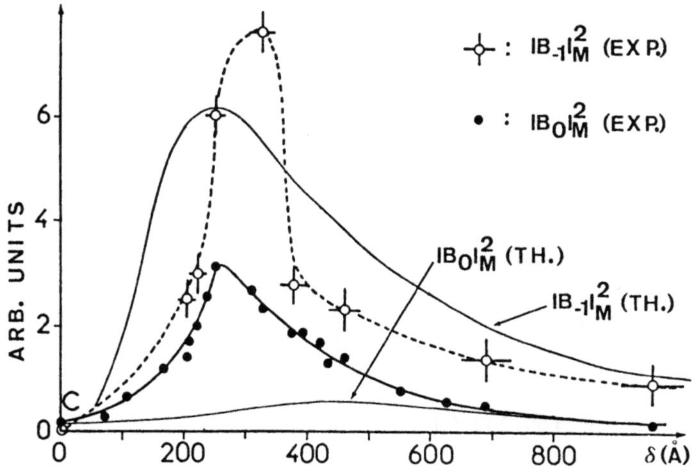

Fig 7. Peak intensity, $|B_0|_M^2$ and $|B_{-1}|_M^2$ at $2\omega_1$ of respectively the propagating zero and -1 diffracted orders as a function of δ .
$\epsilon_2(\omega_1) = -90.16 + j5.7$; $\epsilon_2(2\omega_1) = -9.8571 + j0.314$

and 36 as compared to the flat interface case (point C of fig 7), iiii) there is a good agreement with the corresponding numerical results concerning : the shape of the $|B_0|_M^2$ and $|B_{-1}|_M^2$ curves versus δ , the existence of δ_{opt}, the fact that δ_{opt} depends on n, the enhancement E_n : the computed values of E_0 and E_{-1} are respectively $E_0 = 2.7$, $E_{-1} = 31.3$

This general agreement can be considered as a check of the theory of diffraction in NLO (and also of the study of $\vec{P}^{NL}$ in metals).

Two important results must be emphasized :

(1) Each time the associated smooth structure supports guided normal modes, then, through a modulation of this structure, the possible EM resonances will lead to the enhancement of nonlinear effects taking place in this structure. In other words, depending on the structure which is used, the enhancement of nonlinear effects can be achieved by using surface plasmon-polariton or guided waves. Thus the surface plasmon-polariton resonance is not the only one,and probably not the more suitable EM resonance to be used when one is interested in enhanced nonlinear effects [18].

(2) The contribution of these EM resonances to the enhancement of nonlinear effects can be optimized. It is worth noticing that this optimization, and thus the strongest enhancement, is achieved for low modulation, of the order of 0.03.

5. Surface Enhanced Kerr Effect : Optical Bistability

We are dealing with a NLO process of the kind : $\omega = \omega + \omega - \omega$
The corresponding NL polarization is written as :

$$P_j^{NL}(\omega = \omega + \omega - \omega) = \chi_{j,i,k,l} E_i(\omega) E_k(\omega) E_l^*(\omega) \quad (34)$$

In order to determine the electromagnetic field, we assume that the nonlinear medium is cubic. Then, the 3rd order susceptibility tensor which appears in the expression (34) of the nonlinear polarization is such that it creates nonlinear effects presenting the same character (TE or TM) as the excitation. This allows us to distinguish between these fundamental cases of polarization TE or TM. We deal here with the TE case.

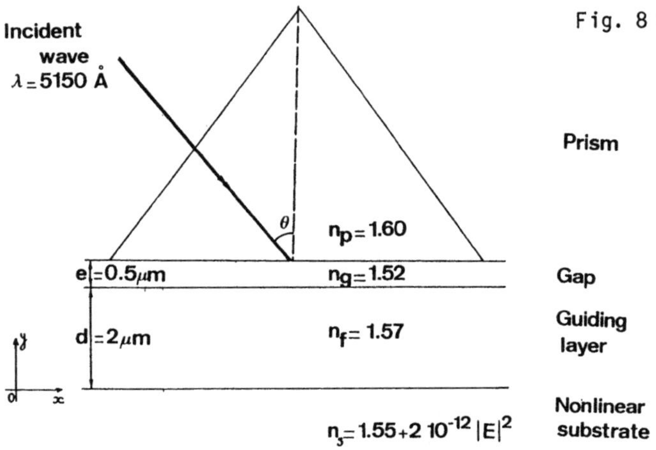

Fig. 8 Geometry of BOD

We consider the device depicted in fig. 8. Its main feature is the existence of a NL medium constituting the substrate. As a result, we deal with a NL waveguide. This structure supports NLGM. According to fig. 8, these NLGM are excited through an ATR process (OTTO geometry) which, in fact, is a NL ATR process. Such a geometry allows a simpler analysis than the NL guiding layer case[49].
The incident plane wave is TE polarized : $E = E_i e^{-j(\omega t - k_x x)}$ (incident power P_i). The index of refraction of the Kerr medium is written as : $n = n_0 + \delta |E|^2$. The numerical values are reported on fig. 8.

The calculation is performed in the following way. i) Write Maxwell equation in the media of indices n_g, n_f and n_s. ii) Solve the corresponding propagation equations in each of these media together with the boundary conditions at the following interfaces : prism - gap, gap - guiding layer, guiding layer-substrate. This allows getting the electromagnetic field in the guiding layer. Points i) and ii) can be easily dealt with in the linear media. However, in the nonlinear medium we have to be careful. Indeed, one has to take into account the y dependence of the index of refraction in the Kerr layer where we deal with a nonlinear Maxwell equation. As the substrate extends to infinity ($y \rightarrow -\infty$), the Kerr substrate configuration can be solved using Kaplan's results[50].

5.1. Homogeneous NL problem

Let E_{GK} be the electric field at the NL substrate guiding layer interface. The method outlined above allows studying the associated homogeneous NL problem ($e \rightarrow \infty$) and yields the curves fig.9 which are a plot of E_{GK} versus the effective index β of the NLGM.

5.2. Inhomogeneous NL problem : BO behaviour and resonant excitation of NLGM

We get the following relation between the incident field amplitude E_i and the modulus of the electric field at the NL substrate/guiding layer interface E_{GK} :

$$E_i^2 = -aE_{GK}^2 + bE_{GK}^2 \sqrt{\Delta_0 - (\Delta_E/2)E_{GK}^2} + cE_{GK}^2 \quad (35)$$

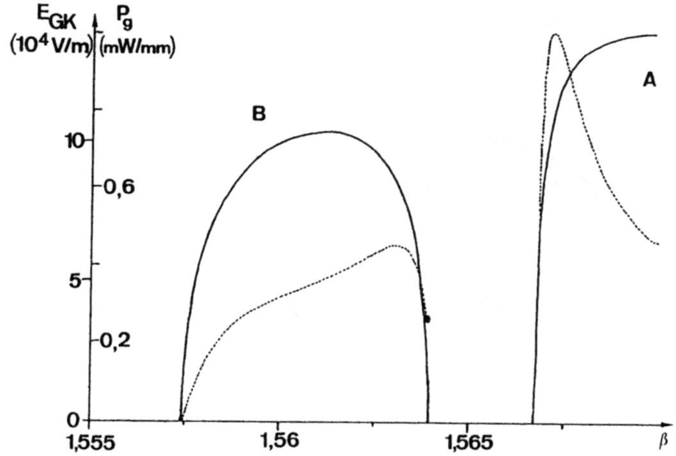

Fig. 9 Homogeneous problem : full line : effective index β of the NLGM versus the electric field E_{GK} at the NL substrate/guiding layer interface. Dotted line : effective index β of the NLGM versus the guided power. A : TE0 NLGM, B : TE_1 NLGM

Δ_0, Δ_E, a, b, c are known quantities. Their expressions are quite intricate and of no interest for the following discussion.

We first notice that P_i and θ are two independant parameters. Thus using (35), we can plot E_{GK} versus θ for different values of P_i (fig.10a) or E_{GK} versus P_i for different values of θ (fig.10b). We shall consider successively fig. 10a and fig. 10b.

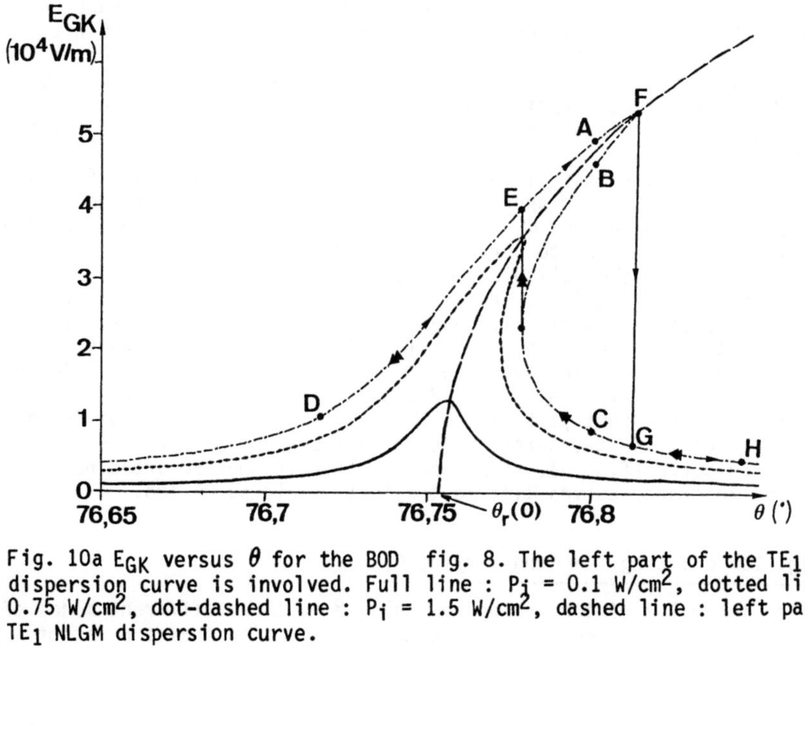

Fig. 10a E_{GK} versus θ for the BOD fig. 8. The left part of the TE_1 NLGM dispersion curve is involved. Full line : P_i = 0.1 W/cm^2, dotted line : P_i = 0.75 W/cm^2, dot-dashed line : P_i = 1.5 W/cm^2, dashed line : left part of the TE_1 NLGM dispersion curve.

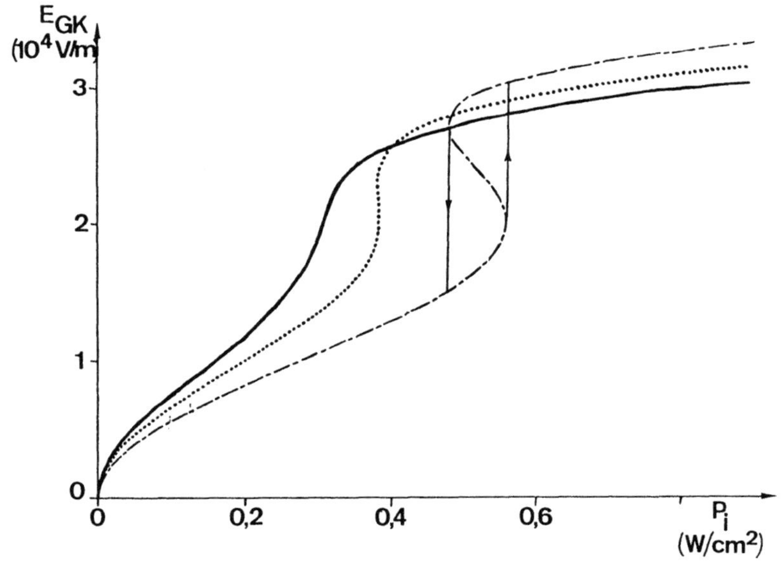

Fig.10b E_{GK} versus P_i for the BOD fig.8. Full line θ =76.765°, dotted line θ = θ_S =76.767°, dot-dashed line θ =76.77°

5.2.1 Figs 10a : fixing P_i and varying θ .

For a low value of P_i everything happens as if one had a linear ATR configuration (the NL term $\delta|E|^2$ is negligible) and, indeed, we get the usual linear resonance curve. But when P_i increases, the peak of the resonance curves tilts to the right and it becomes possible to get 3 operating points : A and C being stable, B unstable. This occurs provided $P_i > P_{is}$. (P_{is} = 0.4 W/cm²). Obviously when $P_i > P_{is}$, the transmitted GW, which is closely related to E_{GK}, experiences optical bistability (OB) (hysteresis cycle DEFGHE in fig. 10a). It is worth noticing that this is a new type of OB, namely an angular OB.

To get some physical insight in the origin of the OB behaviour of E_{GK}, we reported on fig. 10a the dispersion curve of the TE_1 NLGM (i.e. solution of the associated homogenenous NL problem). It is seen that the locus of the peak positions of the resonance curves and the dispersion curve are superimposed. This result has the following important meaning : OB is closely related to the resonant excitation of a NLGM (here the TE_1 NLGM) with a given value of β . Therefore, it clearly appears that this kind of OB is not related to the fact that, for a given E_{GK}, the TE_1 NLGM dispersion curve yields two values of the effective index β . Stated in another way, OB involves the resonant excitation of a NLGM whose value of β belongs either to the left or to the right part of the TE_1 NL dispersion curve. But there is no jump between these two parts. Concerning the right part of the TE_1 NLGM dispersion curve (fig. 9), as P_i goes up, results not reported here show that the resonance curves tilt to the left, yielding OB too.

5.2.2 Fig. 10b : fixing θ and varying P_i

By adjusting the value of the incidence angle θ , three different modes of operation come out : OB appears provided $\theta > \theta_S$. Notice that the set of curves fig. 10b merely corresponds to another way of reading fig. 10a. When $\theta \lesssim \theta_S$, the device fig. 8 works as an intrinsic optical triode. The optical limiter is obtained when $\theta < \theta_S$. As for the angular OB, OB in power and resonant excitation of NLGM are closely linked. This can be checked by noting that when OB takes place the value of $E_{GK}(\theta)$, when reporterd in fig. 9, yields a value of β close to $k_x = n_p(\omega/c)\sin\theta$.

5.2.3 Discussion

It is worth noting that the resonant excitation of a NLGM is a necessary condition to observe OB, but it is not a sufficient one. This can be best seen considering fig. 10b : the incidence angle θ has to be greater than θ_S . But fig. 10a shows that θ_S is greater than $\theta_r(0)$ ($\theta_r(0)$: angle of incidence leading to the resonant excitation of the linear guided mode). Thus OB appears provided the resonance takes place when the value of the offset on angle $\Delta\theta = \theta - \theta_r(0)$ exceeds the threshold value $\Delta\theta_S = \theta_S - \theta_r(0)$. The same conclusion applies to the right part of the TE_1 NL dispersion curve. The physical menaning of $\Delta\theta$ is the following : OB is obtained provided i) the NLGM is resonantly excited and ii) the electric field amplityde in the Kerr medium is high enough in order to allow the NL term, $\delta|E|^2$,to become sufficiently large.

If instead of considering the numerical values of fig. 8, we use : $\lambda = 1.5\,\mu m$, $n_q = 1.5$, $e = 1.5\,\mu m$, $\delta = 0.65\ 10^{-20}\ m^2V^{-2}$ (CS_2) (the other values unchanged ; in this case, the NL waveguide is monomodal), it is found that the critical switching intensity required to get OB is 800 times smaller than when no EM resonance is used[32] and 30 times smaller than when SP resonance is involved[23].

These results clearly demonstrate that EM resonances, especially of the GM type, leads to an important lowering od the threshold of OB.

5.3. Grating induced SEKE : OB in NL integrated optics

Let us now consider the grating excitation of NLGM of the type considered in sections 5.1 and 5.2 (fig. 11a).

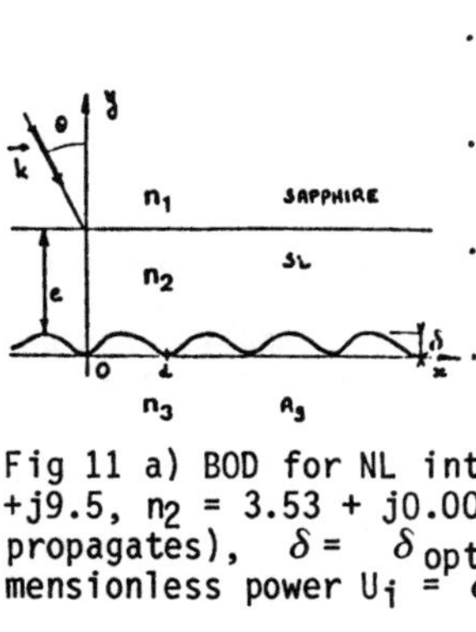

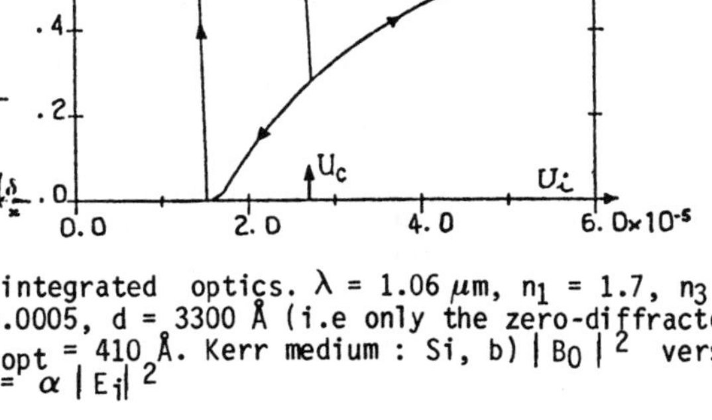

Fig 11 a) BOD for NL integrated optics. $\lambda = 1.06\,\mu m$, $n_1 = 1.7$, $n_3 = 0.3 + j9.5$, $n_2 = 3.53 + j0.0005$, d = 3300 Å (i.e only the zero-diffracted order propagates), $\delta = \delta_{opt} = 410$ Å. Kerr medium : Si, b) $|B_0|^2$ versus the dimensionless power $U_i = \alpha|E_i|^2$

The ω pump fiel is still TE polarized. The starting point for the study of the device fig. 11a is, as usual, Maxwell equations (4a,4b,4c) written at frequency ω. We get

$$\Delta E + (\omega^2/c^2) n_0^2 (1 + \alpha |E|^2) E = 0 \quad (36)$$

with $\alpha = [\chi(x,y)/\epsilon_0 \epsilon(x,y)]$, χ is defined (see(34)) by $P_z^{NL} = \chi E E E^*$. Use has been made of the fact that $\partial/\partial z = 0 \longrightarrow \vec{\nabla}.\vec{E} = 0$. The nonlinear equation (36) is solved using an iterative technique described in ref (27).

Fig. 11b shows the OB behaviour of the zero-order reflected efficiency. The sapphire - Si - Ag device is optimized i.e. $\delta = \delta_{opt} = 410$ Å and the BO cycle arises from the resonant excitation of the fundamental GM. The offset on angle is: $\theta - \theta_{res}(0) = 0.13°$

6. Conclusion

To enhance the efficiency of NLO effects requires an increase of the NL polarization : that is to say an increase of the elements of the NL susceptibility tensor and/or an increase of the EM field in the NL medium (media). In this chapter, we have shown the importance of the EM resonances in this kind of phenomena. These EM resonances can be of various types. We have only considered SP and linear or NL GM. When the excitation of SP or GW is achieved through a grating, it appears that SENLO effects are nothing but a special case of diffraction in NLO.

One may wonder about the interest of such a complicated formalism of diffraction in NLO. The best way to answer this question is a) to consider the optimization of the NL grating and b) to look at the enhancement of the SH efficiency.

We have found that $\delta_{opt}(\omega_1) \neq \delta_{opt,0}(2\omega_1) \neq \delta_{opt,-1}(2\omega_1)$. This has the following meaning : the groove depth for which the enchancement of the SHG is the strongest does not correspond to the optimum groove depth at the pump frequency, i.e. to the groove depth for which the SP resonance intensity is the greatest. Concerning the enhancement, let us first introduce the concept of elementary NL polarization which is the quantity $\vec{p}_{1,p-1}$ (equations 28 and 29). A given $\vec{p}_{1,p-1}$ arises from the NL interaction between two diffracted orders l and m at ω_1 such that l+m = p. The +1 diffracted order at ω_1 being resonantly excited; it is the p = 1 + 1 = 2 elementary NL polarization which will be the most enhanced. It could be then tempting to perform a "quasi-linear" theory of SESHG i.e. to assume that the enhancement of the SH efficiency is equal to the increase of $|P_{1,2-1}|^2$ as compared to the flat case. Results presented elswhere[21], show that proceeding along these lines leads to a contribution of the EM resonance which is nearly 2 10^3 times smaller than that derived from the complete analysis.

This clearly demonstrates the need of performing a full theory of diffraction in NLO. Otherwise we would a) get erroneous results concerning the groove depth for which the SH efficiency is the greatest and b) strongly underevaluate the increase of this SH efficiency.

Let us emphasize that these enhancements are obtained with very shallow modulations. Hence the need of a theory which does not consider the groove depth of the grating as a perturbative parameter. That such a small perturbation of the entrance face of the NL medium leads to so dramatic modifications in the NL response is amazing.

Acknowledegment

The topics covered in this chapter constitute the collective research work of P. Arlot*, G. Chartier*, J.L. Coutaz*, A. Koster***, D. Maystre**, N. Paraire***, E. Pic*, P. Vincent** and G. Vitrant*.

* : LEMO, U.A. 833 du CNRS, Génie Physique, ENSIEG , BP 46
38 402 SAINT MARTIN D'HERES, FRANCE

** : Laboratoire d'Optique Electromagnétique, U.A. 843 du CNRS, Faculté des Sciences et Techniques, Centre de St Jérôme,
13 397 MARSEILLE CEDEX 13, FRANCE

*** : Institut d'Electronique Fondamentale, L.A. 22 du CNRS, Bâtiment 220, Université de Paris XI, 91 405 CEDEX, FRANCE

References

1 E. Burstein, C.Y. Chen, S. Lundqvist : light scattering in solids, Proceeding of the second Joint USA-USSR Symposium (Plenum, New York 1979)
2 S.G. Shultz, M. Janik-Czachor, R. Van Duyne : Surf. Sci. 104, 419 (1981)
3 A. Otto : Appl. Surf. Sci. 6, 309 (1980)
4 T.E. Furtak, J. Reyes : Surf. Sci 93, 351 (1980)
5 M. Fleischman, P.J. Hendra, A.J. Mc. Quillan : Chem. Phys. Lett. 26, 163 (1974)
6 S. Ushioda, A. Aziza, J.B. Valdez, G. Mattei : Phys. Rev. B19, 4012 (1979)
7 R.K. Chang, T.E. Furtak : Surface Enhanced Raman Scattering (Plenum, New York 1982)
8 C.K. Chen, A.R.B de Castro, Y.R. Shen : Phys. Rev. Lett 46, 145 (1981)
9 V.M. Agranovich, D.L. Mills : Surface Polaritons, Electromagnetic waves at surfaces and interfaces (North Holland, Amsterdam 1982)
10 A.D. Boardman : Electromagnetic Surface Modes (John Wiley and Sons, New York 1982)
11 A. Wokaun, J.G. Bergman, J.P. Heritage, A. Glass, P.F. Liao, D.H. Olson : Phys. Rev B24, 849 (1981)
12 M. Yamashita, M. Tsuji : J. of the Phys. Soc. of Jap. 52, 2462 (1983)
13 D.L. Mills, M. Weber : Phys. Rev. B26, 1075 (1982) ; M. Weber, D.L. Mills Phys. B27, 2698 (1983)
14 R. Reinisch, M. Neviere : Opt. Engn. 20, 629 (1981)
15 R. Reinisch, M. Neviere : J. Optics (Paris) : 13, 81 (1982)
16 R. Reinisch, M. Neviere : Phys. Rev. B24, 4392 (1981)
17 M. Neviere, R. Reinisch : Phys. Rev. B26, 5403 (1982)
18 R. Reinisch, M. Neviere : Phys. Rev. B26, 5987 (1982)
19 R. Reinisch, M. Neviere : Phys. Rev. B28, 1870 (1983)
20 G.A. Farias, A.A. Maradudin : Phys. Rev. B30, 3002 (1984)
21 M. Neviere, R. Reinisch, D. Maystre : Phys. Rev. B to be published
22 J.L. Coutaz, M. Neviere, E. Pic, R. Reinisch : Phys. Rev. B to be published
23 G.M. Wysin, H.J. Simon, R.T. Deck : Opt. Lett. 6, 30 (1981)
24 P. Martinot, A. Koster, S. Laval, W. Carvalho : Appl. Phys. B29, 172 (1982) ; P. Martinot : Thèse 3ème cycle, Univ. Paris XI Orsay (30.6.1983)
25 G. Vitrant, P. Arlot, D. Michel, E. Pic, R. Reinisch : 1984 IEEE International Workshop on Integrated Optical and Related Technologies for Signal Processing, Florence (ITALY 1984)
26 P. Arlot, E. Pic, R. Reinisch, G. Vitrant : JOG N°6, Issy les Mx (1985)
27 P. Vincent, N. Paraire, M. Nevière, A. Koster, R. Reinisch : JOSA to be published

28 R. Petit : The Electromagnetic Theory of Gratings (Springer Verlag 1980) and references cited therein
29 N.E. Glass, A.A. Maradudin, V. Celli : Phys. Rev B27, 5150 (1983)
30 R.W. Wood : Phylos. Mag. 4, 396 (1902)
31 C.M. Bowden, M. Ciftan, H.R. Robl : Optical Bistability (Plenum Press, New York 1981) and references cited therein
32 P.W. Smith, J.P. Hermann, W.J. Tomlinson, P.J. Maloney : Appl. Phys. Lett 35, 846 (1979)
33 P.W. Smith, I.P. Kaminow, P.J. Maloney, L.W. Stulz : Appl. Phys. Lett. 34, 62 (1979) ; P.W. Smith : Opt. Engn. 19, 456 (1980)
34 K.M. Leung : Phys. Rev. A31, 1189 (1985) and references cited therein
35 U. Langbein, F. Lederer, H.E. Ponath : Opt. Commun. 46, 167 (1983)
D.J. Robbins : Opt. Commun. 47, 309 (1983)
A.D. Boardman, P. Egan : J. Phys. Colloq. (in press).
G.I. Stegeman, C.T. Seaton, John Childwell, D. Smith : Appl. Phys. Lett. 44, 830 (1984)
36 A. Girlando, M.R. Philpott, D. Heitman, J.D. Swalen, R. Santo : J. Chem Phys. 72, 5187 (1980)
37 F. de Martini, G. Giulani, P. Mataloni, E. Palange, Y.R. Shen : Phys. Rev. Lett 37, 440 (1976) ; F. de Martini, M. Colocci, S.E. Kohn, Y.R. Shen : Phys. Rev. Lett. 38, 1223 (1977)
38 M. Bouix : Les Discontinuités du Rayonnement Electromagnetique (Dunod Paris 1966)
39 S. Biraud-Laval : Thèse d'Etat, Univ. Paris XI Orsay (1973)
40 N. Bloembergen : Nonlinear Optics (Benjamin, New York 1965)
41 N. Marshall , B. Fischer : Phys. Rev. Lett. 28, 811 (1972)
42 W.L Faust, C.H. Henry : Phys. Rev. Lett. 25, 1265 (1966)
43 N. Bloembergen, R.K. Chang, S.S. Jha, C.H. Lee : Phys. Rev. 174, 813 (1968)
44 J.E. Sipe, G.I. Stegeman : see ref. 9 p. 661
45 J.E. Sipe, V.C.Y.So, M. Fukui, G.I. Stegeman : Phys. Rev. B21, 4389 (1980)
46 L.S. Schwartz : Methodes Mathematiques pour les Sciences Physiques (Herman, Paris 1965)
47 R. Reinisch, E. Pic, D. Maystre, M. Nevière : J. de Phys. submitted for publication
48 D. Maystre, M. Nevière, R. Reinisch : Appl. Phys. submitted for publication
49 G. Vitrant, E. Pic, R. Reinisch : submitted to IEEE JQE
50 A.E. Kaplan : Sov. Phys. JETP 45, 896 (1978).

Nonlinear Guided Waves

G.I. Stegeman[1], C.T. Seaton[1], W.M. Hetherington III[2], A.D. Boardman[3], and P. Egan[3]

[1] Optical Sciences Center and Arizona Research Laboratories, University of Arizona, Tucson AZ 85721, USA
[2] Optical Sciences Center and Department of Chemistry, University of Arizona, Tucson, AZ 85721, USA
[3] Department of Physics, University of Salford, Salford, M5 4WT, U.K.

1. Introduction

Since its inception in the 1960s, nonlinear optics has led to a rich variety of wave-mixing interactions that have applications to basic materials research [1-2], to the generation of new frequencies [1], and most recently to all-optical signal processing [3]. In general, nonlinear optical interactions occur whenever the optical fields associated with one or more laser beams propagating in a material are large enough to produce polarization fields proportional to the product of two or more of the incident fields. These nonlinear polarization fields radiate electric fields at the nonlinear frequency. For some interactions, the generated fields grow linearly with propagation distance under optimum conditions of phase-matching. Typically, the efficiency of any nonlinear optical interaction depends on (1) the product of the power densities of the input and output waves, raised to some power, and (2) the interaction distance raised to some power greater than or equal to unity. Since power density is power per unit area, the efficiency of any nonlinear interaction can be enhanced by reducing the cross-sectional area of the interacting beams. For plane waves this can be achieved by focusing with a lens. There is a tradeoff, of course, because the high power density can be maintained only over the depth of focus of the lens, which limits the effective interaction length.

Electromagnetic waves can be guided by the interface between two semi-infinite media, or by single or multiple films bounded by two semi-infinite media. The key feature is that the fields decay exponentially away from the boundaries into both semi-infinite bounding media with 1/e distances of typically a fraction of the wavelength of the radiation being guided. Therefore the effective beam dimension along the direction normal to the surface can be of the order of the wavelength of light, which corresponds to the minimum beam cross-sectional area, and hence maximum power density for a given input power level. These attractive features were recognized at an early stage and have, by now, been applied to an impressive number of nonlinear guided wave phenomena.

Guided waves usually result from a coupling between an electromagnetic field and some resonance. The resonances can be geometric, for example as occurs in a thin-film integrated-optics waveguide [4], where constructive interferences in the film result in waveguide modes. The resonances can also be related to material properties. For example, the plasma resonance associated with the electron gas in a metal is coupled to an electromagnetic field via the interaction between the field and the charges, and leads to surface plasmons [5]. Such material resonances are usually accompanied by large losses and hence the propagation distances attained are limited in these cases. For this reason, most work in guided wave nonlinear optics for device applications has been limited to integrated optics with freely propagating guided waves. In this chapter we will concentrate solely on nonlinear interactions involving freely propagating waves.

The nonlinear guided wave phenomena studied to date have consisted of (1) guided wave analogs of plane wave nonlinear interactions, and (2) a limited number of mixing processes that are unique to guided wave geometries. Certainly the most developed area involves second-order interactions [6]. Second harmonic generation has been demonstrated for waves guided by dielectric films and channels. Other phenomena, such as sum and difference frequency generation, and parametric amplification and oscillation, have also been reported with integrated optics waveguides. Finally, the mixing of oppositely propagating guided waves to produce a second harmonic radiated normal to the guiding surface, a phenomenon unique to waveguiding geometries, has also been observed.

New developments in nonlinear optics in the last few years have centered on phenomena depending on $\chi^{(3)}$, the third-order susceptibility. This case deals with the mixing of up to three separate incident optical fields, for example at the frequencies ω_a, ω_b, and ω_c. Therefore, the nonlinear polarization, and hence the radiated fields, can have the frequency components $\omega_a \pm \omega_b \pm \omega_c$, which leads to a large range of phenomena. In contrast to efficient second-order phenomena that are difficult to phase-match, there are third-order processes such as degenerate four-wave mixing and optical bistability that are automatically phase-matched [3]. There are interactions in which tuning the difference frequency between two of the laser fields through characteristic molecular vibrational frequencies leads to resonant enhancements in the signal and hence can be used for spectroscopy [2]. Here we discuss guided-wave versions of these phenomena [7]. In addition, there are third-order interactions that are unique to integrated optics and have no plane-wave analogs. In particular, there is the nonlinear coherent coupler, power-dependent coupling into waveguides via distributed couplers such as prism and gratings with possible applications to switching, and a new class of guided waves whose properties change dramatically with guided wave power. In this chapter we discuss nonlinear phenomena in waveguides that utilize the third-order susceptibility.

2. Guided Waves

We start our discussion of nonlinear guided waves with a brief review of the properties of guided waves [4,8,9]. In this section we assume that the fields are normal modes and that their properties do not depend on the power of the guided wave.

Guided waves are electromagnetic fields that satisfy both Maxwell's equations in every medium into which the fields penetrate, and continuity of the tangential boundary conditions at every interface. For the most general case, which includes a nonlinear polarization field, the wave equation for fields and polarization sources at the frequency ω takes the form

$$\nabla^2\mathbf{E}(\mathbf{r},t) + \omega^2 \frac{n_\gamma(z)^2}{c^2} \mathbf{E}(\mathbf{r},t) = -\mu_0 \omega^2 \mathbf{P}^{NL}(\mathbf{r},t) , \tag{1}$$

where $n_\gamma(z)$ is the refractive index of the γ'th medium. For the simplest case of an isotropic waveguide media with z normal to the guiding surface(s) and propagation along the x-axis, the normal mode solutions to Eq. (1) with $\mathbf{P}^{NL}(\mathbf{r},t) = 0$ separate into TE (s-polarized) waves with field components E_y, H_x, and H_z, and TM (p-polarized) waves with field components H_y, E_x, and E_z. For s-polarized (TE) waves, the electric field of frequency ω is given by

$$E_y(\mathbf{r},t) = \sum_m E_y^{(m)}(\mathbf{r},t) + \int_* E_y^{(\nu)}(\mathbf{r},t)\, d\nu, \qquad (2)$$

where the summation is taken over the guided wave modes, and the integral is taken over the radiation fields $E_y^{(\nu)}(\mathbf{r},t)$. Guided wave fields are characterized by exponential field decay into both semi-infinite bounding media, in contrast to radiation fields that exhibit oscillatory behavior in at least one bounding medium and extend to infinity in that medium (in the absence of loss). For the m'th order guided wave mode,

$$E_y^{(m)}(\mathbf{r},t) = \frac{1}{2} E_y^{(m)}(z) a^{(m)}(x) e^{i(\omega t - \beta^{(m)} kx)} + \text{c.c.} \qquad (3)$$

The detailed depth dependence of the guided wave fields is contained in the $E_y^{(m)}$ term, which is normalized so the $|a^{(m)}(x)|^2$ is the guided wave power in watts per meter of wavefront (along the y-axis). Since $\omega = kc$, the term $\beta^{(m)}$ plays the role of a refractive index for propagation along the x-axis and is called "the effective index": Its value is obtained from an eigenvalue equation or dispersion relation obtained by satisfying the boundary conditions across each interface. The radiation fields have the same basic structure as Eq. (3) with the mode index m replaced by ν, and $a^{(m)}(x)$ replaced by $b^{(\nu)}(x)$. Since these solutions are normal modes, they satisfy orthogonality relations:

$$\frac{\beta}{2\mu_0 c} \int_{-\infty}^{\infty} dz\, E_y^{(m)}(z) E^*{}_y^{(r)}(z) = \delta_{m,r} \,. \qquad (4)$$

These orthogonality relations will prove very useful later for evaluating the amplitudes of nonlinearly generated fields.

It is useful to consider the simple geometry of a film of thickness "h" and refractive index n_f bounded by a cladding medium ($0 \geq z$, n_c) and a substrate ($z \geq h$, n_s), as illustrated in Fig. 1. A finite number of discrete modes occur, and the field distributions associated with the first few TE_m waves shown in Fig. 2 exhibit oscillatory behavior in the film, and decay exponentially with distance into the cladding and substrate. Therefore nonlinear interactions can take place in any one of the three media. One of the unique features of thin film guided waves is that the $\beta^{(m)}$ for each mode depend on film thickness, as illustrated in Fig. 3.

3. Nonlinear Polarization Fields

The mixing of multiple optical beams leads to nonlinear polarization fields that are usually expanded in products of the mixing fields [1]. A nonlinear polarization field of frequency ω_s can be generated in each of the media in which the guided wave fields exist. It is usually written as

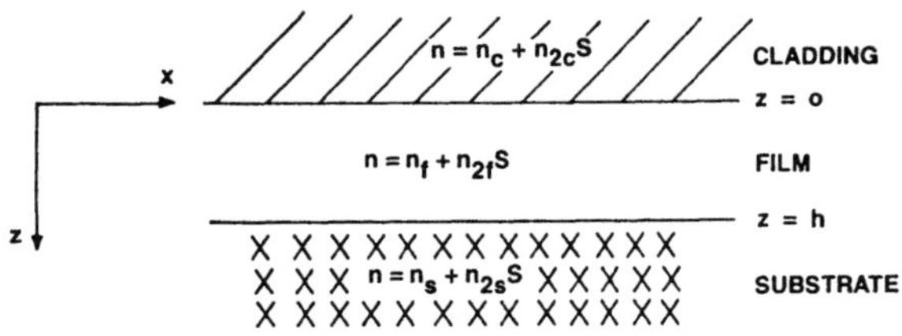

Fig. 1. A thin film dielectric waveguide of thickness "h."

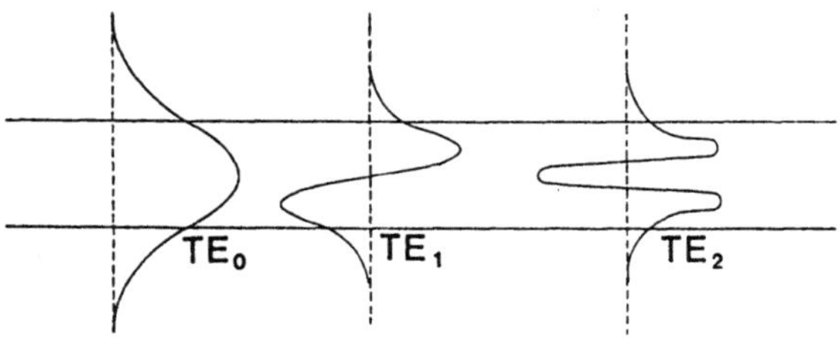

Fig. 2. Typical electric field distributions for TE_m modes.

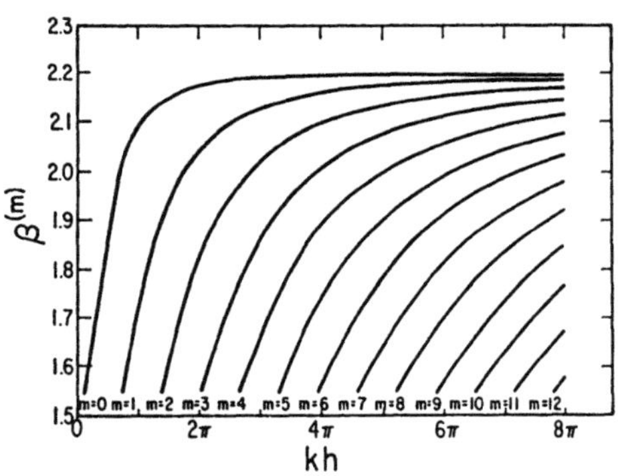

Fig. 3. Dispersion in effective index β with product of film thickness and wavevector.

$$\mathbf{P}^{NL}(\mathbf{r},t) = \frac{1}{2} P^{NL}(z,\omega_s) e^{i(\omega_s t - \beta_p \mathbf{k}_s \cdot \mathbf{r})} + \text{c.c.} \tag{5a}$$

If we now assume up to three input guided-wave fields of frequency ω_a, ω_b, and ω_c (of which two or three may be degenerate)

$$\begin{aligned} \mathbf{P}^{NL}(z,\omega_s) &= \varepsilon_0 \boldsymbol{\chi}^{(2)}(-\omega_s;\omega_a,\pm\omega_b):\mathbf{E}^{(m,a)}(z)\mathbf{E}^{(m',b)}(z)a^{(m,a)}(x)a^{(m',b)}(x) \\ &+ \varepsilon_0 \boldsymbol{\chi}^{(3)}(-\omega_s;\omega_a,\pm\omega_b,\pm\omega_c):\mathbf{E}^{(m,a)}(z)\mathbf{E}^{(m',b)}(z)\mathbf{E}^{(m'',c)}(z) \\ &\times a^{(m,a)}(x)a^{(m',b)}(x)a^{(m'',c)}(x), \end{aligned} \tag{5b}$$

where $\omega_a = k_a c$, $\omega_b = k_b c$, etc., $\chi^{(2)}$ and $\boldsymbol{\chi}^{(3)}$ are the second- and third-order susceptibilities, and a minus sign for a frequency corresponds to taking the complex conjugate of the appropriate field. Note that $\beta_p \mathbf{k}_s = \beta^{(m,a)}\mathbf{k}_a \pm \beta^{(m',b)}\mathbf{k}_b \pm \beta^{(m'',c)}\mathbf{k}_c$ is the wavevector associated with the nonlinear polarization source field, and it is <u>not necessarily</u> equal to $\beta^{(n,s)}\mathbf{k}_s$ which is the value appropriate to a propagating field of frequency ω_s. The case $\beta_p k_s = \beta^{(n,s)} k_s$ corresponds to phase-matching, as will be discussed later. For the second-order processes there are two possible input waves with frequencies ω_a and ω_b, which produce polarization and signal fields at $\omega_s = \omega_a \pm \omega_b$. For third-order processes, the nonlinear polarizations can occur at the frequencies $\omega_s = \omega_a \pm \omega_b \pm \omega_c$. Most of these interactions have now been demonstrated in nonlinear guided-wave experiments.

In addition to the nonlinear polarization that occurs inside a material due to the product of electric fields, nonlinear source terms can also occur near surfaces where the field gradients are large [10]. It can easily be shown that this nonlinear polarization term is zero for TE polarized waves. However, for TM modes, there are contributions because there exists an electric field component parallel to the propagation wavevector, and because the E_z TM field component is discontinuous across the boundary. These contributions to the nonlinear polarization are typically less than the usual $(\chi^{(2)})$ terms by a factor of 10 to 10^4, depending on the material and mode. One would therefore expect these terms to be important only for cases in which $\chi^{(2)}$ is uniquely zero because of symmetry reasons. To date, these terms have only played a role in nonlinear surface plasmon interactions [10].

4. Coupled Mode Theory and Nonlinear Wave Generation

It is important to note that the existence of the nonlinear polarization field does not ensure the generation of strong signal fields that must satisfy the driven wave equation, Eq. (1), with $\mathbf{P}^{NL} \neq 0$. With the exception of phenomena based on an intensity-dependent refractive index, the generation of the nonlinearly produced signal waves at frequency ω_s can be treated in the slowly varying amplitude approximation using well-known guided-wave coupled-mode theory [8,9]. For example, for an s-polarized nonlinear polarization source field that generates guided-wave fields of mode order "n", the general form of the solution field [Eq. (2)] is substituted into Eq. (1). As already explicitly assumed in Eq. (3), the amplitudes of the waves are allowed to vary weakly with propagation distance x, that is $(d^2/dx^2)\, a^{(n',s)}(x) \ll \beta^{(n',s)} k_s (d/dx)\, a^{(n',s)}(x)$, which leads to

$$\sum_{n'} 2i\beta^{(n',s)} k_s \frac{d}{dx} a^{(n',s)}(x)\, E_y^{(n',s)}(z)\, e^{-ik_s\beta^{(n',s)}x}$$

$$+ \int_* d\nu'\, 2i\beta^{(\nu',s)} k_s \frac{d}{dx}\, b^{(\nu',s)}(x)\, E_y^{(\nu',s)}(z)\, e^{-ik_s\beta^{(\nu',s)}x}$$

$$= -\omega_s^2 \mu_0 P_y^{NL}(z,\omega_s) e^{-ik_s\beta_p x} \,. \tag{6}$$

We now make use of the orthogonality relations, [Eqs. (4)], by multiplying both sides of Eq. (6) by $E^*{}_y^{(n,s)}(z)$ and integrating over z. Note that by choosing a guided-wave field distribution as the multiplier, we are explicitly evaluating the guided-wave field amplitude coefficients. This results in

$$\frac{d}{dx} a^{(n,s)}(x) = i \frac{\omega_s}{4} \int_{-\infty}^{\infty} dz \; \mathbf{P}^{NL}(z,\omega_s)\cdot\mathbf{E}^{*(n,s)}(z) \; e^{-i(\beta_p - \beta^{(n,s)})k_s x} \;, \quad (7)$$

where we have generalized the result to also include TM guided-wave fields.

The same formalism can be used to calculate the generation of radiation fields due to nonlinear polarization sources in the vicinity of the waveguide. Multiplying Eq. (6) by $E^{*(\nu,s)}(z)$, integrating over z, and invoking the orthogonality relations leads to the equivalent of Eq. (7) with $a^{(n,s)}(x)$ replaced by $b^{(\nu,s)}(x)$ and the superscript n replaced by its value for radiation fields, namely ν.

5. Second Harmonic Generation (SHG)

SHG is the only nonlinear interaction that has been studied extensively in optical waveguides. Second harmonic signals can be generated by a single guided wave, by the mixing of two co-directional guided waves of different polarization, or by the interaction of two oppositely propagating guided waves. The discussion in each case will be separated into two parts, theory and experiment.

(a) SHG by Co-directional Guided Waves

The simplest case is a single fundamental beam producing a SHG wave via the material nonlinearity. Hence there are two waves present, the fundamental ($\omega_a = \omega$, $k_a = k$) and the second harmonic ($\omega_s = 2\omega$). In terms of the d_{ijk} tensor (which is usually used instead of the χ_{ijk} tensor), the nonlinear polarization fields are

$$P_i^{NL}(z,2\omega) = \varepsilon_0 d_{ijk} E_j^{(m,\omega)}(z) E_k^{(m,\omega)}(z) a^{2(m,\omega)} \;, \quad (8a)$$

$$P_i^{NL}(z,\omega) = 2\varepsilon_0 d_{ijk} E^{*}{}_j^{(m,\omega)}(z) E_k^{(n,2\omega)}(z) a^{(n,2\omega)} a^{*(m,\omega)} \;. \quad (8b)$$

Substituting into Eq. (7) gives

$$\frac{d}{dx} a^{(n,2\omega)}(x)$$

$$= \frac{i\omega\varepsilon_0}{2} \int_{-\infty}^{\infty} dz \; d_{ijk} E^{*}{}_i^{(n,2\omega)}(z) E_j^{(m,\omega)}(z) E_k^{(m,\omega)}(z) \; a^{(m,\omega)2}(x) e^{-2i\Delta kx} \;, \quad (9a)$$

$$\frac{d}{dx} a^{(m,\omega)}(x) = \frac{i\omega\varepsilon_0}{2} \int_{-\infty}^{\infty} dz \; d_{ijk} E^{*}{}_i^{(m,\omega)}(z) E_j^{(n,2\omega)}(z) E^{*}{}_k^{(m,\omega)}(z)$$

$$\times \; a^{(n,2\omega)}(x) a^{*(m,\omega)}(x) \; e^{2i\Delta kx} \;, \quad (9b)$$

which are the two equations that must be solved with $\Delta k = (\beta^{(m,\omega)} - \beta^{(n,2\omega)})k$. At frequencies far from any material resonances, the three indices ijk can be interchanged so that the integrals in Eqs. (9), called the overlap integral, yield the same value, to within a phase factor.

It is very useful to first examine the case in which the fundamental beam is undepleted and the conversion to the harmonic power after an interaction distance L is small. In that case

$$|a^{(n,2\omega)}(L)|^2 = (kL)^2 |K|^2 \frac{\sin^2(\Delta kL)}{(\Delta kL)^2} |a^{(m,\omega)}(0)|^4 . \quad (10a)$$

$$K = \frac{c\epsilon_0}{2} \int_{-\infty}^{\infty} dz\, d_{ijk}\, E_j^{(m,\omega)}(z) E_k^{(m,\omega)}(z) E^*{}_i^{(n,2\omega)}(z) . \quad (10b)$$

The $\sin^2\phi/\phi^2$ term describes the effect of phase-mismatch, that is, $\Delta k \neq 0$. Efficient conversion can be obtained only for $\Delta kL < \pi/4$. The optimum clearly occurs for $\Delta k = 0$, the phase-matched case.

In the limit $\Delta kL = 0$, Eqs. (9) can be solved in closed form. The results are

$$a^{(m,\omega)}(x) = a^{(m,\omega)}(0) \operatorname{sech}\frac{x}{\ell_{SH}} , \quad (11a)$$

$$a^{(n,2\omega)}(x) = i \frac{a^{(m,\omega)2}(0)}{|a^{(m,\omega)}(0)|} \tanh\frac{x}{\ell_{SH}} , \quad (11b)$$

$$\frac{1}{\ell_{SH}} = k\, K|a^{(m,\omega)}(0)| . \quad (11c)$$

Therefore the conversion efficiency is given simply by $\eta = \tanh^2 (L)/(\ell_{SH})$.

Although it initially appears that SHG in waveguides should be simpler to achieve than for the plane wave case, in practice this is not true. For example, the existence of multiple modes of the same polarization at any frequency should offer many more possibilities for phase-matching, that is the superscripts n, m and m'(for two orthogonally polarized input waves) can take multiple values. However, there is dispersion in β with film thickness so that $\beta^{(n,2\omega)} > \beta^{(n,\omega)}$. This dispersion reinforces the usual dispersion in material refractive index with wavelength so that SHG with n = m (= m') is not possible with modes of the same polarization. It is, however, possible to produce SHG with n > m,m'. This leads to a greatly reduced cross-section, because the overlap integral involves interference effects when products of fields are integrated over the depth coordinate. This is clear from Fig. 2. Since $\beta_{TE}{}^{(m,\omega)} > \beta_{TM}{}^{(m,\omega)}$, the situation improves when the fundamental wave is a TE mode and the harmonic is a TM wave. Nevertheless, it has still proven very difficult to find combinations in which there are no interference effects in the overlap integral.

A number of different approaches for facilitating phase-matching have been proposed and implemented. Temperature has been used to tune the refractive indices [11]. Surface gratings have been used to add additional wavevectors to the interaction [12]. One very elegant solution to this interference effect has been demonstrated by Ito and Inaba [13]. They overcoated the nonlinear waveguide with a second film chosen so that no interference effects occurred in the field overlap inside the nonlinear film, as shown schematically in Fig. 4. In fact, this experiment was the only one in which relatively efficient SHG was generated in a slab waveguide geometry.

The second problem with SHG in thin film waveguides is the very strong dispersion with film thickness that occurs, see Fig. 3. Any variation in film thickness in the film fabrication process leads to changes in $\beta^{(r)}$ where r = n, m, m'. Since the slopes of the dispersion curves are different for different mode numbers, the phase-matching

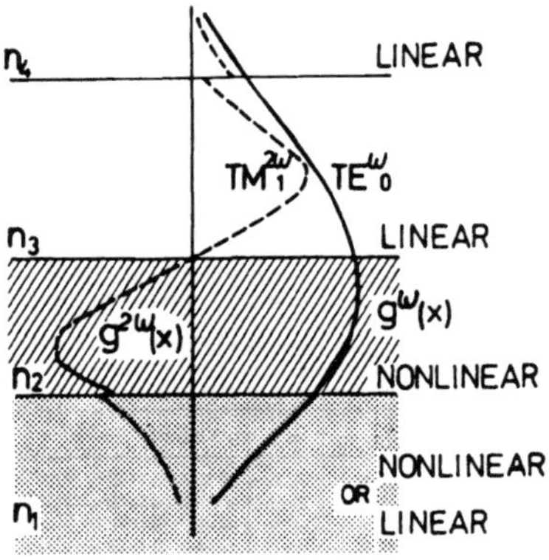

Fig. 4. Guided-wave field overlap in a two-film waveguide for harmonic generation with $TE_0 + TE_0 \rightarrow TM_1$ [13].

tolerances can vary with propagation distance. The larger the total dispersion in β with film thickness, the more sensitive are the phase-matching conditions to changes in the film thickness. In almost all slab waveguide experiments to date, the film thickness was tapered and the beams were coupled in at different points along the taper until phase-matching was achieved [14].

These problems, along with the small values of nonlinear coefficients available for most dielectric materials that can be easily fabricated in thin film form, stopped development of SHG in nonlinear slab waveguides for a number of years. Interest has been revived recently by the development of new, highly nonlinear, organic materials with high damage thresholds. The nonlinear coefficients $\chi^{(2)}$ are one to two orders of magnitude larger than those for materials such as $LiNbO_3$, thus opening new possibilities for efficient harmonic generation at low laser powers. Preliminary experiments have been reported on para-chloro-phenylurea and MNA with very encouraging results [15,16]. However, much more work is still required for finding better materials and film fabrication techniques.

The current state-of-the-art waveguide SHG was obtained with in-diffused $LiNbO_3$ waveguides. Typically, the waveguides are fabricated by diffusing Ti ions into the surface of single crystal $LiNbO_3$, thus increasing the refractive index in the surface region to create a waveguiding region [17]. The net refractive index change in this case is small, less than 0.01 so that the waveguides are quite deep (a few micrometers) and the conditions on maintaining phase-matching are relaxed relative to slab waveguides. Furthermore, the $LiNbO_3$ birefringence can be used to facilitate phase-matching, even between orthogonally polarized modes of the same order. Because of the nature of the in-diffusion process, it has proven possible to fabricate high-quality channel waveguides by depositing the titanium for the diffusion process through a mask. Propagation distances as long as 50 mm have been obtained. Phase-matching has been achieved by tuning the temperature or wavelength, or by applying electric fields to make use of the electro-optic effect in $LiNbO_3$. Such waveguides currently provide the best results reported to date for SHG in waveguides [18,19].

SHG in channel $LiNbO_3$ waveguides has been enhanced by making a resonator out of the channel waveguide [20]. High-reflectivity (0.96 at the fundamental wavelength) mirrors were coated onto the end faces of the channels to increase the fundamental power in a low-loss (0.15 dB/cm) waveguide. Although the structure was not optimized, 1.5 mW of HeNe radiation at 1.15 μm was doubled with an efficiency of 10^{-3} by tuning the temperature until phase-matching was obtained. For a matched resonator, an order of magnitude improvement in the conversion efficiency has been predicted. This requires that the fundamental be trapped in the waveguide until it is either absorbed or converted to the harmonic field.

(b) SHG By Contradirectional Waves

This particular phenomenon is unique to guided-wave geometries because it relies crucially on the strong confinement of the guided-wave fields. For TE incident waves, the nonlinear polarization field has components at $\omega_s = 0$ and $\beta_p = 2\beta^{(m,\omega)}$, and at $\omega_s = 2\omega_a = 2\omega$ and $\beta_p = 0$. The radiative nonlinear polarization field created by the mixing via the d_{yyy} tensor component of guided waves propagating along the ±x axis, $a_+(x)$ and $a_-(x)$ respectively, is given by

$$P_y^{NL}(z,2\omega) = 2\varepsilon_0 d_{yyy} E_y^{2(m,\omega)}(z)\, a_+^{(m,\omega)}(x) a_-^{(m,\omega)}(x) \,. \tag{12}$$

Note that other components of the **d** tensor, which have either x or y as the first index, will also lead to the same phenomenon. The source field has no spatial periodicity parallel to the surface, and hence the harmonic fields are radiated in directions normal to the waveguide surfaces. When this polarization is substituted into the radiative version of Eq. (7), $\beta_p = \beta^{(\nu)} = 0$, and integration over the depth coordinate gives the amplitudes of the fields radiated along the ±z directions.

The second harmonic signal has been observed for pulses of 1.06-μm that are coupled by a radiation prism into opposite ends of a Ti:in-diffused $LiNbO_3$ waveguide. The experimental and theoretical results were in good agreement [21].

This interaction is of special interest, because it can be used for all-optical signal processing or as a picosecond transient digitizer. This application is summarized in Fig. 5. For two incident optical waveforms of the form $U_+(t - x/v)$ and $U_-(t + x/v)$, the field radiated normal to the surface is proportional directly to the instantaneous overlap of the two waveforms. The resulting waveform is the real-time convolution of the incoming waveforms, as shown in Fig. 5d. This outgoing signal profile can be captured by placing a CCD array above and parallel to the waveguide surface. The convolution aspect of this interaction has been demonstrated experimentally [22]. If one of the input beams is effectively a δ-function in time, the radiated signal is exactly the temporal envelope of the second signal. Hence this device can operate as a picosecond transient digitizer for analyzing picosecond laser pulses.

6. Parametric Processes

There are a number of other nonlinear processes that utilize the second-order susceptibility. Those implemented to date in waveguide formats include sum and difference frequency generation, parametric amplification, and parametric oscillation. The only waveguides of sufficient quality for producing parametric amplification and oscillation have been Ti:in-diffused $LiNbO_3$ waveguides.

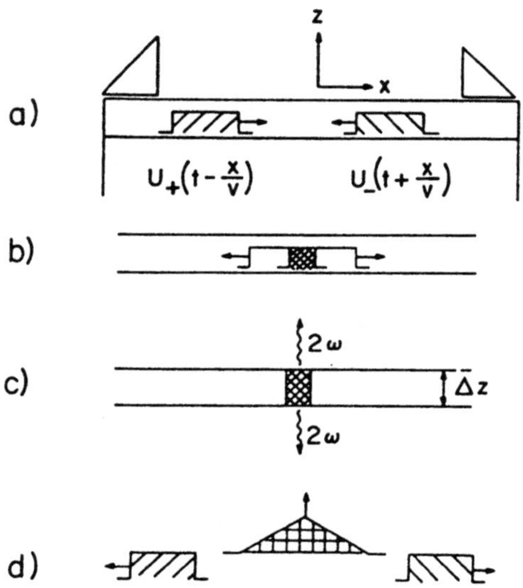

Fig. 5. Schematic diagram of signal convolution by the nonlinear mixing of two counter-propagating waves in an optical waveguide.

(a) Sum and Difference Frequency Generation

This case is essentially a generalization of the preceding SHG discussion. There are two separate incident guided waves at the frequencies ω_a and ω_b ($\omega_a \geq \omega_b$) that mix via the second-order susceptibility to produce nonlinear polarization fields at the frequencies $\omega_s = \omega_a \pm \omega_b$. The pertinent nonlinear polarization fields are

$$P_i^{NL}(z,\omega_a+\omega_b) = 2\varepsilon_0 d_{ijk} E_j^{(m,a)}(z) E_k^{(m',b)}(z) a^{(m,a)}(x) a^{(m',b)}(x) \ , \tag{13a}$$

$$P_i^{NL}(z,\omega_a-\omega_b) = 2\varepsilon_0 d_{ijk} E_j^{(m,a)}(z) E^*{}_k^{(m',b)}(z) a^{(m,a)}(x) a^{*(m',b)}(x) \ , \tag{13b}$$

$$P_i^{NL}(z,\omega_a) = 2\varepsilon_0 d_{ijk} E_j^{(n,s)}(z) E^*{}_k^{(m',b)}(z) a^{(n,s)}(x) a^{*(m',b)}(x) \ , \tag{13c}$$

$$P_i^{NL}(z,\omega_b) = 2\varepsilon_0 d_{ijk} E_j^{(n,s)}(z) E^*{}_k^{(m,a)}(z) a^{(n,s)}(x) a^{*(m,a)}(x) \ . \tag{13d}$$

It is of course possible to derive a perfectly general formalism for this case, similar to that just discussed for SHG. Since the reported conversion efficiencies in waveguides have been small, only the weak generation case is developed here. The difference frequency case has been studied in the context of parametric amplification,and will be addressed in the next section. For sum frequency generation with phase-matching over a distance L

$$|a^{(n,s)}(L)|^2 = (\omega_a + \omega_b)^2 L^2 \ |K|^2 \ |a^{(m,a)}(0)|^2 \ |a^{(m',b)}(0)|^2 \ , \tag{14a}$$

$$K = \frac{\varepsilon_0}{2} \int_{-\infty}^{\infty} dz \ d_{ijk} \ E_j^{(m,a)}(z) E_k^{(m',b)}(z) E^*{}_i^{(n,s)}(z) \ . \tag{14b}$$

The same comments about phase-matching, the overlap integral, etc. made for the SHG case are also valid here.

Two experiments have been reported to date, both in $LiNbO_3$ waveguides. Sum frequency generation has been reported by Uesugi et al. [23] in in-diffused $LiNbO_3$ channel waveguides and by Reutov and Tarashchenko [24] in thin $LiNbO_3$ platelets. They (Uesugi) used either a HeNe (1.19 μm) or a Nd:YAG laser in conjunction with an optical parametric oscillator operating in the near infrared to produce tunable sum-frequency radiation from 0.532 μm to 0.545 μm. Their tuning characteristics are shown in Fig. 6.

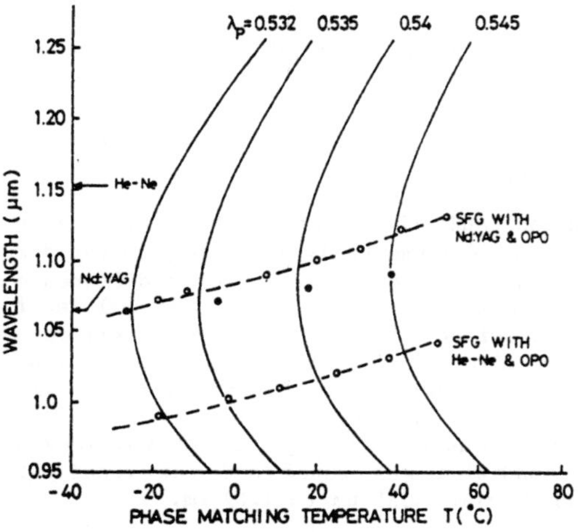

Fig. 6. Tuning characteristics for second harmonic generation (SHG) and sum frequency generation (SFG) in Ti:in-diffused $LiNbO_3$ channel waveguides. Closed and open circles are experimental data for SHG and SFG respectively. The solid lines correspond to tuning curves. The broken lines are theoretical curves for SFG [23].

(b) Parametric Amplification and Oscillation

Difference frequency generation, usually in the form of parametric amplification or oscillation, is a very useful technique for producing tunable radiation in the infrared. In this case, a whole new nomenclature unique to parametric processes has been developed. The signal, "pump," and "idler" waves have the frequencies ω_s (ω_b), ω_p (ω_c) and ω_i (ω_a), respectively.

In parametric amplification, a weak input signal at ω_s is amplified by the phase-matched conversion of the pump wave into both a signal and an idler wave, that is, $\omega_p = \omega_s + \omega_i$ and $\beta^{(m,p)}k_p = \beta^{(n,s)}k_s + \beta^{(m',i)}k_i$. For parametric oscillation, only the pump beam is incident, and the signal and idler build up from noise in the resonator cavity.

The appropriate nonlinear polarization fields are given by Eqs. (13c) and (13d). Substituting into Eq. (7), assuming phase-matching and no significant depletion of the pump beam,

$$a^{(m',b)}(x) = a^{(m',b)}(0) \cosh\frac{x}{\ell_{OPA}} , \quad (15a)$$

$$a^{(m,a)}(x) = i\sqrt{\frac{\beta^{(m',b)}\omega_b}{\beta^{(m,a)}\omega_a}}\; a^{*(m',b)}(0)\; \frac{a^{(n,s)}(0)}{|a^{(n,s)}(0)|}\; \sinh\frac{x}{\ell_{OPA}} , \quad (15b)$$

$$\frac{1}{\ell_{OPA}} = \sqrt{k_a k_b}\; |K|\; |a^{(n,s)}(0)| . \quad (15c)$$

$$K = \frac{c\varepsilon_0}{2}\int_{-\infty}^{\infty} dz\; d_{ijk} E^{*}_{i}{}^{(m,a)}(z) E_j{}^{(n,s)}(z) E_k{}^{(m',b)}(z) . \quad (15d)$$

In the limit $x \gg \ell_{OPA}$, the hyperbolic functions degenerate into exponentials with gain coefficient $1/\ell_{OPA}$.

Optical parametric amplification and the corresponding difference frequency generation have been studied experimentally in $LiNbO_3$ waveguides by two groups [11,20]. Uesugi [11] reported the generation of a difference frequency signal when an idler wave from a cw Nd:YAG ($\lambda \simeq 1.318$ µm) laser and a dye laser ($\lambda \simeq 0.58$ µm), which served as the pump, were mixed in a Ti:in-diffused $LiNbO_3$ channel waveguide. Using the mode combination $TE_0 + TM_0 \rightarrow TM_0$ with pump and idler powers of 70 mW and 1.7 mW respectively, a generation efficiency of 0.014% was obtained at $\lambda \simeq 1.035$ µm. By tuning the temperature and adjusting the dye laser wavelength appropriately, the phase-matching condition was tuned and the resulting difference frequency signal was varied from 1.035 µm to 1.19 µm.

Sohler and Suche managed to reduce their waveguide losses sufficiently so that they were able to measure signal gain. In the initial experiments [20], the signal source was a cw HeNe laser (λ = 1.15 µm) and the pump was derived from a tunable pulsed dye-laser operating in the

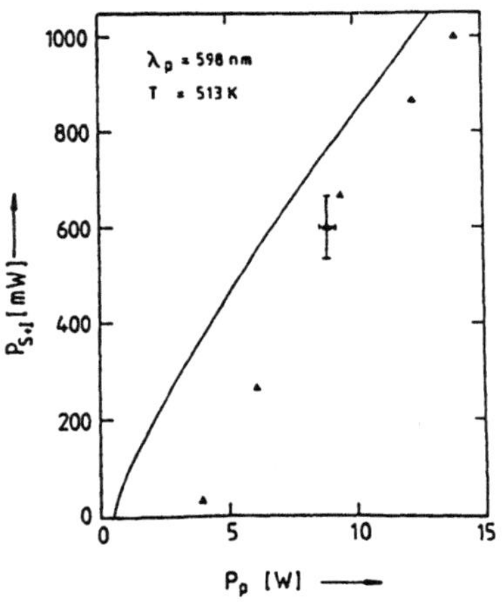

Fig. 7. Sum of the signal and idler powers versus pump wavelength for optical parametric oscillation in Ti:in-diffused $LiNbO_3$ channel waveguides. The solid line is the theoretical curve [20].

visible ($\lambda \simeq 0.65$ μm). The best single-pass parametric gain observed was 1.75 obtained with a peak pump power of about 200 W and a TE_{10}(pump) → TM_{00}(signal, idler) coupling. Much improved performance was obtained in a later experiment [20] with a 20-μm-wide channel, 48-mm long. With 150 W of peak power, a gain of 16 dB, which corresponds to a signal amplification of 43, was achieved.

The quality of the 10-μm-wide channel waveguides made by Sohler and Suche was so good that parametric oscillation was obtained when the ends of the channels were coated to a reflectivity of 0.96 at the signal wavelength [20]. At high enough pump powers, the signal and idler gain surpasses the resonator losses, and simultaneous oscillation at the signal and idler frequencies will start from noise. This will, of course, happen only under phase-matching conditions. The results for the signal and idler power generated are shown in Fig. 7 as a function of pump power (at λ = 0.598 μm). For a pump power of 14 W, 1 W is converted into the sum of signal and idler power. Varying the input pump wavelength results in tuning of the signal idler wavelengths as phase-matching is maintained. This is illustrated in Fig. 8. By also varying temperature, the signal and idler wavelengths can be tuned from 0.587 μm to 0.616 μm.

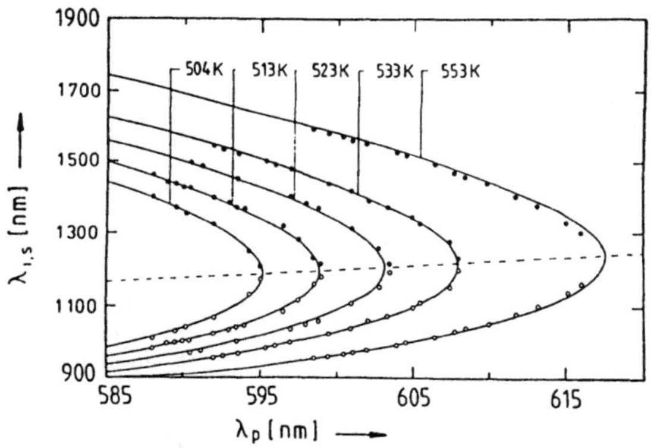

Fig. 8. The signal and idler wavelengths versus pump wavelength for optical parametric oscillation in a Ti:in-diffused $LiNbO_3$ channel waveguide at a series of temperatures [20].

7. Degenerate Four-Wave Mixing

Degenerate four-wave mixing involves three input beams and one output beam, all at the same frequency. It has been studied extensively with plane waves, as well as in fibers [3]. It is unique because it is automatically phase-matched in this interaction geometry, which is shown for two guided-wave cases in Fig. 9. Wavevector conservation is given by $\beta^{(m,1)}\mathbf{k}_1 + \beta^{(m',2)}\mathbf{k}_2 + \beta^{(m'',3)}\mathbf{k}_3 + \beta^{(n,4)}\mathbf{k}_4 = 0$, with $k_1 = k_2 = k_3 = k_4$. If the two input beams 1 and 2 are exactly contra-directional and have the same mode number $m = m'$, $\beta^{(m,1)}\mathbf{k}_1 = -\beta^{(m,2)}\mathbf{k}_2$. Therefore, if $m'' = n$, then $\beta^{(n,4)}\mathbf{k}_4 = -\beta^{(n,3)}\mathbf{k}_3$ and the signal beam 4 is generated backward along the direction of beam 3, independent of the incidence direction of beam 3. Note that this process requires that beams 1 and 2 have the same mode order m, and that beam 4 is radiated into the same mode index n as beam 3, and m and n are not necessarily equal [25].

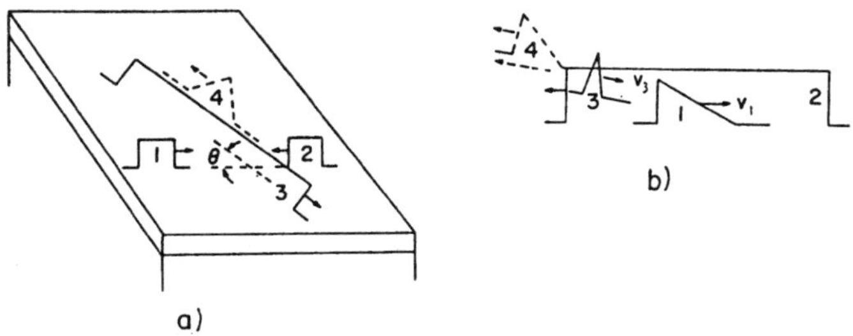

Fig. 9. The application of degenerate four-wave mixing to (a) convolution of pulses 1 and 2, and (b) the time inversion of beam 1.

The nonlinear polarization field corresponding to degenerate four-wave mixing can be obtained from Eqs. (5). For an isotropic material with equal mode indices for all of the input beams,

$$P^{NL}{}_i(z,\omega) = 4\varepsilon_0\chi^{(3)}{}_{1122}(-\omega,\omega,-\omega,\omega)\ E_i{}^{(n,3)}(z)E^*{}_j{}^{(n,2)}(z)E_j{}^{(n,1)}(z) + 2\varepsilon_0\chi^{(3)}{}_{1221}(-\omega,\omega,-\omega,\omega)\ E^*{}_i{}^{(n,1)}(z)E_j{}^{(n,2)}(z)E_j{}^{(n,3)}(z)\ . \quad (16)$$

This expression simplifies far from any resonant behavior in the $\chi^{(3)}$ terms and $\chi^{(3)}{}_{1122} = \chi^{(3)}{}_{1221} = \chi^{(3)}{}_{eff}$. For guided wave fields,

$$P^{NL}{}_{\gamma i}(z,\omega) = 2c\varepsilon_0{}^2 n_\gamma{}^2 n_{2\gamma}[\tfrac{2}{3}E_{\gamma i}{}^{(n,3)}(z)E_{\gamma j}{}^{(n,1)}(z)E^*{}_{\gamma j}{}^{(n,2)}(z) + \tfrac{1}{3}\ E_{\gamma j}{}^{(n,3)}(z)E_{\gamma j}{}^{(n,2)}(z)E^*{}_{\gamma i}{}^{(n,1)}(z)]\ , \quad (17)$$

in each medium (labeled γ) where $n_{2\gamma} \neq 0$. Here we have defined $3\chi^{(3)}{}_{eff} = n_{2\gamma}n_\gamma{}^2\varepsilon_0 c$, the justification for which will be discussed later. Assuming that TE beams 1 and 2 propagate along x', which is oriented at an angle θ to the x-axis along which TE beams 3 and 4 are traveling,

$$\mathbf{E}^{(n,1\text{ or }2)}(z) = E^{(n,1)}(z)\ [\sin\theta,\cos\theta,0]\ ;\ E^{(n,3\text{ or }4)}(z) = E^{(n,3)}(z)\ [0,1,0] \quad (18)$$

In the limit of no attenuation or depletion for the input beams, and assuming that the beams 1, 2, and 3 are introduced into the interaction region at $x' = 0$, $x' = L'$, and $x = L$, respectively, substituting into Eq. (7) and integrating the signal beam from $0 \rightarrow L$ gives

$$a^{(n,4)}(L) = (kL)\ A\ a^{(n,1)}(0)a^{(n,2)}(L')a^{(n,3)}(L)\ , \quad (19)$$

$$A = \frac{ic^2\varepsilon_0{}^2}{2}\ \frac{2+\cos^2\theta}{3}\int_{-\infty}^{\infty} dz n_\gamma(z)^2 n_{2\gamma}(z) \times E_\gamma{}^{(n,1)}(z)E_\gamma{}^{(n,2)}(z)^*E_\gamma{}^{(n,3)}(z)E^*{}_\gamma{}^{(n,4)}(z). \quad (20)$$

In the general case, it is necessary to take incident beam depletion into account, which leads to a series of coupled mode equations between the amplitudes of the various beams.

One experiment has been reported to date [26]. The nonlinear medium was CS_2 used as a cladding medium, inside a cell optically contacted to the surface of a thin-film glass waveguide, as shown in Fig. 10. Three

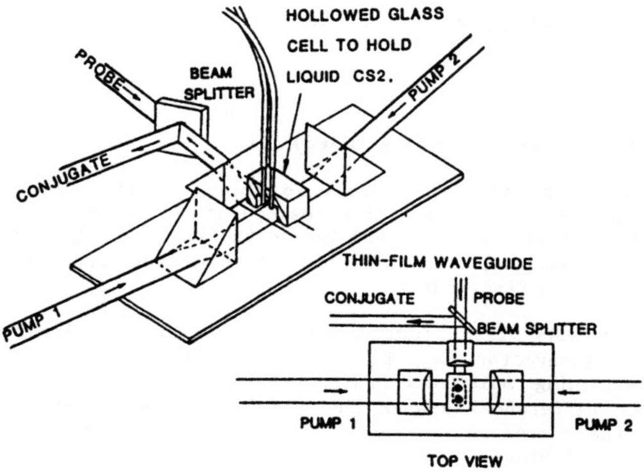

Fig. 10. Sample and coupling degenerate four-wave mixing geometry for CS_2 on a thin-film glass waveguide. The probe beam intersects the pump beams at 90° under the liquid cell [26].

coupling prisms were required for the four beams needed in the interaction, leading to difficult alignment problems. Since the nonlinear mixing occurred via the evanescent tail of the guided waves, and because the nonlinearity of CS_2 is small ($n_{2c} = 3\times10^{-18}$ m^2/W), the reflectivity (fractional conversion of beam 3 into beam 4) was only 10^{-9}. Nevertheless, the cubic dependence of the four-wave mixing signal was verified experimentally for 15-ns-long pulses from a Q-switched, doubled Nd:YAG laser.

This interaction has some possible applications to real-time all-optical signal processing [27], two of which are illustrated in Fig. 9. The two input signals to be convolved are beams 1 and 2, which have pulse envelopes $U_1(t - x/v)$ and $U_2(t + x/v)$. Whether the convolution process occurs or not can be controlled by beam 3, which is assumed to have a constant amplitude during the overlap of pulses 1 and 2 inside the waveguide. Assuming that $\theta = \pi/2$ for this case and that L is small relative to the characteristic pulse size (to preserve detail in the convolved signal), the 4'th beam signal radiated is directly proportional to the instantaneous overlap of beams 1 and 2. Mathematically, the time evolution of the signal beam can be written as $U_4(t) \propto \int U_1(t-\tau)U_2(\tau)U_3\,d\tau$, where it has been explicitly assumed that the control beam is "turned on." For this application, the input signal length cannot be larger than the waveguide dimension, which limits the process to pulses of 100 ps or less.

Another potential application of four-wave mixing to real-time processing is time reversal of an optical waveform, as illustrated in Fig. 9b. Here, $\theta \approx 0$ and beams 1 and 2 are the waveform to be reversed and a control beam of constant amplitude, respectively. Beam 3 is a δ-function in time (very short pulse) and 4 is the time-reversed signal. A crucial factor here is the fact that beams 1 and 3 can be propagated in different modes. If the mode number of beam 3 is larger than that of beam 1, beam 3 can overtake beam 1 from "behind." Thus beam 4 is "read out" from trailing edge to leading edge, which corresponds to a time reversal of the pulse envelope.

8. Coherent Anti-Stokes Raman Scattering (CARS)

This interaction has been used with plane waves to obtain Raman spectra and has recently been applied to the Raman spectroscopy of thin films [28] and monolayers deposited on thin films. It has the potential of becoming a very powerful, albeit specialized, experimental technique for studying surface species on a picosecond time scale.

The basic concept of CARS [2] is summarized in Fig. 11a. Two laser beams of frequency and wavevector (ω_a, $\mathbf{k}_a$) and (ω_b, $\mathbf{k}_b$) with $\omega_a \geq \omega_b$ illuminate a sample containing molecules characterized by a vibrational frequency Ω_r. When the difference between the two laser frequencies is tuned to approximately the resonance frequency, that is $\omega_a - \omega_b \simeq \Omega_r$, the molecules are excited to vibrate at this difference frequency. This results in a coherent (in space and time) excitation of the vibrational normal mode, which can be expressed as a contribution to the linear susceptibility of the material of the form

$$\chi_{ij}(\mathbf{r},t) = \frac{1}{2}\,\frac{\eta\,\epsilon_0}{4m[\omega_a - \omega_b - \Omega_r - i\Gamma_r]}\,\alpha_{ik}\alpha^*_{j\ell}\,E_k^{(m,a)}(z)E^*_\ell{}^{(m',b)}(z)$$
$$\times\, e^{i[(\omega_a - \omega_b)t - (\beta^{(m,a)}\mathbf{k}_a - \beta^{(m',b)}\mathbf{k}_b).\mathbf{r}]} + \text{c.c.}\,, \qquad (21)$$

where η is the density of molecules, m is the electron mass, and α_{ik} is the Raman susceptibility tensor. Information about the orientation of the molecules relative to the coordinate system is contained in the α_{ik} terms. Light incident onto the molecules is now scattered via the usual Raman process, as shown in Fig. 11c. Since the induced nonlinear polarization is just the product of the susceptibility with the incident fields, nonlinear polarization fields oscillating at the frequencies $\omega_s = 2\omega_a - \omega_b$ (CARS) and $\omega_s = 2\omega_b - \omega_a$ (CSRS - Coherent Stokes Raman Scattering) are created. For CARS the spatial periodicity is given by $\beta_p\mathbf{k}_s = 2\beta^{(m,a)}\mathbf{k}_a - \beta^{(m',b)}\mathbf{k}_b$. The corresponding CARS nonlinear polarization field is

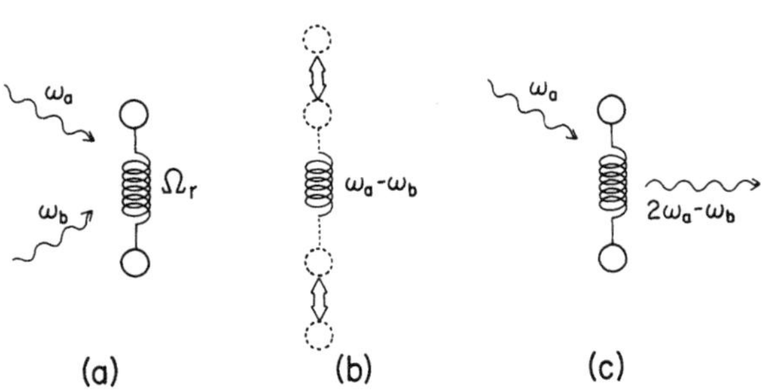

Fig. 11. Schematic of the CARS interaction. (a) Two lasers of frequency ω_a and ω_b mix at a molecule to excite it at the difference frequency $\omega_a - \omega_b$. (b) Laser "a" is Raman scattered by the excited molecule to produce a signal at $2\omega_a - \omega_b$.

$$P_i^{NL}(z,2\omega_a-\omega_b) = \epsilon_0\left[\chi_b{}^{(3)}{}_{ijk\ell} + \sum_r \chi_r{}^{(3)}{}_{ijk\ell}\frac{\Gamma_r/\pi}{[\omega_a - \omega_b - \Omega_r - i\Gamma_r]}\right]$$

$$\times\, E^{2(m,a)}(z)E^{*(m',b)}(z)\; a^{2(m,a)}(x)a^{(m',b)}(x)\;. \qquad (22)$$

The $\chi_b{}^{(3)}$ term corresponds to the background third-order susceptibility term due to the electronic degrees of freedom and can be considered to be independent of frequency. The details of the resonant $\chi_r{}^{(3)}$ terms can be obtained from Eq. (21).

The amplitude of the guided waves generated at the frequencies $2\omega_a - \omega_b$ and $2\omega_b - \omega_a$ are calculated from coupled mode theory via Eq. (7). Assuming an interaction distance L, a CARS guided-wave signal field (mode number n), which starts with zero amplitude at x = 0, and assuming no depletion of the incident beams

$$|a^{(n,s)}(L)|^2 = |\,k_sL\,|^2\;|K_{2a-b}|^2\;\frac{\sin^2\phi}{\phi^2}\;|a^{2(m,a)}(0)|^2\;|a^{(m',b)}(0)|^2\;, \qquad (23)$$

$$K_{2a-b} = \frac{c\epsilon_0}{4}\int_{-\infty}^{\infty}dz\left[\chi_b{}^{(3)}{}_{ijk\ell} + \sum_r \chi_r{}^{(3)}{}_{ijk\ell}\frac{\Gamma_r/\pi}{[\omega_a-\omega_b-\Omega_r-i\Gamma_r]}\right]$$

$$\times\, E^{2(m,a)}(z)E^{*(m',b)}(z), \qquad (24)$$

with $\phi = (1/2)(\beta^{(n,s)} - \beta_p)\mathbf{k}_s\cdot\mathbf{r}$. Phase-matching occurs for $\phi = 0$ and clearly can happen for only the CARS or SCRS case, but not for both simultaneously. Typically, the angles between the two incident beams are only a few degrees.

How this phenomenon can be used for Raman spectroscopy is clear from Eqs. (23) and (24). The CRS signal, CARS and CSRS, are both proportional to $|K|^2$. If the background signal is negligible, or can be eliminated, then the CRS power is

$$P_s \propto \sum_r |\chi_r((3))|^2\;\frac{(\Gamma_r/\pi)^2}{[(\omega_a-\omega_b-\Omega_r)^2 + \Gamma_r{}^2]}\;. \qquad (25)$$

The signal power is a maximum whenever $\omega_a - \omega_b = \Omega_r$. Therefore, by tuning the frequency difference $\omega_a - \omega_b$ it is possible to map out all of the vibrational transitions. Even when the background term cannot be eliminated, the signal power undergoes dispersion of some kind with frequency difference and the characteristic Ω_r can be identified.

Numerical calculations [28] for thin films based on realistic materials indicate that the signal power on a strong resonance can be very large. For example, for a 2.0-μm-thick polystyrene film deposited on a glass substrate, for the 1000-cm^{-1} ring vibration, and for laser pulse energies of less than 0.1 mJ in pulses of ≈15-ns duration, efficiencies of ≈0.2% are predicted for the conversion of the incident beams into the CARS signal. Such laser energies are easily available, making this approach very promising for nonlinear spectroscopy. The initial experiments were carried out on 2-μm-thick polystyrene waveguides using two tuneable dye lasers with 100 ps long pulses [29]. The power densities in the two beams were 30 and 60 MW/cm^2 which corresponded to pulse energies of hundreds of nanojoules. The conversion efficiency was 0.2%, in excellent agreement with the theoretical calculations.

The remarkable efficiency of this process suggests applications to the investigation of monolayers deposited on film surfaces [28]. The problem is that the background term for the film can lead to a large signal that can mask the desired monolayer signal. In other words, under normal conditions the guiding film can be thousands of Angstroms thick, whereas a monolayer has a thickness of say 5 Å. The problem, therefore, is to eliminate the background term in K_{2a-b} as completely as possible. The solution lies in the coherent nature of this process, and the advantages of using guided waves. As noted before for second-harmonic generation, the overlap integral term is reduced if guided waves with unequal mode numbers are used for the interacting fields. If at least one of the input fields corresponds to a higher order mode, that is TE_m with $m > 0$, interference effects occur, and hence the film contribution can be minimized, leaving the monolayer term as the dominant contribution. This approach has been used under vacuum conditions to observe the bonding of Ti atoms to oxygen bonds on the surface of a Nb_2O_5 film. The result is shown in Fig. 12 in which two TiO_2 vibrational modes were observed.

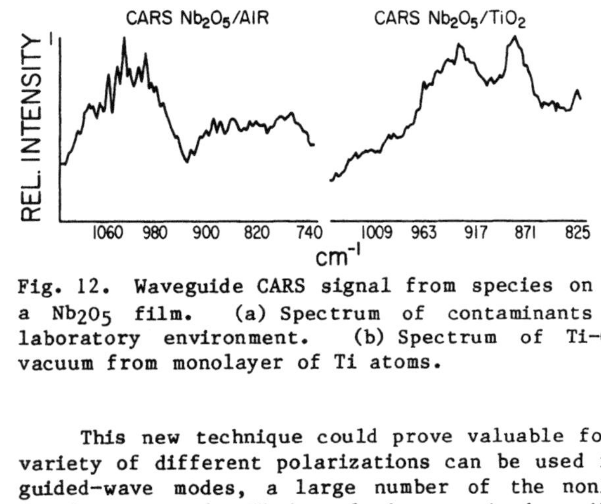

Fig. 12. Waveguide CARS signal from species on the surface of a Nb_2O_5 film. (a) Spectrum of contaminants on surface in laboratory environment. (b) Spectrum of Ti-O vibration in vacuum from monolayer of Ti atoms.

This new technique could prove valuable for surface science. Since a variety of different polarizations can be used for the incident and signal guided-wave modes, a large number of the nonlinear susceptibility terms can be measured. Their relative magnitudes will yield direct information about the orientation of surface species. Furthermore, this remarkable projected sensitivity may also find application in identifying both the nature and location of impurities in thin films. For example, different mode combinations can be used for the input and output guided waves, and the interference effects obtained in K_{2a-b} can then be used to locate the impurity sources as a function of depth into the film.

9. Intensity-Dependent Refractive Index and Dielectric Tensor

The optical properties, specifically the refractive index and relative dielectric constant of a medium, can be changed by the presence of a high-power optical beam. For plane waves, this leads to self-focusing or self-defocusing of the beam, depending on the sign of the nonlinearity. Starting from Eq. (5b) for a single guided wave, the pertinent nonlinear polarization for an isotropic medium is [30,31]

$$P^{NL}_i(z,\omega) = 2\varepsilon_0\chi^{(3)}_{1122}(-\omega,\omega,-\omega,\omega)\, E_i^{(n)}(z)E^*_j{}^{(n)}(z)E_j^{(n)}(z)$$
$$+ \varepsilon_0\chi^{(3)}_{1221}(-\omega,-\omega,\omega,-\omega)\, E^*_i{}^{(n)}(z)E_j^{(n)}(z)E_j^{(n)}(z)\ . \qquad (26)$$

Far from any resonant behavior, Kleinman symmetry is valid and $\chi^{(3)}_{1122} = \chi^{(3)}_{1221} = \chi^{(3)}_{eff}$. For plane waves, Eq. (26) simplifies, and, including the linear susceptibility, the total polarization is

$$P_i(\omega) = \varepsilon_0[\chi^{(1)}_{ii} + 3\chi^{(3)}_{eff}|E_j(\omega)|^2]\, E_i(\omega)\ . \qquad (27)$$

Noting that $\chi^{(1)}_{ii} = n_\gamma^2 - 1$ and expressing $|E_j(\omega)|^2$ in terms of the intensity $S = (1/2)c^2\varepsilon_0 n_\gamma|E_j(\omega)|^2$, the refractive index can be written as

$$n = n_\gamma + n_{2\gamma}S\ ; \quad n_{2\gamma} = \frac{3\chi^{(3)}_{eff}}{n_\gamma^2\varepsilon_0 c}\ . \qquad (28)$$

For guided-wave fields, the expression for the nonlinear polarization is more complex than that for the plane wave case. It is given by [31]

$$P^{NL}_{\gamma i}(z,\omega) = c\varepsilon_0^2 n_\gamma^2 n_{2\gamma}\left[\frac{2}{3}\, E_{\gamma i}^{(n)}(z)|E_{\gamma j}^{(n)}(z)|^2 + \frac{1}{3}\, E^2{}_{\gamma j}^{(n)}(z)E^*{}_{\gamma i}^{(n)}(z)\right]\ . \qquad (29)$$

For TE this simplifies to

$$P^{NL}_{\gamma y}(z,\omega) = c\varepsilon_0^2 n_\gamma^2 n_{2\gamma}|E_{\gamma y}^{(n)}(z,\omega)|^2 E_{\gamma y}^{(n)}(z,\omega)\ . \qquad (30)$$

The situation with TM waves is more complex, because there are two field components E_x and E_z. The difficulty arises from the $E^2{}_{\gamma j}^{(n)}(z)E^*{}_{\gamma i}^{(n)}(z)$ term whose value depends on the relative phase [31] between the fields for products with $i \neq j$. For example, writing

$$E_{\gamma x}^{(n)}(z) = |E_{\gamma x}^{(n)}(z)|\ ; \quad E_{\gamma z}^{(n)}(z) = e^{i\Psi}\,|E_{\gamma z}^{(n)}(z)|\ , \qquad (31)$$

then

$$P^{NL}{}_{\gamma x}(z,\omega) = c\varepsilon_0^2 n_\gamma^2 n_{2\gamma}\left[|E_{\gamma x}^{(n)}(z)|^2 + \left(\frac{2}{3} + \frac{e^{2i\psi}}{3}\right)|E_{\gamma z}^{(n)}(z)|^2\right] E_{\gamma x}^{(n)}(z)\ , \qquad (32a)$$

$$P^{NL}{}_{\gamma z}(z,\omega) = c\varepsilon_0^2 n_\gamma^2 n_{2\gamma}\left[|E_{\gamma z}^{(n)}(z)|^2 + \left(\frac{2}{3} + \frac{e^{-2i\psi}}{3}\right)|E_{\gamma x}^{(n)}(z)|^2\right] E_{\gamma z}^{(n)}(z). \qquad (32b)$$

Fortunately, in media for which $\beta^{(n)} > n_\gamma$, $\Psi = \pi/2$ and Eqs. (32) simplify to

$$P^{NL}{}_{\gamma x}(z,\omega) = \varepsilon_0\alpha_{xx}|E_{\gamma x}^{(n)}(z)|^2 + \varepsilon_0\alpha_{xz}|E_{\gamma z}^{(n)}(z)|^2\ , \qquad (33a)$$

$$P^{NL}{}_{\gamma z}(z,\omega) = \varepsilon_0\alpha_{xz}|E_{\gamma x}^{(n)}(z)|^2 + \varepsilon_0\alpha_{zz}|E_{\gamma z}^{(n)}(z)|^2\ , \qquad (33b)$$

$$\alpha_{ii} = c\varepsilon_0 n_\gamma^2 n_{2\gamma} \qquad \alpha_{ij}(i\neq j) = \frac{1}{3}c\varepsilon_0 n_\gamma^2 n_{2\gamma}\ . \qquad (33c)$$

Before we proceed to guided-wave problems involving nonlinear media, it is useful to consider the effects of nonlinearities on the propagation of plane-wave beams with a finite aperture. The case $n_{2\gamma} > 0$ is called a self-focusing nonlinearity for reasons that are evident in Fig. 13a. Because the phase velocity $v = c/n$ decreases with increasing power, regions

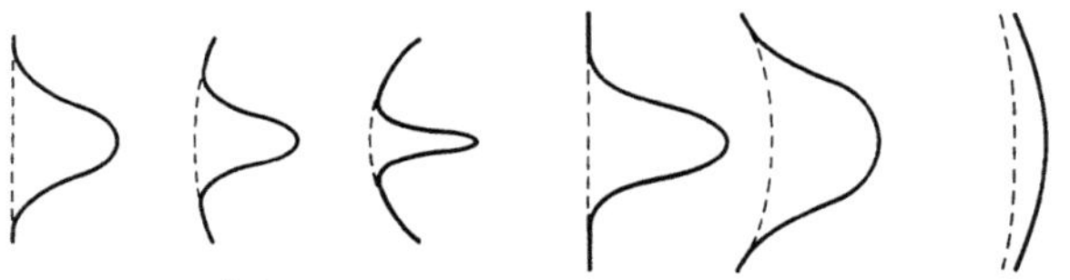

Fig. 13. (a) Evolution of plane-wave field into a self-focused beam in a self-focusing medium. (b) Defocusing of a plane wave in a self-defocusing medium.

of high intensity along a wavefront propagate a shorter distance than regions of low intensity. This leads to curvature in the wavefront equivalent to passing through a focusing lens, and the beam self-focuses under the action of its own high intensity. Conversely, $n_{2\gamma} < 0$ leads to self-defocusing of the wavefront, and hence this case is termed a self-defocusing nonlinearity.

10. Nonlinear Waves - Single Interface

It is well known that waves cannot be guided by the interface between two dielectric media whose refractive indices do not depend on intensity. (TM polarized surface plasmon polaritons can be guided by the interface between a dielectric and a metal by virtue of the negative dielectric constant of the metal: Here we do not consider surface plasmon polaritons because of their high attenuation coefficients.) However, when at least one of the two dielectric media exhibits a self-focusing nonlinearity, a guided wave can exist above a well-defined power threshold [34-36].

a. <u>TE Waves</u>

The single-interface guided-wave geometry is shown in Fig. 1 with h = 0. Including the nonlinearities discussed in the preceding section, the nonlinear wave equation for TE polarized (the only case to be solved exactly to date) waves is [32]

$$\nabla^2 E_{\gamma y}(\mathbf{r},t) + \frac{\omega^2}{c^2}\,[\epsilon_\gamma + \alpha_\gamma |E_{\gamma y}(z)|^2] E_{\gamma y}(\mathbf{r},t) = 0 \;, \tag{34}$$

where $\alpha_\gamma = \alpha_{yy}$ and $\epsilon_\gamma = n_\gamma^2$. The most general form for the field is

$$E_{\gamma y}(\mathbf{r},t) = \frac{1}{2}E_\gamma(z)\,\exp i[\beta kx - \omega t + \phi(z)] + \text{c.c.} \;, \tag{35}$$

where the $E_\gamma(z)$ is real. Substituting into Eq. (34) yields, after equating real and imaginary parts separately to zero,

$$\frac{d^2}{dz^2}E_\gamma(z) - E_\gamma(z)\left[\frac{d\phi}{dz}\right]^2 - q_\gamma^2 k^2 E_\gamma(z) + k^2\alpha_\gamma E^3{}_\gamma(z) = 0 \;; \qquad \frac{d}{dz}\left[E^2{}_\gamma(z)\frac{d\phi}{dz}\right] = 0 \,. \tag{36}$$

The last equation expresses the conservation of energy flux in the nonlinear medium and integrates to a constant K [33]. Therefore the complete equation for $E_\gamma(z)$ is

$$\frac{d^2}{dz^2}E_\gamma(z) - q_\gamma{}^2k^2E_\gamma(z) - \frac{K^2}{E^3{}_\gamma(z)} + k^2\alpha_\gamma E^3{}_\gamma(z) = 0 \ . \tag{37}$$

For the two semi-infinite bounding media, the fields must decay to zero at $\pm\infty$ so that $K = 0$ for this case and the fields are completely real. (For thin film media, or cases in which there is a radiation component that extends to infinity, K need not be zero and this case must be considered carefully.) For a nonlinear upper (cladding, $\gamma \equiv c$, $z \leq 0$) medium, the field solutions are [32]

$$E_c(z) = \sqrt{\frac{2}{\alpha_c}}\ \frac{q_c}{\cosh[q_c k(z_c - z)]} \qquad n_{2c} > 0\ , \tag{38a}$$

$$E_c(z) = \sqrt{\frac{2}{|\alpha_c|}}\ \frac{q_c}{\sinh[q_c k(z_c - z)]} \qquad n_{2c} < 0\ , \tag{38b}$$

where $q_c{}^2 = \beta^2 - n_c{}^2$. For $n_{2\gamma} > 0$, $z_c < 0$ corresponds to a self-focused field maximum in the cladding region. For $n_{2\gamma} < 0$, $z_s < 0$ leads to a divergent field in the cladding, a solution that is rejected as unphysical. Therefore the parameters z_c, which depend directly on the guided wave power, play pivotal roles in defining the field distributions. Similar field solutions are obtained for the substrate medium ($z \geq 0$, $\gamma \equiv s$) with q_c replaced by q_s, and $z_c - z$ by $z + z_s$. For a linear substrate

$$E_s(z) = A_s \exp(-q_s kz) \qquad n_{2s} = 0\ , \tag{39}$$

with $q_s{}^2 = \beta^2 - n_s{}^2$. The parameter z_s, which is related to z_c via the continuity of the tangential electric field, identifies where field maxima occur, and whether the field solutions are physically meaningful.

Matching the tangential electric and magnetic fields across the boundary leads to the dispersion relation [36]

$$q_c \tanh(kq_c z_c) = -q_s \tanh(kq_s z_s)\ ; \quad \alpha_c, \alpha_s > 0\ . \tag{40}$$

Using continuity of the E_y field across $z = 0$ and the dispersion relation, one can show that

$$E_c{}^2(0) = 2\frac{n_s{}^2 - n_c{}^2}{\alpha_c - \alpha_s}\ , \tag{41}$$

which predicts the interesting result that the field at the surface is a constant, independent of guided-wave power. For $n_s > n_c$ (which also implies $q_c > q_s$), the condition $\alpha_c > \alpha_s$ must be satisfied. Furthermore, if $\alpha_\gamma < 0$, then $\tanh(kq_\gamma z_\gamma) \rightarrow \mathrm{cotanh}(kq_\gamma z_\gamma)$ from which all four possibilities are obtained. The limit corresponding to $\alpha_\gamma \rightarrow 0$ (a refractive index that is independent of power) is obtained formally by taking the limit $z_\gamma \rightarrow 0$. Taking into account Eqs. (40) and (41), solutions exist for the following cases (assuming $n_s > n_c$)

$\alpha_c > 0$ and $\alpha_s > 0$ with $\alpha_c > \alpha_s$; $z_c < 0$, $z_s > 0$ <u>or</u> $z_c > 0$, $z_s < 0$

$\alpha_c > 0$ and $\alpha_s < 0$; $z_c < 0$ and $z_s > 0$ $\quad$ $\alpha_c > 0$ and $\alpha_s = 0$; $z_c < 0$, $z_s \rightarrow \infty$.

In each case, at least one of the media must have a self-focusing nonlinearity.

The pertinent experimental quantity is the guided wave power per unit length along the wavefront (y-axis) in Watts per meter or milliwatts per meter. This is obtained from

$$P = \int_{-\infty}^{\infty} \mathbf{ExH}\, dz = P_c + P_s \tag{42}$$

$$P_\gamma = \frac{\beta q_\gamma}{k n_\gamma^2 n_{2\gamma}} \left[1 - [\tanh(k q_\gamma z_\gamma)]^{\pm 1}\right], \tag{43a}$$

where the + and - signs refer to $\alpha_\gamma > 0$ and $\alpha_\gamma < 0$, respectively. If $\alpha_s = 0$ and $\alpha_c > 0$,

$$P_s = \frac{\beta q_c^2}{2 k q_s n_c^2 |n_{2c}| \cosh^2(k q_c z_c)}, \tag{43b}$$

and the $\alpha_c < 0$ case is obtained by replacing cosh with sinh. Note that these expressions contain the parameter z_γ explicitly. Therefore the position of the self-focused peak depends on the guided-wave power.

To date, no exact, analytical, closed-form solution of the nonlinear wave equation which includes loss has been reported. It is, however, very useful as a guideline for experiments to estimate in a simple way the effect of the linear attenuation coefficient on nonlinear guided waves. Assuming that the field distribution obtained in the lossless case will still be valid if the attenuation per wavelength is small, it is possible to calculate the nonlinear guided-wave attenuation coefficient approximately from the imaginary component of the dielectric constant [36]. Expanding $\nabla\cdot(\mathbf{ExH})$ and substituting from Maxwell's equations for $\nabla \mathbf{xE}$ and $\nabla \mathbf{xH}$ leads to

$$\nabla\cdot(\mathbf{ExH}) = \frac{1}{2}\omega\varepsilon_0\varepsilon_{\gamma I}\mathbf{E}(z)\cdot\mathbf{E}^*(z), \tag{44}$$

where $\varepsilon_{\gamma I}$ is the imaginary component of the dielectric constant in the γ'th medium. Integrating over a small volume element, applying the divergence theorem, and assuming small losses gives

$$\beta_I = \Sigma\varepsilon_{\gamma I} P_\gamma / 2P\beta_R, \tag{45}$$

where β_I is the imaginary component of the effective index $\beta = \beta_R - i\beta_I$.

Sample calculations are shown in Figs. 14 and 15 for β versus guided wave power [36]. For a single nonlinear (self-focusing) medium, there is a minimum threshold power required for a guided wave to exist. The larger the difference in indices ($n_s - n_c$) between the two media, the higher the threshold power. This is to be expected, since the index of the cladding near the interface must be increased to the point that it is larger than that of the substrate, which requires higher powers for larger initial values of $n_s - n_c$. As $\beta \to n_s$ (cutoff condition), the field in the substrate degenerates into a plane wave and the substrate power diverges, see Eq. (43b). For two self-focusing media, there are two separate branches, each corresponding to a self-focused field maximum in a different medium. In both figures, the higher the guided-wave power (past the minimum power point), the narrower the self-focused peak.

The key quantity that will limit the experimental utilization of these nonlinear guided waves is Δn_{sat}, the maximum change in refractive index

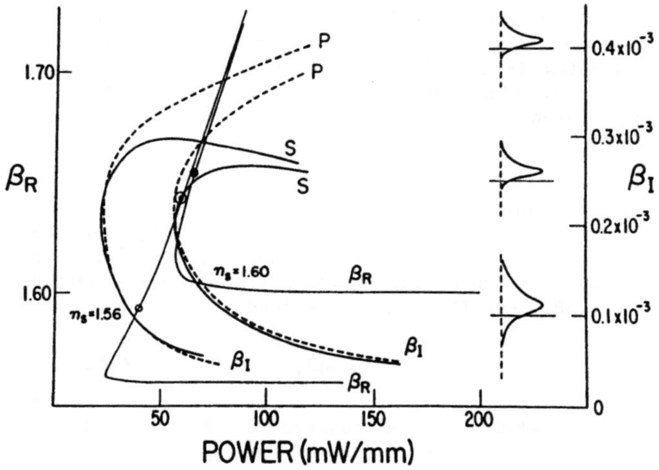

Fig. 14. The effective index $\beta = \beta_R - i\beta_I$ versus power for a wave guided by the interface between a linear and a nonlinear dielectric [36]. Here $\varepsilon_c = 2.4025 - 0.001i$, $n_{2c} = 10^{-9}$ m^2/W, and $n_s = 1.56$ or 1.6 for s- and p-polarized $(\varepsilon_z(E_z^2))$ waves. The evolution in field distribution with β_R is also shown for $n_s = 1.56$. The single and double circles correspond to $\Delta n =$ 0.1 and 0.2 respectively [36].

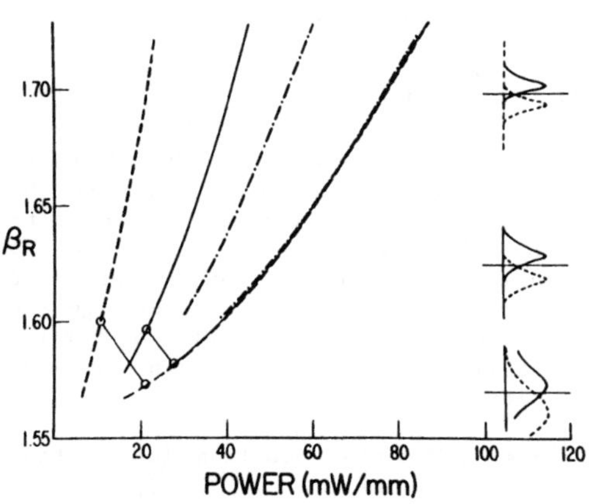

Fig. 15. The effective index versus guided-wave power for TE polarized waves guided by a single dielectric interface separating two self-focusing nonlinear media. Here $n_c = 1.55$, $n_s = 1.56$ and $n_{2s} = 2\times10^{-9}$ m^2/W. The dashed, solid, and dash-dotted curves correspond to $n_{2c} = 4\times10^{-9}$, 2×10^{-9} and 1.5×10^{-9} m^2/W respectively. The separated curves are for $z_c > 0$, and the clustered curves for $z_s > 0$. The field distributions are for the $n_{2c} = 4\times10^{-9}$ m^2/W case [36].

that can be induced optically in a real medium. This quantity is typically less than 0.01, but can be as large as 0.1 for some exceptional cases such as orientational nonlinearities in liquid crystals or semiconductors near their band gaps. As shown in Figs. 14 and 15, this material constraint limits the initial index difference $n_s - n_c$.

b. Tm Waves

The analysis of this case is complicated greatly by the existence of two field components, both of which can contribute to field-dependent dielectric constants and refractive indices. The best starting point for deriving the pertinent nonlinear wave equations are the Maxwell's equations $\nabla\times\mathbf{H} = i\omega\varepsilon_0\varepsilon\mathbf{E}$ and $\nabla\times\mathbf{E} = -i\omega\mu_0\mathbf{H}$. For TM waves, this leads to

$$E_{\gamma x} = \frac{i}{\omega\varepsilon_0\varepsilon_{xx}}\frac{d}{dz}H_{\gamma y} \qquad E_{\gamma z} = -\frac{\beta}{c\varepsilon_0\varepsilon_{zz}}H_{\gamma y} \qquad -i\omega\mu_0 H_{\gamma y} = \frac{d}{dz}E_{\gamma x} + i\beta k E_{\gamma z} \quad (46)$$

from which nonlinear wave equations can be constructed in terms of $E_{\gamma x}$, $E_{\gamma z}$, or $H_{\gamma y}$. However, no analytic solutions have been found yet to these equations without some form of simplifying approximations. The most common one has been to assume $\alpha_{\gamma xz} = \alpha_{\gamma zx} = \alpha_{\gamma zz} = 0$ [37] which we call $\varepsilon_{xx}(|E_x|^2)$. The second has been $\alpha_{\gamma xz} = \alpha_{\gamma zx} = \alpha_{\gamma xx} = 0$ [38], identified as $\varepsilon_{zz}(|E_z|^2)$. Yet a third approach with a different set of approximations has been suggested. Here we will concentrate on the first two, with the purpose of justifying the second one in terms of typical material parameters.

We discuss first the nonlinear wave equation for $E_{\gamma x}$ in the $\varepsilon_{xx}(|E_x|^2)$ approximation [37]. Eliminating $H_{\gamma y}$ and $E_{\gamma z}$ and assuming K = 0,

$$\frac{d^2}{dz^2}E_{\gamma x}(z) - q_\gamma{}^2k^2E_{\gamma x}(z) - \frac{k^2q_\gamma{}^2}{\varepsilon}\alpha_\gamma E^3{}_{\gamma x}(z) = 0 \ , \qquad (47)$$

which has solutions of the form discussed previously for the TE case. There is, however, one very important difference, namely the sign of the nonlinear term is opposite to that of Eq. (37) for TE waves. For a self-focusing nonlinearity ($\alpha_\gamma > 0$) in the cladding, the field solution is proportional to $\sinh[q_ck(z_c - z)]^{-1}$, which cannot lead to self-focusing of a field in the nonlinear medium. If $z_c < 0$, the solution fields diverge and are unphysical. Therefore, the behavior at high powers is quite dissimilar from the TE case, whose interpretation was consistent with previous experience in self-focusing nonlinear optics.

The elimination of E_x and E_z does lead to an equation for H_y that cannot be rigorously solved [38]. In the approximation $\varepsilon_{zz}(|E_z|^2)$,

$$\frac{d^2}{dz^2}H_{\gamma y}(z) - q_\gamma{}^2k^2H_{\gamma y}(z) + \frac{\beta^4}{c^2\varepsilon_0{}^2\varepsilon_{zz}{}^3}\alpha_\gamma|H_{\gamma y}(z)|^2H_{\gamma y}(z) = 0. \qquad (48)$$

Although the sign of the nonlinear term is consistent with that found previously for TE waves, no analytical solutions are possible because the $1/\varepsilon_{zz}$ prefactor depends on the power, and hence on $H_{\gamma y}(z)$. On the other hand, since Δn_{sat} is small for any realistic material, the change in $\varepsilon_{zz}{}^3$ with power will be small, and hence this prefactor constitutes a small correction to a nonlinear term, which is small anyway. Setting $\varepsilon_{zz} = \varepsilon$ and $\beta \simeq \varepsilon$, which is accurate to the same level of approximation, leads for the self-focusing cladding case to

$$H_{\gamma y}(z) = \sqrt{\frac{2c^2\varepsilon_0{}^2 n_c{}^2}{\alpha_c}} \; \frac{q_c}{\cosh[q_c k(z_c - z)]} , \tag{49}$$

and to self-focusing behavior similar to that obtained for the TE case.

Based upon the previous experience with the TE case, it is possible to justify why the $\varepsilon_{zz}(|E_z|^2)$ approximation is the preferable one. From Eqs. (46) and (49)

$$\frac{|E_{\gamma x}|^2}{|E_{\gamma z}|^2} = \frac{1}{\beta^2 k^2} \frac{\left|\frac{dH_{\gamma y}}{dz}\right|^2}{|H_{\gamma y}|^2} = \frac{q_c{}^2}{\beta^2}\tanh^2[q_c k(z_c - z)] \leq \frac{\beta^2 - n_c{}^2}{\beta^2} < \frac{2(\beta - n_c)}{\beta} . \tag{50}$$

It was shown in Figs. 14 and 15 that the value of Δn_{sat} limits the maximum index difference $\beta - n_c$ to less than Δn_{sat}. As mentioned previously, Δn_{sat} is typically 0.01 or less, and can be as large as 0.1 in very special cases only. Therefore $|E_{\gamma z}|^2 >> |E_{\gamma x}|^2$ and the dominant TM nonlinearity will involve $\varepsilon_{zz} = \varepsilon + \alpha_{zz}|E_z|^2$, which suggests that the $\varepsilon_{zz}(|E_z|^2)$ approximation is the most realistic of the two.

The dispersion relations are obtained by matching boundary conditions, and the <u>same</u> relations [36] are obtained for both forms of the uniaxial approximation, namely

$$\frac{q_c}{n_c{}^2} \tanh(kq_c z_c) = -\frac{q_s}{n_s{}^2} \tanh(kq_s z_s) \tag{51}$$

for $\alpha_c > 0$ and $\alpha_s > 0$. If $\alpha_\gamma < 0$, then $\tanh(kq_\gamma z_\gamma) \rightarrow \mathrm{cotanh}(kq_\gamma z_\gamma)$, from which all four possibilities are obtained. The power flow relations for this case can be shown to be exactly the same as for TE waves, and are given by Eqs. (45). Some typical power-β curves are shown in Fig. 14. For small Δn_{sat}, $n_s \approx n_c$ and the dispersion relations Eqs. (40) and (51) are effectively identical.

11. Nonlinear Waves - Thin Film Guided

Thin film waveguides (Fig. 1), in which some combination of the film, the cladding, or the substrate are nonlinear, exhibit many interesting power-dependent characteristics. They are more useful than the single interface nonlinear waves of section 10 because the solutions have low power limits. In addition, self-focused fields are simultaneously possible in more than one medium, which leads to multiple guided-wave branches. This area was first pioneered by Akhemediev [39], and many authors have subsequently contributed [34,40-43], especially for TE cases dealing with self-focusing nonlinearities.

The mathematical analysis for this case follows exactly the procedures outlined for the single interface problem, with the added complication of the thin film. The pertinent nonlinear wave equation for any of the media is Eq. (37). K must be equal to zero for the bounding media, which means that many of the formulas can be used from the preceding section with only small modifications. Since a nonlinear film is a bounded medium, solutions with $K \neq 0$ are also possible: To date only the case $K = 0$ has been analyzed for the nonlinear film case.

(a) Nonlinear Bounding Media

The most general case of two nonlinear bounding media will be treated here since the limits of a vanishingly small nonlinearity can easily be obtained from the most general results. The solutions for the field equations are by now well known [32-43]. The general form was defined previously by Eqs. (1) and (2). It turns out to be convenient to set $a^{(m)}(x) = 1$. Furthermore, since the fields now become power dependent, the solutions are no longer normal modes and the usual orthogonality conditions for normal modes no longer hold. An additional consequence is that the solutions discussed here are valid for only one guided wave present in the waveguide at a time. (This is self-evident, because the presence of any one wave alters the refractive index profile of the waveguide by virtue of the intensity-dependent refractive index.)

Many of the analytical formulas for the field distributions can be taken over directly from the previous section. For example, the fields in the cladding media are given directly by Eqs. (38) with the small change in notation that $E_c(z)$ is replaced by $E_c^{(m)}(z)$ to bring in the dependence on the mode number. Equations (39) describe the field distributions in the substrate with z replaced by z - h to take into account the film thickness, and the superscript (m) added to denote the mode number. The fields inside the film are written in the usual way as a superposition of sine and cosine functions ($n_f > \beta$) or sinh and cosh functions (for $\beta > n_f$, which is now allowed for some cases) with argument $k\kappa z$ where $\kappa^2 = |\beta^{(m)2} - n_f^2|$. For $n_{2c} > 0$ and $\beta^2 < n_f^2$ they are

$$E_f^{(m)}(z) = E_c^{(m)}(0)[\cos(k\kappa z) + \frac{q_c}{\kappa}\tanh(kq_c z_c)\sin(k\kappa z)] \ . \tag{52}$$

For $\beta^2 > n_f^2$, cos → cosh and sin → sinh respectively. When $n_{2c} < 0$, $\tanh(kq_c z_c)$ is replaced by $\mathrm{cotanh}(kq_c z_c)$ with the obvious changes in $E_c^{(m)}(0)$. Continuity of the tangential electric field at z = h relates z_s directly to z_c. The dispersion relations are obtained by matching the tangential magnetic fields at z = h. For $n_f^2 > \beta^2$,

$$\tan(k\kappa h) = \frac{\kappa[q_c\tanh(kq_c z_c) + q_s\tanh(kq_s z_s)]}{\kappa^2 - q_c q_s\tanh(kq_c z_c)\tanh(kq_s z_s)} \ . \tag{53}$$

For $\beta^2 > n_f^2$, $\tan(\kappa kh) \to \tanh(\kappa kh)$ and $\kappa^2 \to -\kappa^2$. For $n_{2\gamma} < 0$, it is straightforward algebra to show that $\mathrm{cotanh}(kq_\gamma z_\gamma)$ replaces $\tanh(kq_\gamma z_\gamma)$ in Eq. (53).

What remains is to determine the constant z_c (and z_s) from the guided wave power, or vice versa. The guided-wave power per unit length along the y-axis is given by Eq. (42) with an additional term P_f to include the power guided by the film. Equation (43a) holds for the power guided by the cladding (P_c) and substrate (P_s). For the film with $n_{2c} > 0$

$$P_f = \frac{\beta^{(m)}q_c^2}{2n_c^2|n_{2c}|}\,\frac{1}{\cosh^2(kq_c z_c)}\,[h(1 \pm \frac{q_c^2}{\kappa^2}\tanh^2(kq_c z_c))$$

$$- \frac{\sin(2k\kappa h)}{2k\kappa}(-1 \pm \frac{q_c^2}{\kappa^2}\tanh^2(kq_c z_c)) \pm \frac{q_c}{\kappa^2}k\tanh(kq_c z_c)(1 - \cos(2k\kappa h))] \ . \tag{54}$$

For $n_f^2 > \beta^2$, the upper sign is used. For $\beta^2 > n_f^2$, the lower sign is appropriate, and, in addition $\sin(2k\kappa h) \to \sinh(2k\kappa h)$ and $\cos(2k\kappa h) \to \cosh(2k\kappa h)$. For $n_{2c} < 0$, one needs to replace $\cosh(kq_c z_c)$ by $\sinh(kq_c z_c)$, and $\tanh(kq_c z_c)$ by $\mathrm{cotanh}(kq_c z_c)$.

The other case of interest here is a substrate whose refractive index is independent of power. This corresponds to taking the limit $z_s \to \infty$ in all of the formulas just given for the two nonlinear media case. For example, in the dispersion relations given by Eqs. (53), $\tanh(kq_s z_s) \to 1$ and $\mathrm{cotanh}(kq_s z_s) \to 1$. The substrate field is given by Eq. (39) with $A_s \to E_f^{(m)}(h)$ and z replaced by z − h. The power flow P_s is given by Eq. (45b) multiplied by the field ratio $|E_f^{(m)}(0)|^2/|E_c^{(m)}(0)|^2$.

Single Nonlinear Bounding Medium

Typical numerical calculations [43] for the effective index β versus the guided-wave power for a self-focusing cladding are shown in Fig. 16. The variation in field distribution with increasing β in Fig. 17 indicates the progressively stronger self-focusing that occurs. For large values of β the TE_0 wave degenerates into a single-interface nonlinear guided wave of the type discussed in the preceding section. A maximum is obtained in the guided-wave power because the power required to sustain a single-interface wave is less than that carried by the film when self-focusing starts to dominate. (For thinner films, no maximum is obtained and the TE_0 wave evolves monotonically into the single-interface solution.) For the TE_1 solution, one of the extrema must remain inside the film, which implies that the solutions in the film are oscillatory, and hence that $n_f > \beta$. Therefore the TE_1 cannot degenerate into a nonlinear single-interface wave, and the branch must terminate for some value β, as seen in Fig. 16. Since there are two possible values for β at some guided wave power levels, the possibility exists of switching, and perhaps bistability under the appropriate conditions.

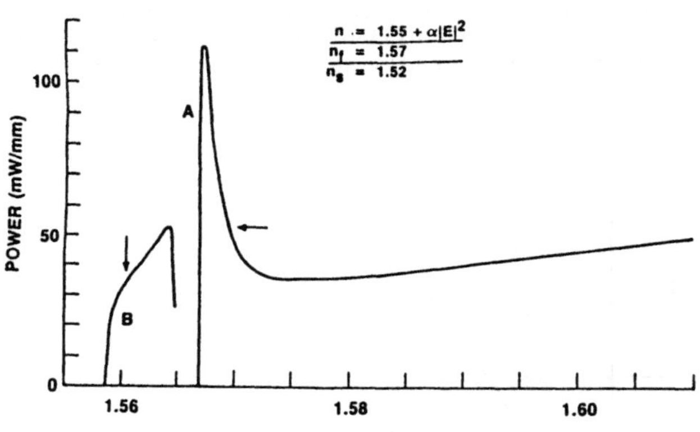

Fig. 16. (a) The guided-wave power versus the effective mode index for TE_0 (A) and TE_1 (B) waves guided by a linear film bounded by a self-focusing cladding and a linear substrate. The waveguide is a 2.0-μm film of Corning 7059 glass on a soda lime glass substrate with a liquid crystal MBBA cladding [43].

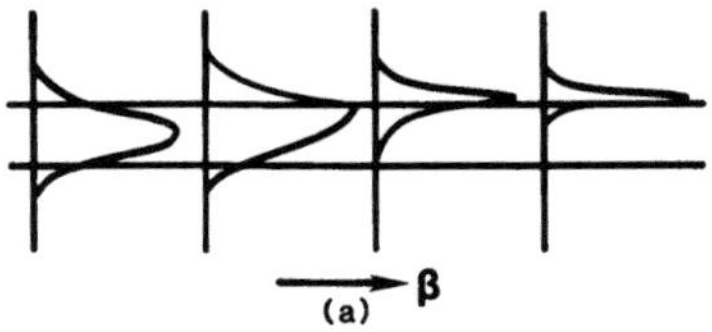

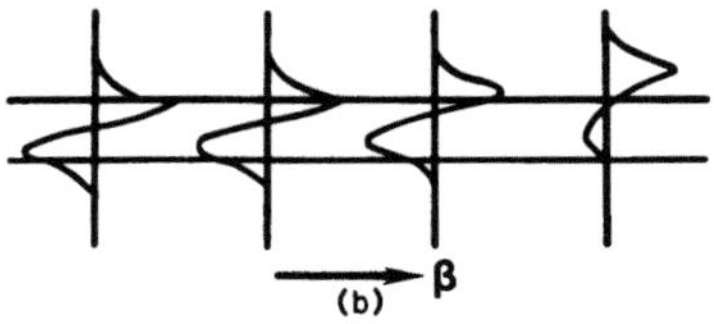

Fig. 17. Field distributions associated with (a) TE_0 and (b) TE_1 nonlinear guided waves for a linear film, linear substrate, and nonlinear cladding. The field evolution with increasing β is shown. Here $n_c = n_s = 1.55$, $n_f = 1.57$, $n_{2c} = 10^{-9}$ m^2/W and $h = 2.0$ μm [49].

The power dependence of the solutions has been verified experimentally [44,45] for the sample geometry shown in Fig. 18. A guided wave was propagated through a transverse boundary from a linear into a nonlinear waveguide region, and then through a second transverse boundary into another linear waveguide. Prisms were used to couple radiation into and out of the linear waveguide sections. The nonlinear medium was liquid crystal MBBA which has a thermal nonlinearity with $n_{2c} \approx 10^{-9}$ m^2/W [45]. With reference to Fig. 16, the guided-wave power available was intermediate between the TE_0 and TE_1 maxima.

The transmitted versus incident power for TE_0 and TE_1 shown in Fig. 19 can be interpreted as follows. The field maximum moves out of the film into the cladding with increasing TE_1 guided-wave power, and hence the transmission coefficient for both transverse boundaries decreases due to field mismatch. Furthermore, the net transmission coefficient decreases as a progressively larger fraction of the guided wave power is propagated in

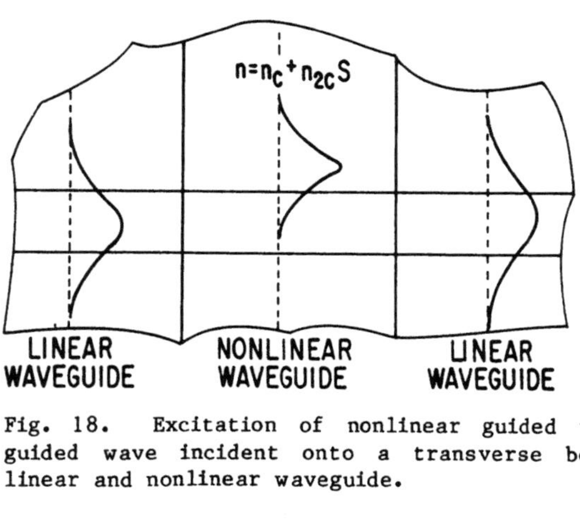

Fig. 18. Excitation of nonlinear guided waves by a linear guided wave incident onto a transverse boundary between a linear and nonlinear waveguide.

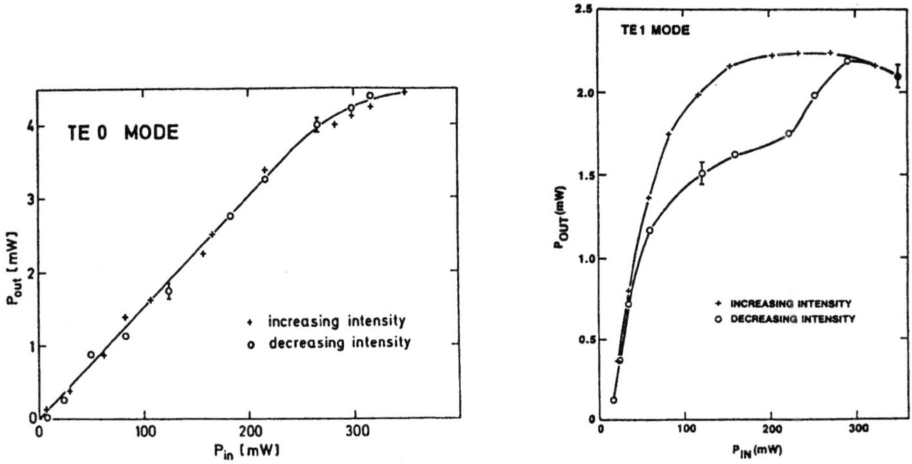

Fig. 19. The guided-wave power for TE_0 (a) and TE_1 (b) waves transmitted through a waveguide with a nonlinear cladding. The waveguide consisted of the liquid crystal MBBA as the cladding (n_c = 1.55, $n_{2c} \simeq 10^{-9}$ m^2/W), a 1.0-μm film of borosilicate glass (n_f = 1.61) and a soda lime glass substrate (n_s = 1.52) [44].

the lossy MBBA cladding. Thus the transmission for the TE_1 wave decreases with increasing power until the transmitted power becomes a constant when the maximum for TE_1 in Fig. 16 is reached. For subsequent decreases in power, both branches of the TE_1 curve are excited. Since the high β branch corresponds to higher losses, hysteresis is expected and observed [44]. For the TE_0 wave, there is insufficient power to significantly move the field extremum towards the boundary, and the transmission remains linear with no hysteresis, as observed [44]. A similar experiment has been carried out with CS_2 as the nonlinear medium [45].

The increase in β with guided-wave power suggests that cutoff may be power dependent for asymmetric ($n_c \neq n_s$) waveguides. That is, for a waveguide of thickness less than that required for low-power TE_0 wave propagation, an optical field can be used to increase the effective index past cutoff. A typical variation [46] in β with guided-wave power is shown in the inset of Fig. 20. The variation in the threshold power with film thickness is shown in Fig. 20. This corresponds to a lower threshold device.

For a self-defocusing cladding medium, β decreases monotonically with guided-wave power [46], see the inset of Fig. 21. If $n_s > n_c$, cutoff occurs when the field in the substrate degenerates into a plane wave and the guided-wave power diverges. However, for $n_c > n_s$, cutoff occurs at a finite power, as shown in Fig. 21. From Eq. (43b), the substrate power remains finite since $q_s \to (n_c^2 - n_s^2)^{1/2} \neq 0$. As $q_c \to 0$, $P_c \to [k^2 z_c n_c |n_{2c}|]^{-1}$ with $z_c > 0$, and the cladding power also remains finite. This phenomenon can be used for upper threshold devices in which the cutoff power can be tuned, for example by tuning the cladding index, see Fig. 21.

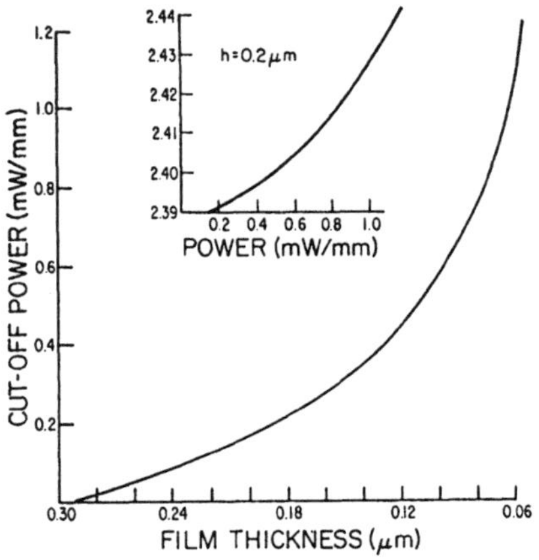

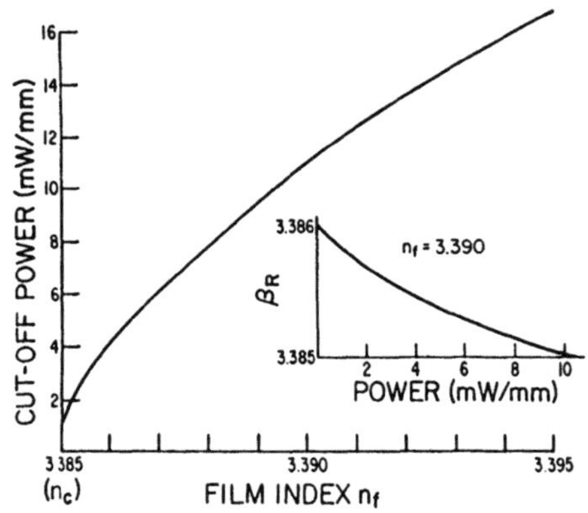

Fig. 20. The cutoff power versus film thickness for the TE_0 wave for a nonlinear self-focusing cladding and a linear film and substrate. Here $n_{2c} = 3x10^{-11}$ m^2/W, $n_c = 2.39$, by n_f = 2.40 and $n_s = 2.38$. The inset shows the variation in effective index β with guided-wave power for a film thickness of 0.2 μm. At low powers, the TE_0 mode for this structure is cut off at a film thickness of ≈0.30 μm [46].

Fig. 21. The maximum TE_0 guided-wave power that can be propagated versus index difference between a linear film and a self-defocusing cladding ($GaAs-GaAl_xAs_{1-x}$. Inset is the effective index versus guided-wave power for n_f = 3.39 and film thickness of 1.07 μm [46].

Two Nonlinear Bounding Media

The most interesting cases here occur for both media with self-focusing nonlinearities [39-43,46-49]. The possibilities are that self-focused fields can occur in either or both of the bounding media. This implies three separate asymptotic curves that degenerate into single-interface waves at high powers.

Numerical results [43,49] are shown in Fig. 22 for a totally symmetric waveguide, that is $n_c = n_s$ and $n_{2c} = n_{2s}$. There are three TE_0 branches, two of which are degenerate with respect to power and have power thresholds. As shown in Fig. 23, the fields associated with branch A remain symmetric with respect to the film center. For large β, this branch degenerates into single-interface surface waves, one guided by each film boundary. On the other hand, for branch B, a single-field maximum remains and moves out into one of the bounding media. Thus the symmetry of the waveguide is broken by the high-power optical field, a rather unique result [39]. Both branches B degenerate at high powers into single-interface surface-guided waves.

There are also multiple branches associated with the TE_1 solutions. Branch C retains symmetry about the film center, and both field extrema move symmetrically into the bounding media and degenerate into two out-of-

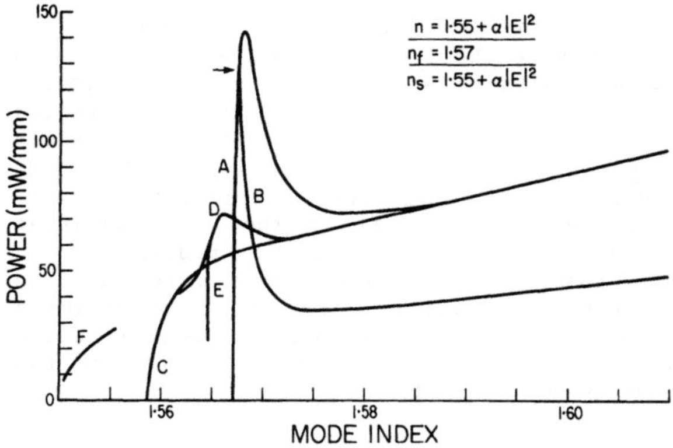

Fig. 22. The mode index β versus guided-wave power for a linear film bounded by identical self-focusing cladding media. There are TE_0-like waves (A,B), TE_1-like waves (C,D,E) and TE_2-like waves (F) for a glass film with n_f = 1.57 with n_c = n_s = 1.55 and $n_{2c} = n_{2s} = 10^{-9} m^2/W$ [44].

phase single-interface nonlinear waves for high β. Branch D has a power threshold and consists of two degenerate curves. The two field extrema both move into the bounding media, but asymmetrically as indicated by the field distributions in Fig. 23. Again, for large β the fields degenerate into two single interface surface waves. Branch E consists of two degenerate curves, each with a field extremum in the film and in one of the self-focusing media. It corresponds to kκh = π so that in Eq. (53) tan(kκh) = 0 and $z_c = -z_s$. For the parameters chosen, the TE_2 solution is cut off at low powers. However, at high powers, the branch F can be excited over a finite range of powers. The corresponding field distributions are shown in Fig. 23.

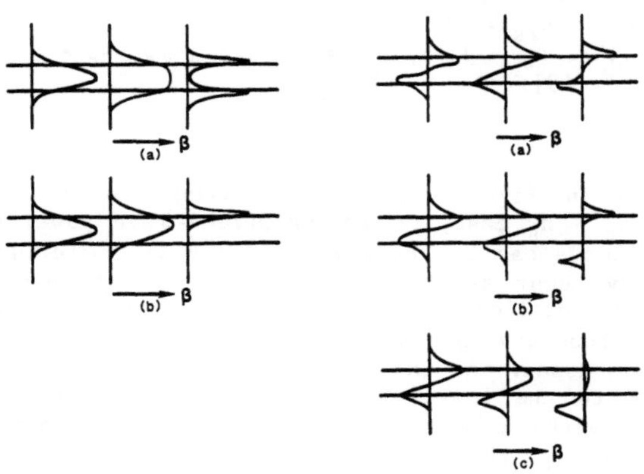

Fig. 23. TE_0 guided-wave field distributions for (a) branch A and (b) branch B. Here $n_c = n_s$ = 1.55, n_f = 1.57, h = 2.0 μm and $n_{2c} = n_{2s} = 10^{-9}$ m^2/W. Field distributions associated with TE_1 guided waves with both bounding media nonlinear; $n_c = n_s$ = 1.55, $n_{2c} = n_{2s} = 10^{-9}$ m^2/W and h = 2.0 μm. Asymmetric branches C, D and E correspond to (a), (b) and (c) respectively [49].

The profusion of solutions indicates that this combination of nonlinear media will be useful only if the film thickness is small enough to ensure only TE_0 solutions. When either the linear or nonlinear symmetry of the sample geometry is broken ($n_c \neq n_s$, $n_{2c} \neq n_{2s}$), two disconnected branches are obtained for TE_0, as shown in Fig. 24 [46,48,49]. We discuss first the case $n_c = n_s$ and $n_{2c} > n_{2s}$. In the limit of large β ($\beta > n_f$), the asymptotic curves in order of increasing power correspond to self-focusing in the most nonlinear medium (cladding), in the least nonlinear medium (substrate), and self-focusing in both bounding media. If in addition $n_c \neq n_s$, it becomes possible to obtain two branches to the dispersion curves that do not overlap in power. That is, there exists a range of powers over which guided waves cannot propagate. Note that the origin of the maxima in the branches is the same as discussed for the single nonlinear cladding in the previous section.

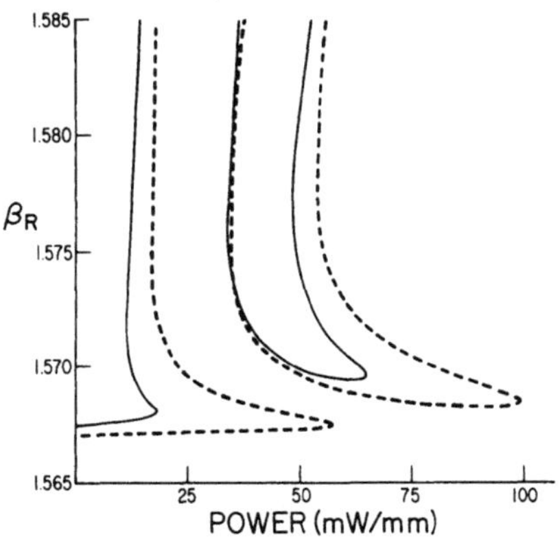

Fig. 24. The effective index versus TE_0 guided-wave power for a linear film bounded by two dissimilar self-focusing media. Solid lines are for h = 2.0 μm, $n_f = 1.57$, $n_c = 1.56$, $n_{2c} = 2\times10^{-9}$ m^2/W, $n_s = 1.55$ and $n_{2s} = 10^{-9}$ m^2/W. Dashed lines are for h = 2.0 μm, $n_f = 1.57$, $n_c = 1.55$, $n_{2c} = 2\times10^{-9}$ m^2/W, $n_s = 1.55$ and $n_{2s} = 10^{-9}$ m^2/W [49].

The curves shown in Fig. 24 appear promising for switching or bistability. For the $n_c = n_s$ case, increasing guided-wave power past the maximum of the lower curve can be accomplished only by switching to the upper curve. This may occur smoothly, or with a discontinuous jump in transmitted power. A detailed calculation of the field transmission through a transverse boundary of the type shown in Fig. 18 will be necessary to settle this issue. As the power is decreased along the upper branch, switching back to the lower branch is required at a much lower power than for the reverse process. This is a prime candidate for bistability. With $n_c \neq n_s$, the gap between the two curves implies a discontinuous jump in power when the guided-wave switches from the lower to the upper branch, and vice versa. Bistability could occur, and these characteristics could be useful for an all-optical switch.

There is one other case of potential interest. If both media have self-defocusing nonlinearities, increasing power always leads to cutoff at finite powers [50].

(b) Nonlinear Film

This case is considerably more complicated than for the nonlinear bounding media, and has not been explored fully yet because of two complicating factors [50-53]. The first is that K is not necessarily zero, since the fields exist only over a bounded region of space. The second is that the fields in the linear film are standing waves, and the generalization of this to the nonlinear case involves Jacobi elliptic integrals. For these reasons we give only a brief summmary, that elucidates the key features for a self-focusing film bounded by linear media.

For a self-focusing nonlinearity ($\alpha_f > 0$), the appropriate field solution to Eq. (37) in the approximation $K = 0$ is [53]

$$E_f^{(m)}(z) = \sqrt{\frac{2(\delta^2 + \kappa^2)}{\alpha_f}}\ \mathrm{cn}[\delta k(z + z_f)|p] , \tag{55a}$$

$$\delta^2 = \sqrt{\kappa^4 + 2\alpha_f E^2_o\left[n_f^2 - n_c^2 + \frac{1}{2}\alpha_f E^2_o\right]} , \tag{55b}$$

and $p = (\delta^2 + \kappa^2)/2\delta^2$ where $\mathrm{cn}(r|p)$ is the Jacobian elliptic function of order p and argument r, and $E_f^{(m)}(0)$ has been abbreviated by E_o. Assuming that the fields in the cladding and substrate are of the form

$$E_c^{(m)}(z) = E_o \exp[kq_c z] \ ; \quad E_s^{(m)}(z) = E_h \exp[-kq_s(z-h)], \tag{56}$$

with $E_h = E_f^{(m)}(h)$, matching the boundary conditions across both film interfaces leads to a dispersion relation of the form

$$\mathrm{cn}(\delta kh|p) = \frac{2E_oE_h(\delta^2 - q_cq_s)}{q_c^2E_o^2 + q_s^2E_h^2 + \delta^2(E_o^2 + E_h^2) + \frac{1}{2}a_f(E_o^2 - E_h^2)} . \tag{57}$$

The power flow is given by

$$P_c = \frac{\beta^{(m)}\varepsilon_0 cE^2_o}{4kq_c} \ ; \qquad P_s = \frac{\beta^{(m)}\varepsilon_0 cE^2_h}{4kq_s} , \tag{58a}$$

$$P_f = \frac{\beta^{(m)}\varepsilon_0 cE^2_o}{2k}\int_0^h dz\ \frac{\delta}{k}\left[\frac{\frac{\delta}{k}\mathrm{cn}\left[\frac{\delta}{k}|p\right] + \frac{q_c}{k}\mathrm{sn}\left[\frac{\delta}{k}|p\right]\mathrm{dn}\left[\frac{\delta}{k}|p\right]}{\frac{\delta^2}{k^2}\mathrm{dn}^2\left[\frac{\delta}{k}|p\right] + \frac{\alpha_f E^2_o}{2}\mathrm{sn}^2\left[\frac{\delta}{k}|p\right]}\right] . \tag{58b}$$

Numerical calculations for TE_0 waves have been performed for thin films with self-focusing nonlinearities. For waveguides with thickness chosen above cutoff for TE_0, the effective index β increases monotonically with power as self-focusing occurs inside the film, as expected [52]. For films whose thickness is below low-power waveguide cutoff, there is a minimum power above which the TE_0 wave can propagate, as shown in Fig. 25 [53]. As β increases past its cutoff value of n_s, the substrate guided-wave field changes from a plane wave to a field progressively more localized near the film-substrate boundary. Therefore the power associated with the substrate field decreases. As the field becomes

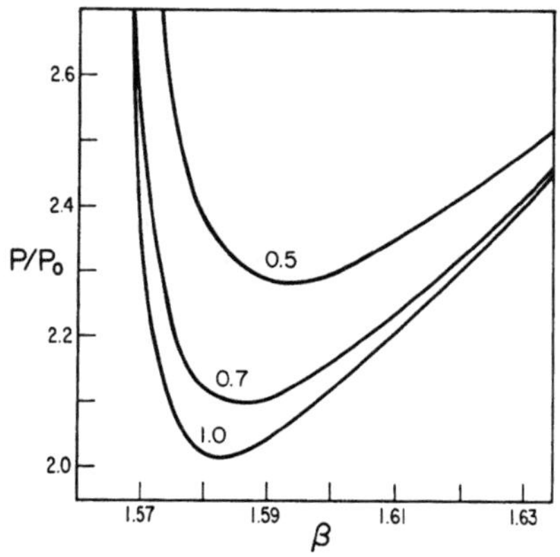

Fig. 25. Normalized power flow for the TE_0 wave guided by a self-focusing thin film versus effective index β. The curves are labelled by $D = kh$ [53].

progressively more self-focused in the film, it presumably degenerates into a single-interface surface wave, and β increases with increasing power. These two effects combine to produce a minimum in the guided-wave power. This phenomenon could find application as a lower threshold device.

12. Intensity-Dependent Wavevector Devices

In the previous section we discussed nonlinear waveguide media excited at high enough power levels to produce power-dependent field distributions. At lower power levels, it can be seen from Figs. 16, 22, and 24 that the variation in effective index with power is linear. (The same result can be obtained from first-order perturbation theory or coupled-mode theory [31].) That is,

$$\beta^{(m)} = \beta_0^{(m)} + \Delta\beta_0^{(m)}|a^{(m)}(x)|^2 \ . \tag{59}$$

Since we are usually dealing with only one guided-wave order m, we drop all the superscripts for simplicity. Such a power-dependent wavevector can affect a synchronous coupling condition for a dual-channel coherent coupler or prism coupler. For a grating imbedded in a nonlinear medium, the Bragg condition becomes power dependent.

(a) Nonlinear Prism Coupling

When a high-index prism is placed at a distance of less than a wavelength above a thin film, waveguide modes can be excited if the projection of the optical wavevector onto the base of the prism matches $\beta^{(m)}k$ for a guided wave. If the guided-wave power increases as the guided wave grows under the base of the prism, β changes, synchronism with the external field is lost, and the coupling efficiency is reduced. This can be expressed mathematically by the coupling equation [54]

$$\frac{d}{dx}a(x) = \hat{t}a_{inc}(x)e^{i(n_p\sin\theta - \beta_0 - \Delta\beta_0|a(x)|^2)kx} - (\ell^{-1} + \alpha)a(x) \ , \tag{60}$$

where $\hat{t}$ is the coupling coefficient, n_p is the prism index, θ is the angle of incidence measured from normal to the surface, ℓ is the characteristic distance for reradiation of the guided-wave field back into the prism and α is the waveguide absorption coefficient. The incident field amplitude varies as $a_{inc}(x)$ when projected onto the base of the prism. If the incidence angle is adjusted for maximum coupling at low powers ($\beta_0 = n_p \sin\theta$), then the guided-wave field falls out of phase with the generating field for high powers and the coupling efficiency is reduced.

This phenomenon has been verified experimentally [55]. The gap between a coupling prism and a thin-film waveguide was filled with the liquid crystal MBBA, which now acts as a nonlinear cladding. Results for the in-coupled versus incident power are shown in Fig. 26. Note that the in-coupled power starts out linear with incident power, and then saturates at high powers. These results are in good agreement with theoretical calculations.

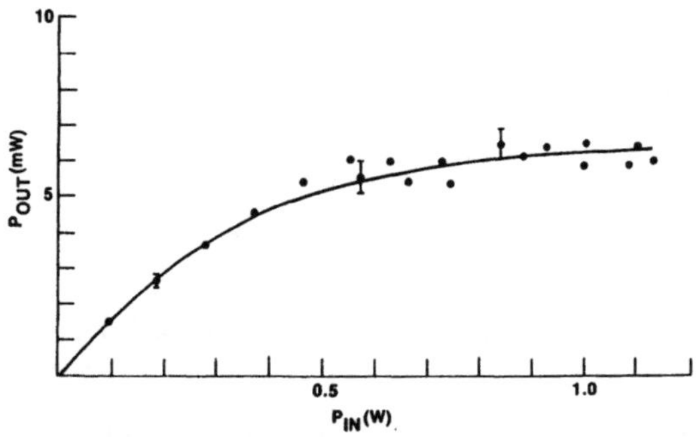

Fig. 26. The power prism-coupled into a nonlinear waveguide versus incident power for the TE_1 wave. The liquid crystal MBBA ($n_c = 1.55$, $n_{2c} = 10^{-9}$ m^2/W) was the nonlinear cladding, the film was 1.7-μm Corning 7059 glass ($n_f = 1.57$) on a Pyrex substrate [55].

(b) Nonlinear Gratings

Guided waves can be deflected efficiently by a grating provided that the Bragg condition is satisfied. For gratings defined either by a surface corrugation or a bulk index modulation with periodicity ℓ, the associated wavevector whose direction is orthogonal to the "grooves" is $\kappa_r = 2\pi r/\ell$ with r = 1 for sinuosoidal gratings. The Bragg condition $\beta^{(n)}\mathbf{k}_s = \beta^{(m)}\mathbf{k}_i + \kappa_r$ is an expression of wavevector conservation, where the subscripts "s" and "i" identify the incident and scattered guided-wave beams. Here it is assumed that m = n and r = 1 since the best coupling coefficients Γ are usually obtained for this case. For simplicity, we drop these subscripts in the subsequent discussion.

Since the effective index of a guided-wave mode depends on power [Eq. (59)], the Bragg condition can be tuned optically by using waveguide media with self-focusing or self-defocusing nonlinearities. Three ways in which this tuning can be effected are shown in Fig. 27. The "control" beams 1, 2, and 3 can change the Bragg condition for the wave incident at an angle to the grating. In addition, beam 1 can tune its own Bragg condition. Most of these cases can be summarized by the coupling equations [56]

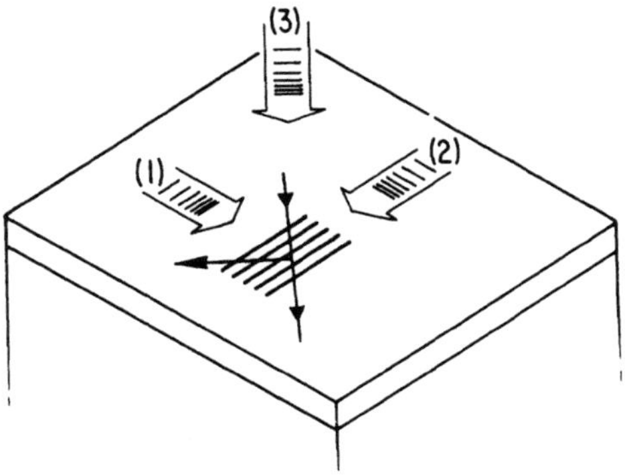

Fig. 27. General nonlinear grating geometry for switching guided waves. The grating can be tuned (in a bistable fashion) by a high-power guided wave incident (for example) along the grating axis, by a control guided wave incident parallel to the lines of the grating, or by illumination from above or below.

$$i\frac{d}{dx}a_s(x) = \Gamma e^{-i\Delta\beta kx}a_i(x) + \Delta\beta_0 k[a^2{}_s(x) + 2a^2{}_i(x)]a_s(x) + 2\Delta\beta_0 ka^2{}_c a_s(x) , \tag{61a}$$

$$-i\frac{d}{dx}a_i(x) = \Gamma e^{i\Delta\beta x}a_s(x) + \Delta\beta_0 k[a^2{}_i(x) + 2a^2{}_s(x)]a_i(x) + 2\Delta\beta_0 ka^2{}_c a_i(x) , \tag{61b}$$

where the initial offset from the Bragg condition is given by $\Delta\beta = 2\beta_0\cos\theta - \kappa$ and θ is the angle between the deflected beam and the grating wavevector. For the surface corrugation $u = u_0\sin(\kappa x)$ centered on the $z = 0$ interface, $\Gamma = \omega\varepsilon_0 u_0/8\ [n_c{}^2 - n_f{}^2]\mathbf{E}_{ic}(z \to 0)\cdot\mathbf{E}_{sf}(z \to 0)$. Here $a^2{}_c$ is the control beam power for cases 2 and 3. (The situation when beam 1 controls the deflection of the second weak beam is more complicated and requires four coupled equations.)

Bistable switching can occur when beam 1 is the only incident field present, that is $a_i(x) \neq 0$, $a_s(x) \neq 0$ and $a_c = 0$. The equations have solutions in terms of Jacobi elliptic functions [56]. Defining $|a_{sw}(0)|^2 = 2\beta_0/3\Delta\beta_0 kL$, $I = |a_i(0)|^2/|a_{sw}(0)|^2$ and $J = |a_i(L)|^2/|a_{sw}(0)|^2$ where I and J are normalized incident and transmitted powers,

$$T = J/I = 2[1 + nd(2\sqrt{\Gamma^2L^2 + J^2}\ |[1 + J^2/\Gamma^2L^2]-1)]^{-1} , \tag{62}$$

where L is the grating length (along $\boldsymbol{\kappa}$). The variation in transmitted versus incident power shown in Fig. 28a is the classic one for optical bistability. Switching powers for InSb and polydiacetylene channel waveguides can be as small as 10's of nanowatts and 100's of milliwatts respectively [57].

Smooth tuning of the grating reflectivity can be achieved with control beams 2 or 3. For a weak incident signal beam, the terms in the [ ] brackets in Eqs. (61) can be neglected and the reflectivity is

$$R = \frac{4\Gamma^2\sinh^2(\mu L)}{4\mu^2\cosh^2(\mu L) + \Delta k^2(a^2{}_c)\sinh^2(\mu L)} , \tag{63a}$$

$$\mu = \sqrt{\Gamma^2 - \Delta k^2(a^2{}_c)} \; ; \qquad \Delta k(a^2{}_c) = k[\Delta\beta_0 + 4\Delta\beta_0|a_c|^2] \; . \tag{63b}$$

Therefore the grating reflectivity can be tuned as shown in Fig. 28b. This property is useful for implementing a variety of all-optical logic functions. For example, if two input beams of power P_s are incident as control beams with the grating initially offset from maximum reflectivity (for a third weak signal beam) by $-P_s$, an XOR operation is obtained for a weak signal beam. Similarly, operations such as AND may also be implemented.

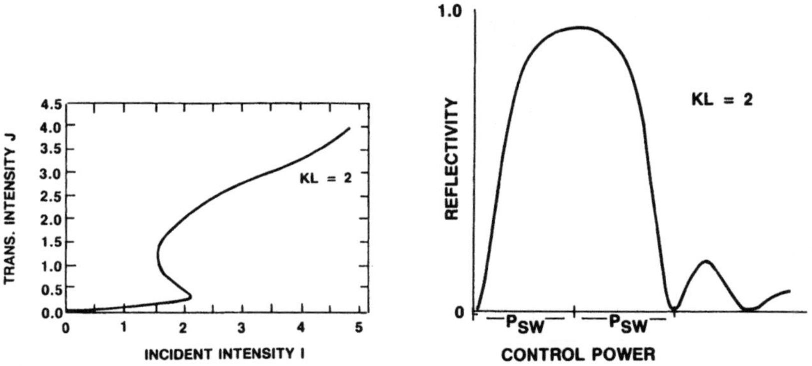

Fig. 28. (a) The transmitted versus incident power for distributed-feedback guided-wave bistability with $\Gamma L = 2$ [56]. (b) Grating reflectivity of a nonlinear grating versus beam power for a control beam incident parallel to the grooves [46].

(c) Nonlinear Coherent Coupler

An intensity-dependent refractive index can be used to alter the phase-matching condition between two coupled channel waveguides. For example, consider the channel waveguide directional coupler shown in Fig. 29. When light is injected into one channel, the overlap of that guided-wave field with the adjacent channel waveguide results in power transfer into the second waveguide. If one of the media in which the guided-wave fields exist is characterized by an intensity-dependent refractive index, the cross-coupling conditions become power dependent and new signal-processing devices become possible [58,59]. That is, the response of the device depends on the intensity of the input waves. Assuming identical waveguides with guided-field amplitudes described by $a_1(x)$ and $a_2(x)$ respectively and neglecting attenuation, the interaction can be described by the coupled-wave equations [58]

$$-i\frac{d}{dx}a_1(x) = \Gamma a_2(x) + [\Delta\beta_0 a_1(x)^2 + 2\Delta\beta_0' a_2(x)^2]a_1(x) \tag{64a}$$

$$-i\frac{d}{dx}a_2(x) = \Gamma a_1(x) + [\Delta\beta_0 a_2(x)^2 + 2\Delta\beta_0' a_1(x)^2]a_2(x). \tag{64b}$$

The parameter $\Delta\beta_0'$ quantifies the nonlinear effect of a strong field in one channel on the propagation characteristics of the neighboring channel.

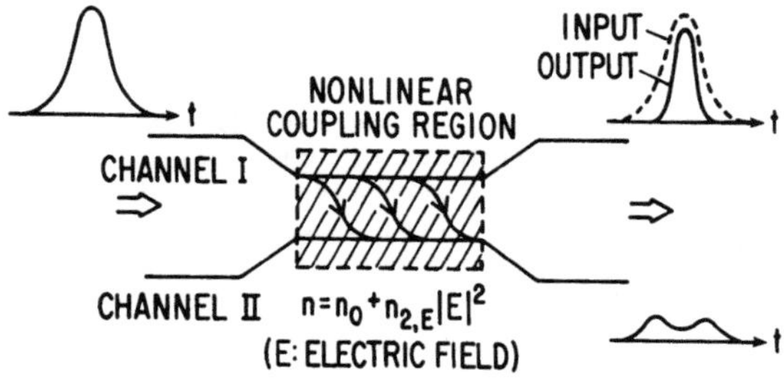

Fig. 29. Schematic of two coupled channel waveguides with a nonlinear material in the coupling region.

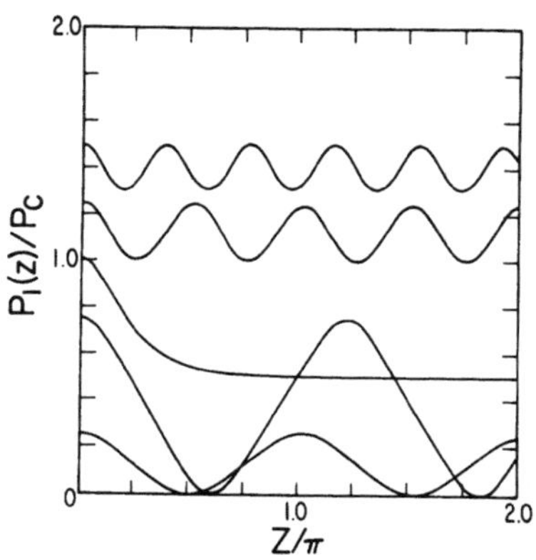

Fig. 30. The amount of power remaining in waveguide 1 as it propagates along a nonlinear coherent coupler. Here z/π is the normalized propagation coordinate, $P_1(z) \rightarrow a_1^2(x)$ and P_c is the typical switching power [58].

The solutions to these equations as obtained by Jensen [58] are shown in Fig. 30 for a single input channel excited at different power levels. They suggest a variety of possible applications. For example, if the device is set to produce complete transfer into the neighboring channel at low powers, at appropriately high-input power levels the transfer can be minimized. This corresponds to an intensity-dependent optical switch. A series of such switches can be used to demultiplex signals that are intensity coded. Alternatively, if the signals are of approximately equal intensity but are wavelength multiplexed, the signals can be demultiplexed if the material nonlinearity is strongly wavelength-dependent.

Intensity-dependent power transfer has recently been demonstrated using in-diffused channel waveguides in $LiNbO_3$ [59], and in GaAs-$Ga_{0.7}Al_{0.3}As$ multiple-quantum-well (MQW) channel waveguides [60]. Although $n_{2\gamma}$ is quite small for $LiNbO_3$ and in fact was measured in this experiment, a measurable transfer with picosecond time resolution has been observed by Haus and colleagues [59]. The recent report [60] of nonlinear switching in the MQW structures demonstrated partial switching of laser diode radiation at waveguide power levels of only 1 mW. In that case the two channel waveguides were strain-induced by depositing two gold stripes on top of a planar MQW waveguide. Although this device operated at a temperature of 180 K, it represents a very new and exciting development in this field.

Acknowledgments

This research was supported by the National Science Foundation (DMR-8300599 and ECS-8304749), the Army Research Office (DAAG-29-85-K-0026), the Air Force Office of Scientific Research (AFOSR-84-0277), the Joint Services Optics Program (ARO, AFOSR and BMSDC) and the NSF-Industry Optical Circuitry Cooperative.

References

1. Y.R. Shen, The Principles of Nonlinear Optics (Wiley, New York, 1984).
2. M.D. Levenson, Introduction To Nonlinear Laser Spectroscopy, (Academic Press, New York, 1982).
3. Reviewed in the book Optical Phase Conjugation, R.A. Fisher, ed. (Academic Press, New York, 1982).
4. P.K. Tien, Rev. Mod. Phys. 49, 361 (1977).
5. In Surface Polaritons, Electromagnetic Waves at Surfaces and Interfaces, V.M. Agranovich and D.L. Mills, eds. (North-Holland, Amsterdam, 1982).
6. W. Sohler, in New Directions in Guided Wave and Coherent Optics, D.B. Ostrowsky, ed. (Plenum Press, New York, 1983).
7. C.T. Seaton, G.I. Stegeman, W.M. Hetherington III and H.G. Winful, in Integrated Optics, H.P. Nolting and R. Ulrich, eds. (Springer-Verlag, New York, 1985) p. 178.
8. H. Kogelnik, Integrated Optics, Vol 7 of Topics in Applied Physics, T. Tamir, ed. (Springer-Verlag, Berlin, 1975), p. 66.
9. D. Marcuse, Theory of Dielectric Optical Waveguides, (Academic Press, New York, 1974).
10. J.E. Sipe and G.I. Stegeman, in Surface Polaritons, D.L. Mills and V.N. Agranovich, eds., (North-Holland, New York, 1982), p. 661.
11. N. Uesugi, Appl. Phys. Lett. 36, 178 (1980).
12. J.P. van der Ziel, M. Ilegems, P.W. Foy, and R.M. Mikulyak, Appl. Phys. Lett. 29, 775 (1976).
13. H. Ito and H. Inaba, Opt. Lett. 2, 139 (1978).
14. Y. Suematsu, Y. Sasaki, K. Furuya, K. Shibata, and S. Ibukuro, IEEE J. Quant. Electron. 10, 222 (1974).
15. G.H. Hewig and K. Jain, Optics Commun. 47, 347 (1983).
16. K. Sasaki, T. Kinoshita, and N. Karasawa, Appl. Phys. Lett. 45, 333 (1984).
17. R.V. Schmidt and I.P. Kaminow, Appl. Phys. Lett. 25, 458 (1974).
18. N. Uesugi and T. Kimura, Appl. Phys. Lett. 29, 572 (1976).
19. W. Sohler and H. Suche, Appl. Phys. Lett. 33, 518 (1978).
20. W. Sohler and H. Suche, in Integrated Optics III, L.D. Hutcheson and D.G. Hall, eds., Proc. SPIE 408, 163 (1983).
21. R. Normandin and G.I. Stegeman, Opt. Lett. 4, 58 (1979).
22. R. Normandin and G.I. Stegeman, Appl. Phys. Lett. 40, 759 (1982).
23. N. Uesugi, K. Daikoku, and M. Fukuma, J. Appl. Phys. 49, 4945 (1978).
24. A.T. Reutov and P. P. Tarashchenko, Opt. Spectrosc. 37, 447 (1974).
25. C. Karaguleff and G.I. Stegeman, IEEE J. Quant. Electron. QE-20, 716 (1984).
26. C. Karaguleff, G.I. Stegeman, R. Zanoni, and C.T. Seaton, Appl. Phys. Lett. 7, 621 (1985).
27. G.I. Stegeman, J. Opt. Commun. 4, 20 (1983).
28. G.I. Stegeman, R. Fortenberry, C. Karaguleff, R. Moshrefzadeh, W.M. Hetherington III, N.E. Van Wyck, and J.E. Sipe, Opt. Lett. 8, 295 (1983).
29. W.M. Hetherington III, N.E. Van Wyck, E.W. Koening, G.I. Stegeman, and R.M. Fortenberry, Opt. Lett. 9, 88 (1984).
30. P.D. Maker and R. Terhune, Phys. Rev. 137, A801 (1964).

31. G.I. Stegeman, IEEE J. Quant. Electron. QE-18, 1610 (1982).
32. A.E. Kaplan, Sov. Phys. JETP 45, 896 (1977).
33. A.E. Kaplan, IEEE J. Quant. Electron. QE-17, 336 (1981).
34. A.A. Maradudin, in Optical and Acoustic Waves in Solids-Modern Topics, M. Borissov, ed. (World Scientific Publ., Singapore, 1983), p. 72.
35. W.J. Tomlinson, Opt. Lett. 5, 323 (1980).
36. J. Ariyasu, C.T. Seaton, G.I. Stegeman, A.A. Maradudin, and R.F. Wallis, J. Appl. Phys., in press.
37. V.M. Agranovich, V.S. Babichenko, and V.Ya. Chernyak, Sov. Phys. JETP Lett. 32, 512 (1981).
38. C.T. Seaton, J.D. Valera, B. Svenson, and G.I. Stegeman, Opt. Lett. 10, 149 (1985).
39. N.N. Akhemediev, Sov. Phys. JETP 56, 299 (1982).
40. A.D. Boardman and P. Egan, J. Physique Colloq. C5, 291 (1984).
41. F. Fedyanin and D. Mihalache, Z. Physik B 47, 167 (1982).
42. F. Lederer, U. Langbein, and H.-E. Ponath, Appl. Phys. B 31, 69 (1983).
43. G.I. Stegeman, C.T. Seaton, J. Chilwell, and S.D. Smith, Appl. Phys. Lett. 44, 830 (1984).
44. H. Vach, C.T. Seaton, G.I. Stegeman, and I.C. Khoo, Opt. Lett. 9, 238 (1984).
45. I. Bennion, M.J. Goodwin, and W.J. Stewart, Electron. Lett. 21, 41 (1985).
46. C.T. Seaton, Xu Mai, G.I. Stegeman, and H.G. Winful, Opt. Eng. 24, 593 (1985).
47. A.D. Boardman and P. Egan, IEEE J. Quant. Electron., in press.
48. A. Boardman and P. Egan, Phil. Trans. Roy. Soc. London A313, 363 (1984).
49. C.T. Seaton, J.D. Valera, R.L. Shoemaker, G.I. Stegeman, J. Chilwell, and S.D. Smith, IEEE J. Quantum Electron. QE-21, 774 (1985).
50. N.N. Akhemediev, K.O. Boltar and V.M. Eleonskii, Opt. Spektrosk. 53, 906 and 1097 (1982).
51. V.K. Fedyanin and D. Mihalache, Z. Phys. B 47, 167, (1984).
52. U. Langbein, F. Lederer, and H.E. Ponath, Opt. Commun. 46, 167 (1983).
53. A.D. Boardman and P. Egan, IEEE J. Quant. Electron., in press.
54. C. Liao, G.I. Stegeman, C.T. Seaton, R.L. Shoemaker, J.D. Valera, and H.G. Winful, J. Opt. Soc. Am. A 2, 590 (1985).
55. J.D. Valera, C.T. Seaton, G.I. Stegeman, R.L. Shoemaker, Xu Mai, and C. Liao, Appl. Phys. Lett. 45, 1013 (1984).
56. H.G. Winful, J.H. Marburger, and E. Garmire, Appl. Phys. Lett. 35, 379 (1979).
57. G.I. Stegeman, C. Liao, and H.G. Winful, in Optical Bistability II, C.M. Bowden, H.M. Gibbs, and S.L.McCall, eds. (Plenum Press, New York, 1984) p. 389.
58. S.M. Jensen, IEEE J. Quant. Electron. QE-18, 1580 (1982).
59. A. Lattes, H.A. Haus, F.J. Leonberger, and E.P. Ippen, IEEE J. Quant. Electron. QE-19, 1718 (1983).
60. P. Li Kam Wa, J.E. Sitch, N.J. Mason, J.S. Roberts, and P.N. Robson, Electron. Lett. 21, 26 (1985).

Guided Nonlinear Waves in Thin Films

A.D. Boardman and P. Egan

Department of Physics, University of Salford, Salford, M5 4WT, U.K.

A few attempts [1-4] to analyse the guiding properties of a nonlinear film have already appeared in the literature. None of the previous papers have exposed any really practical formulae that, say, give any insight into the way in which the wave number depends upon the optical intensity at one of the boundaries or makes the numerical calculations any easier. One of them [3] arbitrarily sets an integration constant to zero,thereby producing a very restricted special solution.

A new treatment is presented here of guided nonlinear TE waves [5] in an optically nonlinear, frequency-independent, dielectric film. The dielectric is cladded by air and sits on a substrate that has a higher linear dielectric constant than the film. New mathematical results are given here that express the properties of the system in terms of its physical parameters. They represent a straightforward way to introduce the Jacobi function field distributions that are necessary to this problem.

By using power levels high enough to make the refractive index dependent upon the guided wave intensity, a threshold intensity is reached beyond which two wave numbers for a single power level become possible. In this calculation the third-order Kerr-type optical nonlinearity is not weak, and an exact theory is developed for TE waves that does not depend upon either perturbation theory or an iterative numerical scheme. Dispersion curves are calculated that are labelled with optical power density at the lower film boundary, and are contrasted to those that are labelled with the total power flowing down the guide. The field shapes are discussed in terms of the power-effective refractive index characteristics, and some comments are made about the device possiblities in the area of optical switching. The stability criteria of the solutions are not given but it is possible, because of the way the nonlinear fields of the film depend upon the variation of the wave index with power flow down the guide, that some kind of threshold switching is possible.

References

1. A.D.Boardman and P.Egan, "Nonlinear surface and guided polaritons of a general layered dielectric structure", in Journ.de Physique,Coll.C5, **45**, 291, (1984)
2. V.K.Fedyanin and D.Mikhalake, "P-polarised nonlinear surface polaritons in layered structures", in Z.Phys.B, **47**, 167, (1982)
3. U.Langbein, F.Lederer and H.-E.Ponath, "A new type of nonlinear slab-guided waves", in Optics Comm., **46**, 167, (1983)
4. N.N.Akhmediev,K.O.Boltar and V.M.Eleonskii, "Dielectric optical waveguide with nonlinear susceptibility.Asymmetric refractive-index profile", in Opt.Spectr.(USSR), **53(b)**, 654, (1982)
5. A.D.Boardman and P.Egan, "Optically nonlinear waves in thin films" in IEEE J.Quant.Elec.(in the press) (1985)

Nonlinear Prism Coupling and Propagation in Nonlinear Waveguides

J.D. Valera, R.B. Svensson, C. Seaton, G. Stegeman, and Xu Mai
Optical Scienes Center, University of Arizona, Tucson, AZ 85721, USA

Nonlinear waveguide theory has been shown to lead to a variety of intriguing phenomena [1]. Two of these effects have been studied experimentally: nonlinear prism coupling and propagation in nonlinear waveguides.

The coupling efficiency versus incident power was studied experimentally for a nonlinear prism coupler. The sample geometry consisted of two strontium titanate coupling prisms and a thin film waveguide of Corning 7059 glass (n=1.57, and 1.7 um thick) deposited onto a Pyrex substrate (n=1.478). The gap between the input prism and the film was filled with the liquid crystal MBBA, whose refractive index is given by $n=n_o + n_2 S$, where S is the local light intensity. When a few milliwatts of 0.5145 um light were transmitted through the waveguide it was seen that, under appropriate conditions, the transmitted intensity was not linear with the input intensity; limiting action was observed as shown in Fig. 1. It was also observed that the position and the angle of incidence of the input beam for optimum coupling at low input powers was different than that at high powers [2]. These effects occur because the waveguide is characterized by an intensity-dependent propagation constant [3]. These results are all in qualitative agreement with theory. [4].

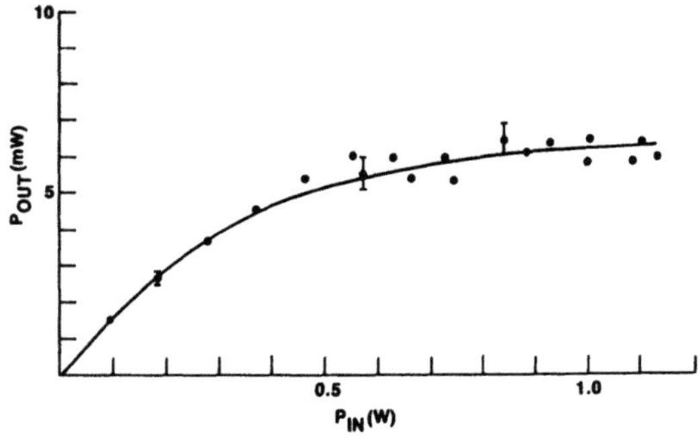

Fig. 1. P_{out} versus P_{in} for TE_o mode

Propagation in nonlinear waveguides was also studied experimentally. The thin film waveguide used was the same as that already described. The liquid crystal was on top of the thin film and was not in contact with any of the prisms. When the liquid crystal used was E7 and the TE_2 mode was observed, switching behavior was obtained. However, when K-15 was the liquid crystal and the TM_o mode was studied, mode cut-off was observed. Modulation of light by light was also demonstrated for the last sample.

Nonlinear waveguide phenomena has proved to be very interesting, as experiments have shown. Other phenomena, like thresholding and light-controlled spatial scanners, are currently under investigation.

References

1. C.T. Seaton, Xu Mai, G.I. Stegeman and H.G. Winful; Optical Engineering 24, 593, (1985)
2. J.D. Valera, C.T. Seaton, G.I. Stegeman, R.L. Shoemaker, Xu Mai, and C. Liao: Appl. Phys. Lett. 45, 1013 (1984)
3. G.I. Stegeman: IEEE J. Quantum Electron. QE-18, 1610 (1982)
4. C. Liao and G.I. Stegeman: Appl. Phys. Lett. 44, 164 (1984)

Index of Contributors

MIX
Papier aus verantwortungsvollen Quellen
Paper from responsible sources
FSC® C105338

If you have any concerns about our products,
you can contact us on
ProductSafety@springernature.com

In case Publisher is established outside the EU,
the EU authorized representative is:
Springer Nature Customer Service Center GmbH
Europaplatz 3, 69115 Heidelberg, Germany

Printed by Libri Plureos GmbH
in Hamburg, Germany